Explainable, Interpretable, and Transparent AI Systems

Transparent Artificial Intelligence (AI) systems facilitate understanding of the decision-making process and provide opportunities in various aspects of explaining AI models. This book provides up-to-date information on the latest advancements in the field of explainable AI, which is a critical requirement of AI, Machine Learning (ML), and Deep Learning (DL) models. It provides examples, case studies, latest techniques, and applications from domains such as healthcare, finance, and network security. It also covers open-source interpretable tool kits so that practitioners can use them in their domains.

Features:

- Presents a clear focus on the application of explainable AI systems while tackling important issues of "interpretability" and "transparency".
- Reviews adept handling with respect to existing software and evaluation issues of interpretability.
- Provides insights into simple interpretable models such as decision trees, decision rules, and linear regression.
- Focuses on interpreting black box models like feature importance and accumulated local effects.
- Discusses capabilities of explainability and interpretability.

This book is aimed at graduate students and professionals in computer engineering and networking communications.

Explainable, Interpretable, and Transparent AI Systems

Edited by

B. K. Tripathy and Hari Seetha

CRC Press is an imprint of the
Taylor & Francis Group, an **informa** business

Designed cover image: B. K. Tripathy

First edition published 2025
by CRC Press
2385 NW Executive Center Drive, Suite 320, Boca Raton FL 33431

and by CRC Press
4 Park Square, Milton Park, Abingdon, Oxon, OX14 4RN

CRC Press is an imprint of Taylor & Francis Group, LLC

Library of Congress Cataloging-in-Publication Data
Names: Tripathy, B. K., 1957- editor. | Seetha, Hari, 1970- editor.
Title: Explainable, interpretable, and transparent AI systems / edited by
B. K. Tripathy and Hari Seetha.
Description: First edition. | Boca Raton : CRC Press, 2024. | Includes
bibliographical references and index. | Identifiers: LCCN 2024008454 (print) |
LCCN 2024008455 (ebook) | ISBN 9781032528564 (hbk) | ISBN 9781032580920 (pbk) |
ISBN 9781003442509 (ebk)
Subjects: LCSH: Artificial intelligence–Industrial applications.
Classification: LCC TA345 .E984 2024 (print) | LCC TA345 (ebook) |
DDC 670.285/63–dc23/eng/20240404
LC record available at https://lccn.loc.gov/2024008454
LC ebook record available at https://lccn.loc.gov/2024008455

ISBN: 978-1-032-52856-4 (hbk)
ISBN: 978-1-032-58092-0 (pbk)
ISBN: 978-1-003-44250-9 (ebk)

DOI: 10.1201/9781003442509

Typeset in Times
by codeMantra

The editors would like to dedicate this volume to Dr. G. Viswanathan, the Honourable Chancellor and Founder of Vellore Institute of Technology, for his constant support, for providing a congenial atmosphere in the institute, and for his encouragement throughout the preparation of this piece of work.

-B. K. Tripathy

-Hari Seetha

Contents

Preface

As stated by Stephen Hawking, intelligence is the ability to adapt to changes. Artificial intelligence (AI) enables computers to execute tasks that are easy for people to perform but difficult to describe formally. It is one of the most-discussed technology trends in research and practice today and is estimated to deliver an additional global economic output of around USD 13 trillion by the year 2030. AI algorithms can be classified into two categories, namely, white box and black box. Machine learning algorithms that provide results that are understandable to users are white box models. An expert in the domain also does not understand black box models, creating ambiguity about how they function and, ultimately, how they make predictions.

There is an imminent need for explainable AI (XAI) techniques that are accountable, fair, transparent, and trustworthy without compromising the performance of the models, which would lead to better interaction with machines by users. It becomes essential to explain why and how decisions are made by machines. Machine learning models must have the ability to provide reasoning, describe their merits and demerits, and convey an understanding of how they will behave. To achieve this, new or modified machine-learning techniques that provide the required explanations must be developed. These models, in combination with human-computer interface techniques, must be able to provide users with a greater understanding of why, how, and when a machine model or a deep learning model will perform well. It is important for any business to understand the process used to make decisions by AI models.

Transparent AI systems facilitate understanding of the decision-making process and also provide opportunities in various aspects of explaining AI models. AI systems often depend on machine learning and deep learning-based opaque algorithms. Interpreting them is a pressing need to magnify their implementation in various sectors of organizations, overcome failures, strengthen trust, and use them in accordance with relevant (inter)national policies. Without making opaque systems transparent, explaining the inner tricks of derivation or interpreting the outcomes is like seeing magic with astonishing eyes without knowing the internal tricks, which may be simple to understand.

Under the above background, this book aims to provide readers with up-to-date information on the latest advancements in the field of explainable AI, which is a critical requirement of AI, machine learning (ML), or deep learning (DL) models. It provides examples, case studies, latest techniques, and applications from the domains of healthcare, finance, network security, etc. It also covers open-source interpretable tool kits so that practitioners can use them in their domains. This book comprises of 17 chapters. We briefly characterize the contents of the chapters for comprehension and to provide readers with an overall holistic view.

XAI is a rapidly growing field that has the potential to make ML models more transparent, understandable, and accountable. XAI techniques can help build trust between users and ML models and can help identify and correct biases in these models. There are several applications of XAI; in the healthcare domain, it can diagnose diseases, predict patient outcomes, and organize treatment plans; in the financial sector, it can detect fraud, assess credit risk, provide investment recommendations, and develop trading strategies; in the field of ML algorithms, it can provide insights to improve content moderation, personalize recommendations, target ads, and analyse user behaviour; in the area of decision-making, it can assist in autonomous driving, predictive maintenance, fault detection, and customer service applications; in the judicial system, it can help judges, lawyers, and defendants make more informed decisions and prevent bias and discrimination in the legal system. XAI has the potential to make ML models more transparent, understandable, and accountable. In **Chapter 1** of this book, the authors present these application areas and many more fields.

In the creative computing scenario, explainability and evaluation remain two major challenges for researchers. Additionally, the evaluation of creativity remains a challenging affair. Real-life creative environments are dynamic and partially observable. Another interesting aspect of creativity is its contextual nature and high temporal significance. Any single direction overriding the whole scenario imposes constraints on the freedom and flexibility demanded by creativity. Multidirectional semi-supervised reinforcement with multiple context-driven fitness functions definitely appears as a promising option. While deliberating evaluation and exploration, we can utilize event-based temporal measurements or episodic responses to gain insights regarding explainability. **Chapter 2** of this collection works through these pieces to strike a balance between evaluation and explainability, highlighting a multitude of possibilities.

As stated above, XAI models play a major role in the decision-making process, a vital component of applications in various sectors like healthcare, education, media, and entertainment. These models act as a white box between the input data and the output decision. In **Chapter 3** of this edited volume, a detailed study of the impact of XAI on modern automotive, financial, and manufacturing sectors is explained. The problems associated with using AI models in these applications are discussed, and appropriate solutions are suggested with the help of XAI tools and techniques. The wide range of XAI usability in each of the mentioned sectors is highlighted, guiding the improvement of existing methods and promising to build trustable systems. Specific applications target the level of XAI and its role in each of these sectors for the modern industry and society.

In this digital world, the creation and use of online social networks generate a huge amount of data in an ever-increasing rate. The security of these data is of utmost importance. Also, the prevalence of cyber-attacks has necessitated their prevention and detection. Despite the development of several intrusion detection systems, IT organizations still face the threat of these attacks, leading to financial losses. Obtaining optimal dimensionality is a crucial task in modelling an effective detection mechanism. The emergence of AI has brought enormous models for detecting a variety of cyber-attacks. However, some of these models are black box models with complex underlying algorithms that offer less transparency, raising concerns about their trustworthiness. In this regard, XAI-based methods would be more powerful in not only providing trustworthiness but also determining the contribution of the features towards prediction. **Chapter 4** of this compilation presents various techniques and XAI methods used in obtaining specific features that could distinguish cyber-attacks.

Although ML and DL techniques under AI have provided very effective AI applications and are highly accurate and robust, these models do not articulate human-understandable explanations for the outcomes generated in the form of results and decisions. **Chapter 5** of this volume aims to create a suite of ML or deep learning techniques that produce a more explainable model, with a high level of learning performance for decision-making and trained AI models. It aims to develop a novel explainable AI framework that is data-driven with feature quantities indication using SHapley Additive exPlanations (SHAP) for feature reduction using Random Forest and Explain Like I'm 5 (ELI5) for model interpretation. XAI-based methodologies help visualize the important features for crop suitability analysis. It is shown how the process of selecting suitable crops helps in developing user-friendly models to improve crop suitability prediction with cost-effectiveness and to reduce the time taken for making decisions to find the suitable crop to be cultivated in agricultural land.

XAI has provided successful applications in real-life scenarios with respect to innovations, risk mitigation, ethical issues, and logical values to the users. In **Chapter 6** of this book, the focus is to deliberate on the principles of XAI and its possible applications in terms of methodologies and trust, along with future directions. Experimentation is carried out over other models like Local Interpretable Model-agnostic Explanations (LIME) in addition to the SHAP model on the publicly available Diabetes dataset.

The business world has been using AI and XAI to large extents, and the growth in use can be termed as exponential. In fact, XAI defines model precision, fairness, transparency, and AI-supported decision-making, leading to confidence and trust in the act of AI being used by an organization. However, privacy concern grows at the same pace as the growth in the use of AI and alleged programs. Out of the available attributes, XAI selects those which should be visible. This is an attractive prospect as it attracts business aspects with a minimum visibility of sensitive information to the end users and customers. While transforming the black-box regression model into XAI, there are chances that some part of the sensitive information can be exposed in the absence of proper rules based on feature selection for explainability. **Chapter 7** of this work provides a detailed real-time discussion of AI, XAI, and their pros and cons. It is focused on discussions on the affluence of AI and XAI in numerous applications and their associated risks. Most importantly, the social and legal phenomena of AI and XAI-enabled structures are strategically discussed here.

AI has revolutionized the world with its applications in modern devices like self-driving cars to smart-device assistants. However, the level of trust in such applications used in systems is an important factor. In recent times, there is a great deal of concern about the trustworthiness and dependability of AI systems. In **Chapter 8** of this edited volume, it is substantiated how the various methods and practices lead to the development of unbiased or fair AI and XAI systems. Different approaches are followed by different systems for making systems explainable. In this chapter, a generic approach to be followed by stakeholders while developing any kind of systems, leading to the development of fair and explainable AI solutions that would put them in better positions in making decisions, is discussed.

Deep learning models are black box models as the user cannot directly interpret them to provide solutions to the problems experienced in model awareness regarding data model effect, data relation, quality, and risk to increase the explainability of applications. In recent years, studies with image, signal, and textual data have been investigated to explain the validity of prediction decisions and the behaviour of deep learning models. LIME is one of the solution aspects for black box models. It has been developed to solve applications that use various data types such as image, signal, and unstructured text. It makes an essential contribution to the fact that models developed for AI could be explained in an automatic structure and transparent in the life cycle with modern information architecture. In **Chapter 9** of this volume, a comprehensive review of deep neural networks (DNNs) is provided as black box models for various data models.

Explainable techniques (ET) have been developed to incorporate in many sectors where AI is used for application. ET is supposed to enable users, especially novices, to understand the system's decision-making. At the same time, the necessity of trust from a socio-technical perspective pinpoints its ability to both foster transparent and long-term relationships and enable the adoption and proliferation of emerging novel technologies in society. **Chapter 10** of this book discusses the importance of trust in any application domain and explores social trust and bias in the context of intelligent and computable models. Two real-life datasets are used to outline parametric evaluation measures related to social networks, and the findings are presented graphically. Real-life data are also analysed to examine the core concept of explainable parameters using a Python-based narration to explain the concepts in detail. The chapter outlines a trust prediction model that uses a trust score and the Diverse Counterfactual Explanations (DiCE) explainable technique. The findings indicate that this approach may have significant practical expansion not only for predictions derived from data but also for graph-based investigations. The contents of this chapter emphasize the significance of XAI in fostering social trust in social networks and offer a comprehensive framework for achieving this objective.

Real-time analysis is a growing trend, and the explainability of learning about the environment derived from the data is equally important for understanding the application. Due to the dynamic, continuous, and drifting nature of streaming data, every learning model should be capable of discerning the properties of the environment in real time with adaptability. Because concept drift and the emergence of new concepts are very common in a streaming environment, fuzzy clustering seems advantageous for consideration, as it provides the feature where each data point will have a dedicated member for each cluster, and due to the drifting nature of the environment, this membership will play a critical role in making the clustering algorithm adaptive. In **Chapter 11,** an algorithm is proposed which determines the number of clusters and the fuzzifier value depending on the data entropy. The membership of each data point changes as concepts drift, and the algorithm detects this. The novelty of the algorithm lies in the detection of a number of clusters through learning with data and adaptability to the environment. The experimental analysis demonstrates the efficacy of the proposed method on real-world and benchmark synthetic data.

XAI helps make complex systems more transparent, providing explanations for the decisions to some level of detail, which is important to identify potential bias in data and algorithmic fairness and also to ensure the proper functioning of the system. Another aspect of XAI is to understand which input features contribute to the predictions, allowing humans to verify whether the decision is based on plausible features rather than spurious correlations. **Chapter 12** of this work highlights the importance of AI decisions and the explanatory approaches that currently exist. Additionally, this chapter deals with the common methods to explain the decisions of AI systems like LIME, SHAP, Saliency Maps, and other relevant techniques.

The popularity of online services has increased manifold since the recent pandemic, as these platforms give customers the privilege to stay at home and shop in their comfortable way. These websites and platforms can be aided with support systems that help customers in many shapes and forms. Explainable recommendations bring trust to customers in the industrial world and are essential for their satisfaction. **Chapter 13** presents an overview of existing ML and DL-based approaches, also highlighting the position of an explainable product recommendation system that assists users in purchasing from and using the platform. A two-dimensional taxonomy to distinguish existing explainable recommendation studies is drawn, which consists of model-intrinsic and model-agnostic approaches. As a result, it contributes to the visibility, trustworthiness, effectiveness, and customer satisfaction of recommendation systems. It also makes it easier for system designers to examine, debug, and enhance the recommendation algorithm.

We have become so dependent on AI that it is now impossible to think about our lives without it. However, there has been a spate of accidents involving AI systems, leading to serious losses including loss of lives, financial fines, and loss of reputation for the relevant AI providers. All these accidents typically lead to a lack of trust in AI, resulting in gaps in the implementation of trustworthy AI.

The primary root cause of these accidents is the lack of sufficient testing of AI systems, by the relevant developers and quality professionals involved in the development of AI solutions. In **Chapter 14** of this volume, the challenges involved in the testing of AI and solutions are highlighted. The focus is primarily on those powered by ML due to the lack of a test oracle. Also, a solution to the lack of a test oracle in AI systems is presented through the use of a metamorphic testing-based approach to test AI systems. The different metamorphic testing-based approaches to testing AI systems are elaborated.

The rapid growth of DNNs has led to their application in various industries, including healthcare, finance, transportation, etc. However, the adoption and trustworthiness of these DNN models are challenged by their need for more transparency and interpretability. In **Chapter 15** of this work, state-of-the-art methods, techniques, and tools developers use to implement explainable AI systems to ensure transparency while adopting AI models are discussed. This chapter also discusses implementing different explainable AI algorithms with popular DNNs. In addition, it compares different explainable AI algorithms and the open-source software packages used to implement those algorithms. Furthermore, the challenges of implementing different explainable AI algorithms due to the complexity of other neural network structures and the limitations of the available open-source software packages are presented.

Visualization of AI is highly essential in the context of its explainability and interpretability. **Chapter 16** discusses various types of visualization metrics and techniques, followed by the importance and need for visualization in the context of AI. The available evaluation strategies are also discussed. Different metrics in terms of complexity, fidelity, understandability, coverage, and consistency are analysed with respect to quantification. The solo and combined applications of visualization and interpretable AI are systematically presented. Additionally, some challenges and corresponding future prospects pertaining to visualization and interpretable AI are elaborated.

Recommender systems are becoming increasingly ubiquitous, powering the way we discover new products, services, and information. However, the black box nature of these systems can make it difficult for users to understand why they are being recommended certain items. So, transparency, trustworthiness, scrutability, and fairness in the recommendations generated by the recommender system are crucial for the users to consider. The tool to make the recommendations trustable, explainable, and scrutable is the transparent recommender systems. The final chapter (**Chapter 17**) of this book provides a detailed study on transparent AI, its benefits and challenges, and open-source packages available to make the ML models understandable. It gives an outline of transparent recommender systems and includes the most recent work in this domain.

AI has become a buzzword in modern society and has become inevitable in the advancement of technology. However, because of its black box nature, it is losing its trust among the users. The three concepts, namely, explainability, interpretability, and transparency, are useful in bringing back trust among the users in a big way.

If the contributions in different chapters of this volume attract the eyes of researchers in the field of AI; make them able to understand the basic concepts involved in the field of explainable, interpretable, and transparent AI; provide them with the existing technologies in this field; and inspire them to come up with new solutions to tackle the problems elaborated in this volume, the editors will have the satisfaction that their tedious effort is successfully rewarded. Also, we are hopeful that beginners in this highly attractive field will find this book to be an invaluable support to quench their thirst to explore the future of AI as a whole.

B. K. Tripathy
Hari Seetha
Vellore Institute of Technology

Acknowledgments

The editors are grateful to CRC Press, Taylor & Francis Group for permitting them to edit this volume, *Explainable, Interpretable, and Transparent AI Systems*. Special thanks are due to Dr. Gagandeep Singh, Senior Editor, Acquisitions (Engineering/Environmental Sciences), for his constant support, encouragement, and suggestions.

We are thankful to the authorities of Vellore Institute of Technology for providing a congenial atmosphere and moral support to carry out this work smoothly.

Also, we are thankful to all the contributors for their insightful contributions and the reviewers for their timely support and constructive suggestions to improve the quality of the content substantially.

Our family members, friends, and colleagues helped us in one way or the other. We wish to thank them all.

About the Editors

Dr. B. K. Tripathy is a distinguished researcher in the fields of Computer Science and Mathematics and is working as a professor (Higher Academic Grade) in the SCORE School of VIT, Vellore. He received his Ph.D. degree in 1983. During his student career, he received three gold medals for securing first position at the graduation level, securing first position at the postgraduate level, and being adjudged as the best postgraduate of the year from Berhampur University, Odisha. He has the distinction of receiving a national scholarship at the PG level, UGC (Govt. of India) fellowship for pursuing his research, DST (Govt. of India) fellowship for pursuing M.Tech. (Computer Science) at Pune University, and the SERC fellowship (DOE, Govt. India) for joining IIT Kharagpur as a visiting fellow. He has published more than 740 articles in international journals, proceedings of international conferences of repute, and chapters in edited research volumes. Also, he has edited 11 research volumes and written two books and two monographs. He has acted as a member of the International Advisory Committee/Technical Program Committee of more than 140 international conferences and has delivered the keynote addresses at some of them. During his career so far, he has supervised candidates for 32 Ph.D. degrees, 13 M. Phil degrees, and 5 M.S and M. Tech by research degrees. Also, he has guided numerous students with their major projects at B. Tech., M. Tech., and MCA levels. Dr. Tripathy is a member of more than 25 professional bodies such as a senior member of IEEE, a senior member of ACM, a senior member of IRSS, and a senior life member of CSI, Indian Academy of Mathematics, IEEE Communication Society, Indian Mathematical Society, ACM Compute News group, Indian Science Congress Association, International Science and Technology group, and IEEE fuzzy logic society. Recently, Dr. Tripathy has been selected as Fellows of The Institution of Electronics and Telecommunication Engineers (IETE) and Institution of Engineers India (IEI). As recognition of his distinguished research in the field of Mathematics, he was included by the American Mathematical Society as a reviewer for Mathematical Reviews in 1982 and was later honoured with an honorary membership of the American Mathematical Society in 1992. In 1983, he was selected as a reviewer for Zentralblutt fur Mathematik (Germany). At present, Dr. Tripathy is a reviewer for over 100 international journals from all top publishing houses all over the world. Some of the journals of repute where he has published articles include *Information Sciences, Applied Soft Computing, IEEE Access, Journal of Mathematical Analysis and Applications, Transactions on Data Privacy, Bulletin of the Malaysian*

Mathematical Society, Indian Journal of Pure and Applied Mathematics, Journal of Educational Technology and Society, International Journal of Earth Sciences and Engineering, Kybernetes, World Applied Sciences Journal, International Journal of Communication Systems, Journal of Web Engineering and Evolutionary Intelligence, International Journal of Uncertainty, Fuzziness and Knowledge-Based Systems, Frontiers in Public Health, Sensors (Basel), Evolutionary Intelligence, only to name a few. His book, *Soft Computing- Advances and Applications,* is recommended as a textbook in many universities and several programmes. He has evaluated the Ph.D. theses of approximately 25 university students from all over India. He has been sanctioned with three research projects from central agencies of India. Also, Dr. Tripathy has served in several administrative positions, such as head of the department programme manager of different academic programmes, division leader of research groups, chairperson of the disciplinary committee, member of faculty selection committees, member of research admission committees, and Dean of SCOPE School in 2015 and the Dean of the SITE School from 2018 to 2021 at VIT University, Vellore. During his tenure as Dean of the SITE School, he has organized three international conferences.

Dr. Hari Seetha obtained her master's degree from the National Institute of Technology (formerly R.E.C.) Warangal and obtained her Ph.D. from the School of Computer Science and Engineering, VIT University, Vellore, India. She worked on Large Data Classification during her Ph.D. She has research interests in the fields of pattern recognition, data mining, text mining, soft computing, XAI, IDS, and machine learning. She received the Best Paper Award for the paper entitled "On improving the generalization of SVM Classifier" at the Fifth International Conference on Information Processing held in Bangalore. She has published several research papers in national and international journals of repute. She has been one of the editors for the edited volume, *Modern Technologies for Big Data Classification and Clustering* published in 2017. She is a member of the editorial board for various international journals. She guided six Ph.D. students, and several scholars are working under her guidance. She had been a co-investigator for a major research project sponsored by the Department of Science and Technology, Government of India. She served as a Division Chair for the Software Systems Division and Program Chair for B.Tech. (Computer Science and Engineering) programme in the School of Computer Science and Engineering at VIT University, Vellore, India. She had been Assistant Director (Ranking and Accreditation) at VIT University, Vellore, Tamil Nadu, India. She also served as Dean of the School of Computer Science and Engineering at VIT-AP University, Near Vijayawada, Andhra Pradesh, India. She is currently working as Professor and Director of the Center of Excellence in Artificial Intelligence and Robotics at VIT-AP University, Near Vijayawada, Andhra Pradesh, India.

Contributors

A. Anitha received her Ph.D. in Information Technology from VIT University, India. At present, she is working as an Associate Professor (Sr.) in the School of Information Technology, VIT University, Vellore, India. She has authored more than 35 international and national journal papers for Elsevier, Inderscience, and Springer journals. She has presented research work at various international conferences. She is an associate editor for several international journals and a reviewer for international journals such as *Applied Soft Computing, Soft Computing, Tech Science Press,* and *MDPI.* She is associated with professional bodies such as CSI, Indian Science Congress, and IAENG. Her research interests include explainable AI, neural networks, artificial intelligence, soft hybrid techniques, rough computing, business intelligence, machine learning, deep learning, blockchain, and Industrial Revolution 4.0.

Raj Kumar Batchu received his Ph.D. in Computer Science and Engineering from VIT-AP University, Near Vijayawada, Andhra Pradesh. He completed his master's degree in Computer Science and Engineering from KL University, AP, in 2013, and B.E in Computer Science and Engineering from Anna University, Chennai, in 2011. He has six years of academic and industrial experience He has published several research papers in reputed journals and conferences. His research interests include machine learning, deep learning, intrusion detection, big data analytics, and data science. He is currently working as an Assistant Professor in the Department of Computer Science and Engineering at Amrita Viswa Vidyapeetham, Amaravathi Campus, Andhra Pradesh, India.

Sengul Bayrak received her Ph.D. in Computer Engineering from Istanbul University, Cerrahpasa in 2021. She has been working as an Assistant Professor in Istanbul Sabahattin Zaim University's Software Engineering Department since 2021. Her fields of study include machine learning, artificial intelligence, natural language processing, data mining, and image processing.

Anuhya Bhagavatula has a master's degree in Data Science from the University of Washington, Seattle, and a bachelor's degree in Computer Science Engineering from GITAM (Deemed to be University), Visakhapatnam. With three years of industry experience in master data management, data analysis, and graph data science, her research interests include graph machine learning, data visualization, and explainable AI.

Parth Birthare, holding a Bachelor of Technology in Computer Science from Vellore Institute of Technology, has marked his presence felt through open-source contributions and participations in events and activities where he used his technical skills to solve the problems faced by social-good organizations. He also has exemplary working exposure with clients in the Business and Technology sectors from firms like JP Morgan Chase & Co. Furthermore, he is a dedicated and quick learner, who likes to engage in areas that lead to innovation and welfare. With a strong technological intellect and empathy, he strives to build next-generation solutions to existing problems.

Subhadip Boral received a master's degree in Computer Science from West Bengal State University in 2016. He is pursuing his Ph.D. from the University of Calcutta and working as a Research Fellow at the Technology Innovation Hub at Indian Statistical Institute. His research interests include unsupervised learning, dimensionality reduction, streaming data analysis, and anomaly detection.

V. Lakshmi Chetana received her M.Tech. in Computer Science and Engineering from JNTUK, Kakinada, in 2014, and her master's degree in Computer Applications from ANU, Guntur, in 2009. She is currently pursuing her Ph.D. in Computer Science and Engineering at VIT-AP University, Amaravathi, near Vijayawada, Andhra Pradesh. She had an academic experience of 14 years. Her research interests include recommender systems, machine learning, deep learning, big data analytics, and data science.

Arka De is presently pursuing his Bachelor of Technology in Computer Science and Engineering at Vellore Institute of Technology, Vellore, India. He has been actively researching in the domain of artificial intelligence and Internet of Things (IoT). He has been a consistent grade-point seeker. He has participated and won many hackathons at the state level and organization level. He is an active scholar with dedicated team spirit working presently on developing an automated handwritten script evaluation mechanism.

Shrusti Ghela has a master's in Data Science from the University of Washington, Seattle, and a bachelor's degree from the Vellore Institute of Technology. She is an accomplished Data Scientist and Engineer with a passion for leveraging data to drive insights and solve complex problems. Shrusti has a diverse range of experiences in data analysis, machine learning, and data engineering. Her research interests include explainable AI and natural language processing.

Ashish Ghosh is a Senior Professor at the Indian Statistical Institute. He has published more than 270 research papers in international journals and conferences and is acting as the Principal Investigator of several funded projects. His current research interests include machine and deep learning, data science, image/video analysis, and computational intelligence.

Sharath Kumar Jagannathan is currently working as an Assistant Professor at Saint Peters University, New Jersey, USA. He obtained his bachelor's degree in Computer Science from Madras University, India. He then pursued his master's degree in Software Systems Engineering (M.S.S.E) from the University of Melbourne, Melbourne, Australia. He has 10 years of teaching experience from VIT, Chennai, India, and 5 years of IT experience from Intelligent IP, Perth, Australia. He received his doctoral degree in Social Network Privacy Preservation from Vellore Institute of Technology, Chennai. He has 17 journal publications, one international conference presentation, and a patent. His research interests include machine learning, artificial intelligence, deep learning, and anonymization techniques.

Kaushik K. is currently an undergraduate (UG) student at Vellore Institute of Technology in the School of Computer Science and Engineering at the Chennai campus. He is currently pursuing a B.Tech. in Computer Science and Engineering and will graduate in 2024. His areas of interest include image processing, data science, automata theory, artificial intelligence, and machine learning. He has started a "Mathematics Club" in his college and holds the position of "Founder" of the club. He also serves as the "President" of the Nature Club at his college.

Parag Kulkarni is one of the pioneering contributors to machine learning and AI. He is an entrepreneur and consultant in knowledge innovation, ML, and innovation strategy. Dr. Kulkarni has also authored some best-selling innovation strategy and AI books. He was conferred with a higher doctorate D.Sc. by UGSM monarch, Switzerland. He was also awarded IETE-KR Phadke Award for innovative entrepreneurship and research in 2019.

Dr. Parag has published over 300 research papers/articles in peer-reviewed journals and conferences. He invented over a dozen patents and authored 14 books (with the world's best publishers like Bloomsbury, IEEE, Wiley, Prentice Hall, Springer, OUP, etc.). His book *YD-YearDown* was adapted for a TV serial by Sony TV. His poem collection was appreciated by one of the greatest romantic poets of all time late Mangesh Padgaonkar. Parag's book *Knowledge Innovation Strategy* was listed as a game-changing business book by *Hindustan Times. Niigata Times Japan* mentioned that "it is enlightening experience for readers" – It has foreword and endorsement by Dr. FC Kohli and Ratan Tata. Parag was the first Ph.D. guide in Computer Engineering at COEP, Pune University, and has guided 20 Ph.D. scholars. As an AI consultant, he helped build game-changing products for companies like Envestnet, TechM, etc.

Parag founded startups iKnowlation, India, and Kvinna Ltd, New Zealand, and created social value through innovation. He is a prolific speaker and is associated with many technical and B-schools of repute like IITs, IIMs, Tokyo Int. Uni. Japan, and Masaryk Uni, Brno – Czech. He is a pioneer of concepts of systemic ML, reverse hypothesis ML, and choice computing. His specializations include innovation and business strategy, research in AI and machine learning, knowledge innovation, data and text analytics, and business analytics.

Soumya Ranjan Mahanta completed his M. Tech. in Computer Science and Engineering at Utkal University, Bhubaneswar. His main research interests include computer vision, artificial intelligence, and machine learning.

Maheswari Raja is Professor and Head of the Department, Computer Science and Engineering – Cyber Physical Systems, SCOPE, Centre for Smart Grid and Technologies, Vellore Institute of Technology, Chennai. She has professional experience of over 22 years in the industry and various prestigious institutions. She has published seven patents and 50+ research works in various books, book chapters, international magazines, and journals in her research domains of AI, ML, and IoT.

She has received many awards including Master Award 2023, Outstanding FOSSEE Contributor Award-IIT Bombay & MHRD, Govt. of India, Best Faculty Award, Best Achiever Award, Best Researcher Award, Best Alumni Chapter Award, Best Paper Award, Excellent Paper Award, and Best Club Coordinator Award. Dr. Maheswari has served as a resource person, panel member, chief guest, and guest of honour and has given plenary talks in various industries, international and national institutions as part of training, seminars, workshops, and conferences.

Seshu Bhavani Mallampati received her M.Tech. in Computer Science and Engineering from JNTUH in 2012 and B.Tech. in Computer Science and Engineering at Kakatiya University, Telangana. She obtained her Ph.D degree recently from VIT-AP University, Amaravathi, near Vijayawada, Andhra Pradesh. She had an academic experience of 7 years. Her research interests include machine learning, deep learning, intrusion detection, and data science.

Rajesh Mamilla is working as a Professor and Head of the Department-MBA with more than 17 years of experience in teaching in the field of Management. He specializes in Finance, especially teaching Corporate Finance, Security Analysis and Portfolio Management, Accounting and Financial Management, Basic and Advanced Financial Management and International Financial Management. He has published more than 20 papers in various reputed national and international journals, which are indexed in ABDC, Scopus, and UGC-listed journals. He has published 14 case studies in the area of Finance in the European Case Centre Head (ECCH-UK) and has presented more than 18 papers in various government-funded seminars and conferences like IIMs, GLOGIFT, AICTE, UGC, and DST. Dr. Mamilla has published a textbook in collaboration with German publishers and has attended in numerous workshops and refresher courses conducted by prestigious institutes like Central University of Hyderabad, IIT Bombay, etc. Under his supervision, three scholars completed their Ph.D. in the domain of Finance, and four more research scholars are in the process of completing their Ph.D.

Diptimaya Mishra, an experienced SAP consultant, is currently working as a freelance SAP HANA architect for multiple Levi's global ERP implementations. With two decades of corporate experience, he has been involved in software development, integration, and project management worldwide. Diptimaya has collaborated with renowned companies such as IBM, Capgemini, Mindtree, and others, serving major industry leaders including Nestle, PepsiCo, Philip Morris, Unilever, and Abbott Laboratories. He holds a B.Tech. in Computer Science and Engineering from Berhampur University, a PGDBA in Finance from Symbiosis, Pune, and a postgraduate degree in AI and Machine Learning from Purdue University, Indiana.

Tusar Kanti Mishra is presently working as an Associate Professor in the School of Computer Science and Engineering at Vellore Institute of Technology, Vellore, India. He has a total of more than 18 years of experience in teaching and research. His research interests include artificial intelligence, computer vision, and IoT. He earned his doctorate in the year 2015. He has been a reviewer for many referred journals. He has been the proud mentor for multiple winning teams in different national and state-level hackathons. Presently, he is working on the development of cutting-edge expert systems in the direction of healthcare computing.

Swathi Jamjala Narayanan obtained her Ph.D. from the Vellore Institute of Technology in 2015. She is currently designated as Professor Grade 1 in the School of Computer Science and Engineering, VIT Vellore, India. She has 17 years of teaching experience in Computer Science. Her research interests include soft computing, pattern recognition, machine learning, and data mining. She has been awarded with Best Ph.D. Thesis by Computer Society of India. She currently holds the position of faculty coordinator for the IEEE Computer Society student chapter in VIT. She is a member of the International Association of Engineers and also a lifetime member of Computer Society of India and the Soft Computing Research Society.

Subhadeep Nayak is currently associated with NEC Corporation in London, UK, where he works as an ORAN-Integration & Operations Support Associate. Prior to his current role, he has worked with Altran-Capgemini, UK, Vodafone IoT GDSP, UK, and VIAVI Solutions. Throughout his professional journey, Subhadeep has gained expertise in various areas including Testing in 5G, LTE support, VF IoT GDSP platforms, M2M and b2c portals, VMware vSphere/ESXi, and ORAN-Kubernetes. With 17 years of corporate experience, he holds a master's degree in Computer Application.

Srinivas Padmanabhuni was awarded the title of "Eminent Engineer" by the prestigious and renowned organization "Institution of Engineers" (IEI). The award was given to him in recognition of his experience and contributions to the profession of Computers. His vast experience in the computing field, including the marquee area of Artificial Intelligence (AI), brings a unique blend of academic, research, and industry orientation, a rare combination. He has been the past president of ACM India, the top professional organization of computing professionals. He specializes in areas of software engineering, AI, and machine learning. He is also a guest faculty of Computer Science at IIT Tirupati. He also co-founded a responsible AI start-up testAIng.com and is the CTO. Earlier, he was a Principal Research Scientist and Associate Vice President at Infosys where he overlooked research and innovation group in Infosys Labs. He has won several other awards including the Excellence Award for Innovation at Infosys, the Research Excellence Award at the University of Alberta, and the Merit Award at IIT Kanpur. He earned his Ph.D. in AI from the University of Alberta, Edmonton, Canada, specializing in AI. Prior to Ph.D., he secured his B.Tech. from IIT Kanpur and M.Tech. in Computer Science from IIT Bombay. He has to his credit seven granted patents, one published book by Wiley, 100+ refereed international publications, 400+ invited talks across eight countries, and 15,000+students lectured on AI-related topics, and he is associated with four start-ups.

Koustav Pal received his bachelor's degree in Computer Science from the University of Calcutta in 2021 and a master's degree in Computer Science from Pondicherry University in 2023.

G. K. Panda, currently serving as the Director, holds the position of Professor in the Department of Computer Science at MITS School of Biotechnology, Utkal University in Odisha, India. He completed his M.Tech. in Computer Science from Berhampur University, where he ranked first in his class. He pursued his M.Phil. from VM University in Tamil Nadu and obtained his Ph.D. from Berhampur University in India. He has made significant contributions to the field of research and has authored numerous papers published in national and international conferences and journals. His areas of expertise include medical healthcare analytics, hyperspectral image classifications, social network analysis, sentiment analysis, anonymization techniques, rough set theory, and applications. He actively participates in various professional organizations and affiliations.

Mrutyunjaya Panda holds a Ph.D. in Computer Science from Berhampur University. He obtained his master's degree in Communication System Engineering from the University College of Engineering, Burla, under Sambalpur University; MBA in HRM from IGNOU, New Delhi; and bachelor's degree in Electronics and Tele-Communication Engineering from Utkal University. He has 23 years of teaching and research experience. He is presently working as an Associate Professor and Head, in the P.G. Department of Computer Science and Applications, Utkal University, Vani Vihar, Bhubaneswar, Odisha, India. He is a member of KES (Australia), IAENG (Hong Kong), ACEEE (India), IETE (India), CSI (India), and ISTE (India). He has published about 100 papers in international and national journals and conferences. He has published 20 book chapters, and edited five books in Springer, Taylor Francis, IGI Global, River Publishers, etc., to name a few. He has also authored two textbooks on: (i) *Soft Computing Techniques* by Universal Science Press, New Delhi, and (ii) *Modern Approaches of Data Mining: Theory and Practice* by Narosa Publications, New Delhi. He has published one special issue in the *International Journal of Computational Intelligence Studies*, Inderscience, UK. He serves as an associate editor, editorial board member, and programme committee member of various international journals and conferences. He is an active reviewer of various international journals including *IEEE Access, Applied Soft Computing, Neural Computing and Applications, Wiley WIREs, IJCINI, IGI Global, IJKESDP, Inderscience, IJRSDA, IGI Global,* etc. His active area of research includes applications of data mining, granular computing, big data analytics, IoT, intrusion detection and prevention, social networking, wireless sensor networks, image processing, natural language processing, sign language recognition, software defect prediction, etc.

Prabhavathy Paneer is working as an Associate Professor in the School of Information Technology and Engineering, VIT, Vellore. Her research area includes computational intelligence, data mining, data science, machine learning, and deep learning. She has published around 20 journal papers in her research field. She is a life member of CSI and IEEE. She is also part of various school activity committees. She has published a number of papers in international conferences. She had completed a funded project from ISRO, SAC, Ahmadabad, Gujarat, on "Development of Automated Web based Online Feature Tracking system using Shape Adaptive Curvelet with PCA and Haralick Texture Feature".

L. M. Patnaik obtained his Ph.D. in 1978 in the area of Real-Time Systems, and D.Sc. in 1989 in the areas of Computer Systems and Architecture, both from the Indian Institute of Science, Bangalore. During March 2008 – August 2011, he was the Founding Vice Chancellor, Defence Institute of Advanced Technology, Deemed University, Pune. For more than four decades, he had an illustrious career as a faculty member with the IISc, Bangalore, where starting from the early 1980s, he had initiated several novel research areas in computer science and engineering. Currently, he is a NASI (National Academy of Sciences India) Senior Scientist and Adjunct Professor with the National Institute of Advanced Studies, Bangalore. He is a Life Fellow of the IEEE, and Fellow of The World Academy of Sciences (TWAS), Trieste, Italy, Indian National Science Academy, Indian Academy of Sciences, National Academy of Sciences, Indian National Academy of Engineering, and the Computer Society of India. His research interests span high-performance computing, soft computing, VLSI system design, machine learning, data analytics, social networks, and machine consciousness. In these areas, he has published over 1300 research papers, books/book chapters, and reports. One of his research papers on adaptive genetic algorithms has been cited over 4150 times by Google Scholar. He has significantly contributed to four premier organizations of the country during their formative stages: CDAC (Centre for Development of Advanced Computing), NBRC (National Brain Research Centre), IIIT Allahabad, and DIAT (Defence Institute of Advanced Technology). He had played a pivotal role in initiating key conferences such as HiPC, International Conference on VLSI Design, ADCOMM, and many more. As a recognition of his contributions in the areas of Electronics, Informatics, Telematics, and Automation, he was awarded the Dr. Vikram Sarabhai Research Award. He was awarded more than thirty prestigious awards and honours like the Platinum Jubilee Lecture Award (CS), Indian Science Congress, and the IEEE Computer Society's Technical Achievement Award for contributions to parallel, distributed, and soft computing and high-performance genetic algorithms, the VASVIK Award, and the Pandit Jawaharlal Nehru National Award for Engineering and Technology, to name just a few of them. It needs to be emphasized that his training as well as research work conducted by him are entirely indigenous and his contributions to the Indian higher technical education system are outstanding. His passion has been to improve the quality of research in engineering colleges by organizing international conferences, co-authoring papers with faculty members and students, mentoring students, delivering invited talks, and serving as an advisor on various committees of several engineering colleges across the country.

Pavithra L. K. is currently working as an Assistant Professor Senior at Vellore Institute of Technology in the School of Computer Science and Engineering at the Chennai campus. She finished her Ph.D. from Anna University, Chennai, in 2020.

She has published more than 20 papers in well-renowned international and national journals and conferences. Her areas of interest are image preprocessing, texture analysis, underwater image analysis, machine learning, and deep learning techniques.

Boominathan Perumal has obtained his Ph.D. from the Vellore Institute of Technology. He is currently designated as an Associate Professor Senior in the School of Computer Science and Engineering, VIT Vellore, India. He has 19 years of teaching experience in Computer Science. His research interest includes cloud computing, optimization techniques, and machine learning. Currently, he holds the position of faculty coordinator for the IEEE Computer Society student chapter in VIT. He is a lifetime member of IEEE and Computer Society of India.

Shivam Sakshi is an Assistant Professor of Marketing at VIT University, Vellore. Shivam holds a Ph.D. from the University of Debrecen and Post-Doctoral from Indian Institute of Management Bangalore. His current research interests are consumer behaviour, consumer decision-making styles, and the influence of AI on buying patterns. His research works are published in high-quality peer-reviewed journals listed in SCOPUS, ABDC, ICI, and other indices.

Sangeetha Saman received her B.E degree in Computer Science and Engineering from Anna University of Technology, Tiruchirappalli, India, in 2011 and M.E degree in Computer Science and Engineering from C. Abdul Hakeem College of Engineering & Technology, Melvisharam, Vellore, India, in 2016. She is currently pursuing a Ph.D. in the School of Computer Science and Engineering, VIT University, Vellore, India. Her current research interests include medical image processing, computer vision, and machine learning.

Manthan Sanghavi is an experienced individual with a strong computer science background. He has a master's degree in Computer Science from Arizona State University. Before that, he completed his bachelor's degree in Computer Science from Vellore Institute of Technology (VIT), India. Manthan has valuable industry experience from his machine learning and data science work. He works as a Data and Applied Scientist at Uplinq Inc., where he was instrumental as a founding data science and machine learning team member. At Xpress Technologies Inc. (a U.S. Xpress Company), Manthan was the Engineering Team Lead for the machine learning team and led several successful machine learning projects. He has significantly contributed his extensive experience and machine learning and data science expertise. His skills and knowledge have enabled him to develop innovative solutions that increased efficiency and profitability in the organizations he worked with.

Sameeksha Saraf is presently pursuing her Bachelor of Technology in Computer Science and Engineering at Vellore Institute of Technology, Vellore, India. She has been actively researching in the domain of artificial intelligence and IoT. She has been a consistent grade-point seeker. She has participated and won in many hackathons at the state level and organization level. She is an active scholar with dedicated team spirit working presently on developing an automated handwritten script evaluation mechanism.

Subbulakshmi P. is currently working as an Assistant Professor Senior at Vellore Institute of Technology in the School of Computer Science and Engineering at the Chennai campus. She finished her Ph.D. from Anna University, Chennai, in 2019. She has published more than 30 papers in well-renowned international and national journals and conferences. Her areas of interest are cognitive networks, wireless networks, game theory, and big data.

Nagalakshmi Vallabhaneni is working as an Assistant Professor at VIT University, Vellore. She is pursuing her Ph.D. at the School of Information Technology and Engineering, VIT University, Vellore. She received her M.Tech. in Computer Science and Technology with a specialization in Artificial Intelligence and Robotics from Andhra University in 2010. She received her B.Tech. in Computer Science Engineering from S.S.I.E.T., JNTU Hyderabad, in 2007. Her research interests include artificial intelligence, machine learning, and deep learning.

Venkatesan M. is an Assistant professor in the Department at the National Institute of Technology Puducherry, Karaikal, India. He has more than 20 years of teaching experience. He was an Assistant Professor at the National Institute of Technology Karnataka, Surathkal, from 2016 to 2021. He obtained Ph.D. in Computer Science and Engineering in 2014, Master of Technology in Information Technology in 2006, and Bachelor of Technology in Computer Science and Engineering in 1999. His areas of interest include data science and big data analytics. He was a visiting faculty at Monash University, Clayton, Melbourne, Australia.

Neelima Vobugari is a seasoned entrepreneur with extensive IT experience of over 20 years. She is a certified Data Scientist with a degree in Data Science from John Hopkins University, Maryland, USA. She has an MBA from Symbiosis and a B.Tech. from JNTU. Her entrepreneurial stint is focused on consulting and training in the areas of data science, artificial intelligence, and related technologies. She is the COO of an emerging start-up in the responsible AI space called testAIng.com. She recently received the "Karnataka Mahila Ratna" award for her contributions to society in areas of women's education and empowerment. She has recently been conferred the prestigious "Women in AI" award given to select women leaders in data science and AI. She is a strong innovator, has filed three patents in the areas of AI, and has experience providing innovative solutions to over 20 AI consulting projects at marquee clients. She has the distinction of having trained over 2000 students and professionals. She is an invited speaker at marquee conferences like GHCI, Data Science Congress, International Women Entrepreneur Conference, etc. She was invited by the Chief Minister of Karnataka for the pre-budget session of Karnataka as an expert representing women entrepreneurs of Karnataka.

1 Unveiling the Power of Explainable AI

Real-World Applications and Implications

Shrusti Ghela, Anuhya Bhagavatula, and B. K. Tripathy

1.1 INTRODUCTION

The field of machine learning has rapidly advanced over the past few years, allowing machines to learn complex patterns in data and make predictions or decisions with high accuracy. However, these models are often seen as black boxes, meaning it can be challenging to understand how the model arrived at its predictions [1]. This lack of transparency can be problematic, especially in domains such as healthcare or finance, where the consequences of incorrect predictions can be severe [2]. Therefore, there has been an increasing interest in developing interpretable machine learning (IML) techniques and explainable AI (XAI), which can provide insights into how machine learning models make decisions [3].

In this survey paper, we will discuss various applications of XAI and review recent advances in the field. We will begin by providing an overview of IML and XAI techniques and discussing their importance. We will then explore specific applications of XAI, including healthcare, finance, and social media, and highlight some of the challenges and opportunities associated with each domain. Finally, we will discuss current research trends and potential future directions in the field.

1.2 INTERPRETABLE MACHINE LEARNING AND EXPLAINABLE AI

IML refers to the ability of a machine learning model to provide human-understandable explanations of its predictions or decisions [3]. IML aims to make machine learning models more transparent and understandable so that humans can have confidence in their predictions and better understand the underlying data patterns [4]. XAI is a subset of IML that focuses on developing techniques for providing such explanations [4].

There are several reasons why IML and XAI are essential. First, in many domains, such as healthcare or finance, it is crucial to understand the reasoning behind a model's predictions [5]. For example, suppose a model predicts that a patient is at high risk of developing a particular disease. In that case, healthcare professionals need to know the

DOI: 10.1201/9781003442509-1

reasons behind the prediction so that they can take appropriate actions to prevent the disease. Similarly, in finance, it is essential to understand why a model recommends a particular investment strategy so that investors can make informed decisions. Second, IML and XAI can increase the transparency and accountability of machine learning models, making it easier to identify and correct biases or errors in the models [6]. Finally, IML and XAI can help to build trust between humans and machines, which is critical for the widespread adoption of machine learning in many domains [7].

1.3 IML AND XAI TECHNIQUES

Many techniques exist for developing interpretable machine learning models and providing human-understandable explanations of their predictions. This section will give an overview of some of the most common methods.

1.3.1 Rule-Based Models

Rule-based models are interpretable machine-learning models that are easy for humans to understand. Rule-based models consist of a set of logical rules applied to the input data to make predictions. Humans can easily understand and modify these rules, making rule-based models a popular choice in many domains, such as healthcare and finance [4,7].

1.3.2 Decision Trees

Decision trees are another popular technique for developing interpretable machine-learning models. Decision trees are easy to interpret and can be used for classification and regression tasks. Decision trees consist of a tree-like structure, where each internal node represents a decision based on one or more input features, and each leaf node represents a prediction [8].

1.3.3 LIME

LIME (Local Interpretable Model-Agnostic Explanations) is a technique for explaining the predictions of any classifier [9]. LIME works by generating a set of interpretable models that approximate the behavior of the black box model for a specific instance of input data. These interpretable models can then provide local explanations of the black box model's predictions for that particular instance of input data. LIME has been used in various applications, including image classification and natural language processing [9–11].

1.3.4 SHAP

SHAP (SHapley Additive exPlanations) is another technique for explaining the predictions of machine learning models. SHAP is based on game theory and assigns a score to each feature, indicating how much it contributes to the prediction [12]. SHAP has been used in various applications, including healthcare and finance [13,14].

1.3.5 Model-Agnostic Techniques

Model-agnostic techniques can be applied to any machine learning model, regardless of the specific algorithm used. These techniques include partial dependence plots, permutation feature importance, and global surrogate models. Model-agnostic techniques are often used when the particular details of the model's algorithm are unknown or when the model is complex and difficult to interpret [15–17].

1.3.6 Feature Importance Methods

Feature importance methods are XAI techniques that can identify the essential features in a machine learning model. These methods can help explain why a model is making a particular prediction [18–20].

1.4 APPLICATIONS OF XAI

In this section, we will explore some of the specific applications of XAI, including healthcare, finance, and social media.

1.4.1 Healthcare

Healthcare is an essential domain for XAI, as the consequences of incorrect predictions can be severe. XAI techniques can help healthcare professionals better understand the reasoning behind a model's predictions and make more informed decisions.

One example of the use of XAI in healthcare is in the development of predictive models for diseases such as cancer. XAI techniques can be used to explain the factors contributing to a patient's risk of developing the disease, which can help healthcare professionals develop targeted prevention and treatment strategies [21–25].

Another example is the use of XAI in medical imaging. XAI techniques can be used to explain the features used by the machine learning model to make its predictions. This can help healthcare professionals better understand underlying disease mechanisms and develop more accurate diagnoses [26,27].

1.4.2 Finance

Finance is another critical domain for XAI, as machine learning models are increasingly used to make investment decisions. XAI techniques provide transparency and accountability in these models, allowing investors to make more informed decisions [28].

One example of the use of XAI in finance is in the development of credit risk models. XAI techniques can be used to explain the factors contributing to a borrower's credit risk score, which can help lenders make more accurate lending decisions [29–32].

Another example is the use of XAI in algorithmic trading. XAI techniques can explain the reasons behind machine learning model's trading decisions, which can help traders identify and correct biases or errors in the model [33–35].

1.4.3 Social Media

Social media is another important application area for XAI, as machine learning models are increasingly used to make decisions about content moderation and recommendation systems. XAI techniques provide transparency and accountability, allowing users to better understand why certain content is promoted or suppressed.

One example of the use of XAI in social media is the development of content moderation systems. XAI techniques explain why specific posts are flagged as violating community guidelines, fostering trust between users and the platform [36–38].

Another example is the use of XAI in recommendation systems. XAI techniques explain why certain content is recommended to users, enhancing the relevance and accuracy of recommendations [39,40].

We will discuss applications in more detail in later sections.

1.5 CHALLENGES AND OPPORTUNITIES

While XAI has many potential benefits, several challenges need to be addressed. One of the main challenges is the trade-off between interpretability and accuracy. In some cases, more interpretable models may sacrifice accuracy to provide human-understandable explanations. Therefore, there is a need to develop techniques that can balance these two factors [41–43].

Another challenge is the lack of standardization in XAI techniques. Currently, there are no standardized approaches for evaluating the effectiveness of XAI techniques or for comparing different techniques. This lack of standardization can make it challenging to compare results across various studies and to determine which methods are most effective in different applications [42,43].

Additionally, there is the potential for bias in XAI models. While XAI techniques can help identify biases in machine learning models, they can also introduce their own biases if not used correctly. For example, if the data used to train the XAI model is biased, the model's explanations may also be biased [42,43].

Despite these challenges, there are also many opportunities for further development and application of XAI techniques. One opportunity is the development of more advanced XAI techniques that can provide more detailed and nuanced explanations of machine learning models. Another opportunity is the development of XAI techniques that can be used in real-time applications, such as self-driving cars or robotic systems [42,43].

1.6 HEALTHCARE

The healthcare industry holds immense promise for XAI techniques. Machine learning models can diagnose diseases, predict patient outcomes, and optimize treatment plans. However, these models are often perceived as black boxes, and their predictions are difficult to understand and explain. XAI techniques can help address this issue by providing transparent and interpretable explanations of these models [21–27].

XAI techniques have the potential to revolutionize healthcare by providing transparent and interpretable explanations of machine learning models. These models can predict patient outcomes, identify disease risk factors, and optimize treatment plans. The explanations can help build trust between doctors and machine learning models and enable more informed decision-making [21–27].

1.6.1 Applications of XAI in Healthcare

XAI techniques have diverse applications in healthcare. Some of the most important ones include:

1.6.1.1 Diagnosis of Diseases

XAI techniques can diagnose diseases by analyzing patient data, including medical records and imaging data. These techniques help identify patterns and risk factors that may be challenging for healthcare professionals to identify independently [44–46].

1.6.1.2 Prediction of Patient Outcomes

XAI techniques can predict patient outcomes, such as the likelihood of developing a particular disease or responding to a specific treatment. These predictions help guide treatment decisions and improve patient outcomes [47].

1.6.1.3 Optimization of Treatment Plans

XAI techniques can optimize treatment plans by analyzing patient data and identifying the most effective treatments for individual patients. These techniques contribute to reducing healthcare costs and improving patient outcomes [48,49].

1.6.1.4 Drug Discovery

XAI techniques are used in drug discovery by analyzing extensive datasets of chemical compounds and identifying potential drug candidates. These techniques accelerate the drug discovery process, reducing the time and cost required to bring new drugs to market [50,51].

1.6.1.5 Clinical Trial Design

XAI techniques contribute to designing more effective clinical trials by identifying patient subgroups most likely to benefit from specific treatments. These techniques reduce the cost and time required to conduct clinical trials while also improving the chances of success [52,53].

1.6.2 Challenges and Limitations

Despite the potential benefits of XAI in healthcare, several challenges and limitations must be addressed. Some of the most critical challenges include:

1.6.2.1 Data Quality and Availability

XAI techniques rely on large amounts of high-quality data to be effective. However, healthcare data is often fragmented and of variable quality, which can limit the effectiveness of these techniques [53].

1.6.2.2 Ethics and Bias

XAI techniques can be vulnerable to ethical issues and bias, particularly when making decisions that impact human lives. It is important to ensure that these techniques are used responsibly and ethically [53].

1.6.2.3 Regulatory Issues

XAI techniques in healthcare are subject to regulatory scrutiny, which can slow down their adoption. It is essential to collaborate with regulatory bodies to ensure that these techniques comply with relevant regulations and guidelines [53].

1.6.2.4 Interpretability vs. Accuracy

There is often a trade-off between interpretability and accuracy in machine learning models. XAI techniques can provide interpretable explanations of these models, but they may sacrifice some degree of accuracy [53].

1.6.3 Conclusion

XAI techniques have the potential to revolutionize healthcare by providing transparent and interpretable explanations of machine learning models. These techniques contribute to diagnosing diseases, predicting patient outcomes, and optimizing treatment plans. However, addressing several challenges and limitations is crucial, including data quality and availability, ethics and bias, regulatory issues, and the trade-off between interpretability and accuracy. Despite these challenges, the future of XAI in healthcare appears promising, anticipating more applications of these techniques in the years ahead.

1.7 FINANCE

The finance industry is a data-intensive field that heavily relies on data analytics and machine learning to make informed decisions. However, the complexity of financial data and models often makes it challenging to interpret and understand the outputs of machine learning algorithms. This is where explainable artificial intelligence (XAI) comes in. XAI techniques can provide transparent and interpretable explanations of machine learning models, assisting financial institutions in making better decisions and reducing the risk of errors [28–35].

1.7.1 Applications of XAI in Finance

1.7.1.1 Fraud Detection

XAI techniques can detect fraudulent transactions in real time by analyzing patterns in historical data. These techniques can identify unusual activity that may indicate fraudulent behavior and provide explanations for why a transaction was flagged as fraudulent [54,55].

1.7.1.2 Credit Risk Assessment

XAI techniques can be employed to assess the creditworthiness of borrowers by analyzing a wide range of data, including credit history, income, employment status, and other factors. These techniques can provide transparent and interpretable explanations of the elements used to make a credit decision [29–32].

1.7.1.3 Investment Recommendations

XAI techniques can provide personalized investment recommendations based on a client's financial goals, risk tolerance, and other factors. These techniques can explain why a particular investment was recommended, fostering trust with clients [28,56–58].

1.7.1.4 Trading Strategies

XAI techniques can be used to develop trading strategies based on historical data and market trends. These techniques can provide transparent and interpretable explanations for why a particular trading strategy was selected, aiding traders in understanding the underlying factors driving their decisions [33–35].

1.7.2 Challenges and Limitations

Despite the potential benefits of XAI in finance, several challenges and limitations must be addressed. Some of the most critical challenges include:

1.7.2.1 Data Quality and Availability

XAI techniques rely on large amounts of high-quality data to be effective. However, financial data is often fragmented and of variable quality, which can limit the effectiveness of these techniques [41–43].

1.7.2.2 Ethics and Bias

XAI techniques can be vulnerable to ethical issues and bias, particularly when making decisions that impact human lives. It is important to ensure that these techniques are used responsibly and ethically [41–43].

1.7.2.3 Regulatory Issues

XAI techniques in finance are subject to regulatory scrutiny, which can slow down their adoption. It is important to collaborate with regulatory bodies to ensure compliance with relevant regulations and guidelines [41–43].

1.7.2.4 Interpretability vs. Accuracy

There is often a trade-off between interpretability and accuracy in machine learning models. While XAI techniques can provide interpretable explanations of these models, they may sacrifice some degree of accuracy [41–43].

1.7.3 Conclusion

XAI techniques have the potential revolutionize the finance industry by providing transparent and interpretable explanations of machine learning models. These techniques can detect fraud, assess credit risk, provide investment recommendations, and develop trading strategies. However, several challenges and limitations must be addressed, including data quality and availability, ethics and bias, regulatory issues, and the trade-off between interpretability and accuracy. Despite these challenges, the future of XAI in finance looks bright, anticipating more applications of these techniques in the years to come.

1.8 SOCIAL MEDIA

Social media platforms generate vast amounts of data used for targeted ads, personalized recommendations, and content moderation. Machine learning algorithms analyze this data to extract insights. However, the opacity of these algorithms raises concerns about privacy, accountability, and transparency. XAI techniques offer insight into these algorithms work, enhancing transparency and accountability [36–40].

1.8.1 Applications of XAI in Social Media

1.8.1.1 Content Moderation

Social media platforms employ machine learning algorithms for content moderation, yet the lack of transparency in these algorithms can lead to inconsistent moderation decisions. XAI techniques offer a solution by enhancing transparency and providing explanations for why certain content was flagged or removed [36–38].

1.8.1.2 Personalized Recommendations

Social media platforms use machine learning algorithms to provide personalized recommendations to users. Using XAI techniques can enhance transparency of these recommendations by providing explanations for why specific recommendations were made [39,40].

1.8.1.3 Ad Targeting

Social media platforms leverage machine learning algorithms to target ads to specific users. Employing XAI techniques can enhance the transparency of these algorithms by explaining why certain ads target particular users [59].

1.8.1.4 User Behavior Analysis

Social media platforms use machine learning algorithms to analyze user behavior and identify patterns. Using XAI techniques can enhance the transparency of these algorithms by explaining how user data is analyzed and utilized [36–40].

1.8.2 Challenges and Limitations

Despite the potential benefits of XAI in social media, several challenges and limitations must be addressed. Some of the most critical challenges include:

1.8.2.1 Complexity of Algorithms

Social media platforms use complex machine-learning algorithms that can be challenging to explain fully. XAI techniques may face limitations in fully explaining these algorithms, which can restrict their effectiveness in certain contexts [36–40,60].

1.8.2.2 Privacy Concerns

XAI techniques have the potential to unveil sensitive information about users, which they may prefer to keep private. Hence, it is essential to develop XAI techniques that strike a balance between transparency and privacy [36–40,60].

1.8.2.3 Cost and Resources

XAI techniques require significant computational resources and expertise, which could pose a barrier to adoption for some social media platforms [36–40,60].

1.8.2.4 Adversarial Attacks

XAI techniques can be vulnerable to adversarial attacks, wherein attackers attempt to manipulate the output of a machine learning algorithm. It is essential to develop XAI techniques that are resilient to these attacks [36–40,60].

1.8.3 Conclusion

XAI techniques can provide valuable insights into how machine learning algorithms work in social media, making them more transparent and accountable. These techniques can be used to improve content moderation, personalize recommendations, target ads, and analyze user behavior. The future of XAI in social media looks promising, and we can expect to see many more applications of these techniques. However, several challenges and limitations need to be addressed, including the complexity of algorithms, privacy concerns, cost and resources, and vulnerability to adversarial attacks.

1.9 AUTOMOBILES

The automobile industry is rapidly evolving with technological advancements, including the integration of machine learning algorithms in various domains, such as autonomous driving, predictive maintenance, and fault detection. However, as these systems become increasingly complex, understanding the decision-making processes of these algorithms becomes challenging. This is where XAI comes into play, enabling transparency and interpretability of these algorithms [61–71].

1.9.1 Applications of XAI in the Automobile Industry

1.9.1.1 Autonomous Driving

Autonomous driving stands as one of the most popular applications of XAI in the automobile industry. XAI can offer insight into the decision-making process of the algorithms used in autonomous vehicles. It can explain why the vehicle undertook a particular action, such as slowing down or changing lanes, and rationalize those actions [61–65].

1.9.1.2 Predictive Maintenance

Another area where XAI finds utility in the automobile industry is predictive maintenance. Machine learning algorithms can forecast when a vehicle component is likely to fail. XAI can elucidate why the algorithm predicted a specific element to fail and provide actionable insights for maintenance personnel [66–68].

1.9.1.3 Fault Detection

XAI can also play a crucial role in fault detection systems in vehicles. These systems utilize machine learning algorithms to identify when a vehicle component is malfunctioning. XAI can explain why the algorithm detected a particular fault and provide suggestions for repair or replacement [69–71].

1.9.1.4 Customer Service

XAI can be applied in customer service within the automobile industry. Machine learning algorithms can predict customer preferences and provide personalized recommendations. XAI can be used to explain why a particular recommendation was made and provide transparency in the decision-making process.

1.9.2 Challenges and Limitations

Despite the potential benefits of XAI in the automobile industry, several challenges and limitations must be addressed. Some of the most critical challenges include:

1.9.2.1 Complexity of Algorithms

The algorithms used in the automobile industry are complex, and explaining the decision-making process can be challenging. XAI techniques may require assistance in fully explaining these algorithms, thereby limiting their usefulness in some contexts [61–71].

1.9.2.2 Data Availability

XAI techniques require large amounts of data to train the algorithms, and high-quality data is necessary. In some cases, obtaining data may be challenging, limiting the use of XAI techniques [61–71].

1.9.2.3 Cost and Resources

XAI techniques require significant computational resources and expertise, which may pose a barrier for some companies in the automobile industry [61–71].

1.9.2.4 Safety Concerns

XAI techniques must be developed with safety in mind, especially in applications such as autonomous driving. XAI can provide valuable insights into the decision-making process of these algorithms, but safety considerations must be prioritized [61–71].

1.9.3 Conclusion

XAI techniques can offer valuable insights into the decision-making processes of the algorithms used in the automobile industry, making them more transparent and interpretable. The future of XAI in the automobile industry appears promising, with an anticipation of witnessing numerous additional applications of these techniques in the years to come. XAI finds utility in autonomous driving, predictive maintenance, fault detection, and customer service applications. However, several challenges and limitations need to be addressed, including the complexity of algorithms, data availability, cost and resources, and safety concerns.

1.10 JUDICIAL SYSTEMS

The judicial system is complex, and decisions made by judges and juries can have significant consequences for individuals, society, and the justice system. Therefore, ensuring that these decisions are fair, unbiased, and based on accurate information is crucial. The judicial system has recently explored using XAI to improve decision-making processes, increase transparency, and reduce biases. In this section, we will discuss the applications of XAI in the judicial system, its benefits, challenges, and some use cases [72–81].

1.10.1 Applications of XAI in the Judicial System

1.10.1.1 Sentencing

One of the main applications of XAI in the judicial system is to assist judges in determining appropriate sentences. XAI models can analyze various factors, such as the nature of the crime, the defendant's criminal history, and other relevant information, to provide recommendations for a sentence. These models can reduce the risk of human biases and ensure consistency in sentencing across different cases [72–74].

1.10.1.2 Predictive Analytics

XAI models can also be used for predictive analytics to identify patterns in criminal activities and forecast potential crimes. For instance, law enforcement agencies can use XAI models to analyze crime data, identify high-risk areas, and allocate resources to prevent criminal activities. The models can also predict recidivism rates, which can assist judges in making informed decisions when granting parole [75,76].

1.10.1.3 Case Management

XAI models can assist in managing the judicial system's workload by providing insights into case management. These models can analyze various data, such as the number of cases in a particular court, the judge's workload, and the expected time for a case to be resolved. This information can be used to allocate resources more efficiently and optimize the judicial system's operations [77,78].

1.10.1.4 Legal Research

XAI can assist legal researchers in identifying relevant cases, statutes, and other legal documents. By analyzing large volumes of legal data, an XAI system can help identify meaningful connections and relationships between legal concepts, making it easier for researchers to find the information they need [78,79].

1.10.1.5 Document Analysis

XAI can analyze legal documents, including contracts, pleadings, and other court filings. By analyzing the language and content of these documents, an XAI system can help identify potential legal issues, such as ambiguities or inconsistencies that might otherwise be overlooked [79–81].

1.10.1.6 Compliance Monitoring

XAI can be used to monitor compliance with legal regulations and requirements. By analyzing data from various sources, including financial reports, environmental monitoring systems, and other relevant data sources, an XAI system can help identify potential violations and other compliance issues, allowing regulators to take appropriate action [79].

1.10.2 Benefits of XAI in the Judicial System

1.10.2.1 Improved Decision-making

XAI models can provide judges with more accurate and objective information, reducing the risk of human biases in decision-making. This can result in fairer and more consistent decisions across different cases [72–81].

1.10.2.2 Increased Transparency

XAI models can explain their recommendations, making the decision-making process more transparent. This can help defendants, lawyers, and the general public understand the reasoning behind a decision, increasing trust in the judicial system [72–81].

1.10.2.3 Better Resource Allocation

XAI models can assist in managing the judicial system's workload by providing insights into case management. This can help allocate resources more efficiently and optimize the judicial system's operations, resulting in cost savings and improved efficiency [72–81].

1.10.3 Challenges of XAI in the Judicial System

1.10.3.1 Data Bias

XAI models are only as good as the data they are trained on. If the training data is biased, the model will produce biased recommendations. Therefore, ensuring that the data used to train the models is diverse and representative is crucial [72–81].

1.10.3.2 Privacy Concerns

The judicial system deals with sensitive information, such as criminal records and personal details of defendants. Therefore, there are concerns about the privacy and security of this data when using XAI models [72–81].

1.10.3.3 Legal and Ethical Considerations

Legal and ethical considerations should be considered when using XAI models in the judicial system. For example, there may be concerns about the right to a fair trial when using XAI models to assist judges in making decisions [72–81].

1.10.4 Use Cases

One area of judicial decision-making where XAI has shown significant promise is in the field of criminal justice. In the United States, for example, there have been several cases where judges have relied on machine learning algorithms to predict the likelihood of recidivism or bail violation for defendants. These algorithms have come under fire for various reasons, including bias and lack of transparency. XAI tools can help alleviate some of these concerns by making it easier to understand how these algorithms make their predictions.

One notable example of XAI applied in the criminal justice system is the COMPAS (Correctional Offender Management Profiling for Alternative Sanctions) system. COMPAS is a proprietary software system developed by Northpointe, Inc. to predict recidivism rates for criminal defendants. The system uses machine learning algorithms to analyze various data, including criminal history, age, gender, and education level, to predict the likelihood of future criminal behavior [82].

The use of the COMPAS system has been controversial, with critics arguing that the algorithm is biased against African Americans and not transparent enough to be used in the criminal justice system. In 2016, a group of researchers from the Massachusetts Institute of Technology (MIT) released a study showing that the COMPAS algorithm was more likely to classify black defendants as higher risk than white defendants, even when controlling for factors such as age and prior criminal record [82].

In response to these concerns, researchers have been developing XAI tools to make the COMPAS system more transparent and easier to understand. For example, researchers at the University of California, Berkeley, have developed a "Recidivism Risk Explorer" tool that allows users to explore how the COMPAS algorithm makes its predictions. The tool provides visualizations of the data used by the algorithm and allows users to change various parameters to see how this affects the predicted recidivism rate.

Another example of XAI being applied in the judicial system is the use of machine learning algorithms to predict the outcome of legal cases. Several companies have developed software tools that use machine learning to analyze past legal cases and predict the likelihood of success for a given case. Lawyers can employ these tools to help decide whether to take on a case.

One company that is working on this is Legal Robot, a startup based in San Francisco. Legal Robot has developed a software tool that uses natural language processing and machine learning to analyze legal documents and predict the outcome of legal cases. The tool provides a probability score for the likelihood of success and a breakdown of the critical factors that influence the outcome.

Another example of XAI being applied in the judicial system is using chatbots to help defendants navigate the legal system. Chatbots are computer programs that use natural language processing to communicate with users conversationally. In the legal system context, chatbots can help defendants understand their legal rights, the charges against them, and the legal process in general.

One company working on this is DoNotPay, a startup based in London. DoNotPay has developed a chatbot to help defendants appeal parking tickets, dispute credit card charges, and even apply for asylum. The chatbot uses natural language processing and machine learning to understand the user's situation and provide appropriate advice.

1.10.5 Conclusion

In conclusion, XAI has the potential to revolutionize the judicial system by making it more transparent and easier to understand. XAI tools can assist judges, lawyers, and defendants in making more informed decisions and preventing bias and discrimination in the legal system. While there are still challenges to overcome, such as ensuring that XAI tools are accurate and reliable, the future looks promising for XAI in the judicial system.

1.11 CONCLUSION

XAI is a rapidly growing field that holds the potential to render machine learning models more transparent, understandable, and accountable. XAI techniques can aid in establishing trust between users and machine learning models, as well as in identifying and rectifying biases in these models. While numerous challenges still require addressing, there are also ample opportunities for further development and application of XAI techniques across a diverse array of domains.

REFERENCES

1. Adadi, A., & Berrada, M. (2018). Peeking inside the black-box: A survey on explainable artificial intelligence (XAI). *IEEE Access*, 6, 52138–52160.
2. Carvalho, D. V., Pereira, E. M., & Cardoso, J. S. (2019). Machine learning interpretability: A survey on methods and metrics. *Electronics*, 8(8), 832.
3. Gilpin, L. H., Bau, D., Yuan, B. Z., Bajwa, A., Specter, M., & Kagal, L. (2018, October). Explaining explanations: An overview of interpretability of machine learning. In *2018 IEEE 5th International Conference on Data Science and Advanced Analytics (DSAA)*, Turin, Italy, October 1–3, 2018 (pp. 80–89). IEEE.

4. Molnar, C., Casalicchio, G., & Bischl, B. (2021, February). Interpretable machine learning-a brief history, state-of-the-art and challenges. In *ECML PKDD 2020 Workshops: Workshops of the European Conference on Machine Learning and Knowledge Discovery in Databases (ECML PKDD 2020): SoGood 2020, PDFL 2020, MLCS 2020, NFMCP 2020, DINA 2020, EDML 2020, XKDD 2020 and INRA 2020,* Ghent, Belgium, September 14–18, 2020, Proceedings (pp. 417–431). Cham: Springer International Publishing.
5. Tjoa, E., & Guan, C. (2020). A survey on explainable artificial intelligence (xai): Toward medical xai. *IEEE Transactions on Neural Networks and Learning Systems*, 32(11), 4793–4813.
6. Shin, D. (2021). The effects of explainability and causability on perception, trust, and acceptance: Implications for explainable AI. *International Journal of Human-Computer Studies*, 146, 102551.
7. Kliegr, T., Bahník, Š., & Fürnkranz, J. (2021). A review of possible effects of cognitive biases on interpretation of rule-based machine learning models. *Artificial Intelligence*, 295, 103458.
8. Molnar, C. (2020). Interpretable machine learning. Lulu.com.
9. Mishra, S., Sturm, B. L., & Dixon, S. (2017, October). Local interpretable model-agnostic explanations for music content analysis. In 18th International Society for Music Information Retrieval Conference (ISMIR'17), Suzhou, China, October 23–27, 2017 (Vol. 53, pp. 537–543).
10. Palatnik de Sousa, I., Maria Bernardes Rebuzzi Vellasco, M., & Costa da Silva, E. (2019). Local interpretable model-agnostic explanations for classification of lymph node metastases. *Sensors*, 19(13), 2969.
11. Bramhall, S., Horn, H., Tieu, M., & Lohia, N. (2020). Qlime-a quadratic local interpretable model-agnostic explanation approach. *SMU Data Science Review*, 3(1), 4.
12. Mangalathu, S., Hwang, S. H., & Jeon, J. S. (2020). Failure mode and effects analysis of RC members based on machine-learning-based SHapley Additive exPlanations (SHAP) approach. *Engineering Structures*, 219, 110927.
13. Nohara, Y., Matsumoto, K., Soejima, H., & Nakashima, N. (2022). Explanation of machine learning models using shapley additive explanation and application for real data in hospital. *Computer Methods and Programs in Biomedicine*, 214, 106584.
14. Mokhtari, K. E., Higdon, B. P., & Başar, A. (2019, November). Interpreting financial time series with SHAP values. In *Proceedings of the 29th Annual International Conference on Computer Science and Software Engineering,* Toronto, Ontario, Canada, November 4–6, 2019 (pp. 166–172).
15. Jiarpakdee, J., Tantithamthavorn, C. K., Dam, H. K., & Grundy, J. (2020). An empirical study of model-agnostic techniques for defect prediction models. *IEEE Transactions on Software Engineering*, 48(1), 166–185.
16. Ribeiro, M. T., Singh, S., & Guestrin, C. (2016). Nothing else matters: Model-agnostic explanations by identifying prediction invariance. arXiv preprint arXiv:1611.05817.
17. Zhang, J., Wang, Y., Molino, P., Li, L., & Ebert, D. S. (2018). Manifold: A model-agnostic framework for interpretation and diagnosis of machine learning models. *IEEE Transactions on Visualization and Computer Graphics*, 25(1), 364–373.
18. Altmann, A., Toloşi, L., Sander, O., & Lengauer, T. (2010). Permutation importance: A corrected feature importance measure. *Bioinformatics*, 26(10), 1340–1347.
19. Schlegel, U., Arnout, H., El-Assady, M., Oelke, D., & Keim, D. A. (2019, October). Towards a rigorous evaluation of XAI methods on time series. In *2019 IEEE/CVF International Conference on Computer Vision Workshop (ICCVW),* Seoul, Korea (South), October 27–28, 2019 (pp. 4197–4201). IEEE.

20. Speith, T. (2022, June). A review of taxonomies of explainable artificial intelligence (XAI) methods. In *2022 ACM Conference on Fairness, Accountability, and Transparency*, Seoul, Republic of Korea, June 21–24, 2022 (pp. 2239–2250).
21. Nazar, M., Alam, M. M., Yafi, E., & Su'ud, M. M. (2021). A systematic review of human-computer interaction and explainable artificial intelligence in healthcare with artificial intelligence techniques. *IEEE Access*, 9, 153316–153348.
22. Gerlings, J., Jensen, M. S., & Shollo, A. (2022). Explainable AI, but explainable to whom? An exploratory case study of xAI in healthcare. In Lim, C.-P., Chen, Y.-W., Vaidya, A., Mahorkar, C., Jain, L.C. (Eds.) *Handbook of Artificial Intelligence in Healthcare: Vol 2: Practicalities and Prospects* (pp. 169–198). Cham: Springer International Publishing.
23. Adadi, A., & Berrada, M. (2020). Explainable AI for healthcare: From black box to interpretable models. In *Embedded Systems and Artificial Intelligence: Proceedings of ESAI 2019*, Fez, Morocco (pp. 327–337). Singapore: Springer Singapore.
24. Dauda, O. I., Awotunde, J. B., AbdulRaheem, M., & Salihu, S. A. (2022). Basic issues and challenges on explainable artificial intelligence (XAI) in healthcare systems. *Principles and Methods of Explainable Artificial Intelligence in Healthcare*, ISBN: 9781668437919 (pp. 248–271).
25. Srinivasu, P. N., Sandhya, N., Jhaveri, R. H., & Raut, R. (2022). From blackbox to explainable AI in healthcare: Existing tools and case studies. *Mobile Information Systems*, 2022, 1–20.
26. Van der Velden, B. H., Kuijf, H. J., Gilhuijs, K. G., & Viergever, M. A. (2022). Explainable artificial intelligence (XAI) in deep learning-based medical image analysis. *Medical Image Analysis*, 79, 102470.
27. Jin, W., Li, X., & Hamarneh, G. (2022, June). Evaluating explainable AI on a multi-modal medical imaging task: Can existing algorithms fulfill clinical requirements? In *Proceedings of the AAAI Conference on Artificial Intelligence*, February 20–27, 2024 in Vancouver, Canada (Vol. 36, No. 11, pp. 11945–11953).
28. Ohana, J. J., Ohana, S., Benhamou, E., Saltiel, D., & Guez, B. (2021). Explainable AI (XAI) models applied to the multi-agent environment of financial markets. In *Explainable and Transparent AI and Multi-Agent Systems: Third International Workshop, EXTRAAMAS 2021*, Virtual Event, May 3–7, 2021, Revised Selected Papers 3 (pp. 189–207). Springer International Publishing.
29. Heng, Y. S., & Subramanian, P. (2022, October). A systematic review of machine learning and explainable artificial intelligence (XAI) in credit risk modelling. In *Proceedings of the Future Technologies Conference (FTC) 2022* (Vol. 1, pp. 596–614). Cham: Springer International Publishing.
30. Gramegna, A., & Giudici, P. (2021). SHAP and LIME: An evaluation of discriminative power in credit risk. *Frontiers in Artificial Intelligence*, 4, 752558.
31. Misheva, B. H., Osterrieder, J., Hirsa, A., Kulkarni, O., & Lin, S. F. (2021). Explainable AI in credit risk management. arXiv preprint arXiv:2103.00949.
32. Bussmann, N., Giudici, P., Marinelli, D., & Papenbrock, J. (2021). Explainable machine learning in credit risk management. *Computational Economics*, 57, 203–216.
33. Bianchia, M., & Brièreb, M. (2023). Robo-advising: Less AI and More XAI? Augmenting algorithms with humans-in-the-loop. In *Machine Learning and Data Sciences for Financial Markets: A Guide to Contemporary Practices*. doi: 10.1017/9781009028943.004.
34. Karpagam, G. R., Varma, A., & Samrddhi, M. (2022, March). Understanding, visualizing and explaining XAI through case studies. In *2022 8th International Conference on Advanced Computing and Communication Systems (ICACCS)*, Coimbatore, India (Vol. 1, pp. 647–654). IEEE.

35. Deeks, A. (2019). The judicial demand for explainable artificial intelligence. *Columbia Law Review*, 119(7), 1829–1850.
36. Mehta, H., & Passi, K. (2022). Social media hate speech detection using Explainable Artificial Intelligence (XAI). *Algorithms*, 15(8), 291.
37. Roessner, V., Rothe, J., Kohls, G., Schomerus, G., Ehrlich, S., & Beste, C. (2021). Taming the chaos?! Using eXplainable Artificial Intelligence (XAI) to tackle the complexity in mental health research. *European Child & Adolescent Psychiatry*, 30, 1143–1146.
38. Hardage, D., & Najafirad, P. (2020, December). Hate and toxic speech detection in the context of covid-19 pandemic using xai: Ongoing applied research. In *Proceedings of the 1st Workshop on NLP for COVID-19 (Part 2) at EMNLP 2020.*
39. Samih, A., Adadi, A., & Berrada, M. (2019, October). Towards a knowledge based explainable recommender systems. In *Proceedings of the 4th International Conference on Big Data and Internet of Things (BDIoT'19)*, Tangier-Tetuan, Morocco, October 23–24, 2019 (pp. 1–5).
40. Vultureanu-Albişi, A., & Bădică, C. (2021, August). Recommender systems: An explainable AI perspective. In *2021 International Conference on INnovations in Intelligent SysTems and Applications (INISTA)* (pp. 1–6). IEEE.
41. Caruana, R., Lundberg, S., Ribeiro, M. T., Nori, H., & Jenkins, S. (2020, August). Intelligible and explainable machine learning: Best practices and practical challenges. In *Proceedings of the 26th ACM SIGKDD International Conference on Knowledge Discovery & Data Mining*, CA, USA, July 6–10, 2020 (pp. 3511–3512).
42. Das, A., & Rad, P. (2020). Opportunities and challenges in explainable artificial intelligence (XAI): A survey. arXiv preprint arXiv:2006.11371.
43. Antoniadi, A. M., Du, Y., Guendouz, Y., Wei, L., Mazo, C., Becker, B. A., & Mooney, C. (2021). Current challenges and future opportunities for XAI in machine learning-based clinical decision support systems: A systematic review. *Applied Sciences*, 11(11), 5088.
44. Muneer, S., & Rasool, M. A. (2022). AA systematic review: Explainable Artificial Intelligence (XAI) based disease prediction. *International Journal of Advanced Sciences and Computing*, 1(1), 1–6.
45. Kavya, R., Christopher, J., Panda, S., & Lazarus, Y. B. (2021). Machine learning and XAI approaches for allergy diagnosis. *Biomedical Signal Processing and Control*, 69, 102681.
46. Zhang, Y., Weng, Y., & Lund, J. (2022). Applications of explainable artificial intelligence in diagnosis and surgery. *Diagnostics*, 12(2), 237.
47. Isobe, T., & Okada, Y. (2020). Rehabilitation XAI to predict outcome with optimal therapies. In *Artificial Intelligence and Mobile Services-AIMS 2020: 9th International Conference, Held as Part of the Services Conference Federation, SCF 2020*, Honolulu, HI, USA, September 18–20, 2020, Proceedings 9 (pp. 127–139). Springer International Publishing.
48. Kamal, M. S., Dey, N., Chowdhury, L., Hasan, S. I., & Santosh, K. C. (2022). Explainable AI for glaucoma prediction analysis to understand risk factors in treatment planning. *IEEE Transactions on Instrumentation and Measurement*, 71, 1–9.
49. Vandewinckele, L., Claessens, M., Dinkla, A., Brouwer, C., Crijns, W., Verellen, D., & van Elmpt, W. (2020). Overview of artificial intelligence-based applications in radiotherapy: Recommendations for implementation and quality assurance. *Radiotherapy and Oncology*, 153, 55–66.
50. Jiménez-Luna, J., Grisoni, F., & Schneider, G. (2020). Drug discovery with explainable artificial intelligence. *Nature Machine Intelligence*, 2(10), 573–584.
51. Iswarya, B., & Manimekalai, K. (2022). Drug discovery with XAI using deep learning. In *Principles and Methods of Explainable Artificial Intelligence in Healthcare* (pp. 131–149). IGI Global.

52. Schoonderwoerd, T. A., Jorritsma, W., Neerincx, M. A., & Van Den Bosch, K. (2021). Human-centered XAI: Developing design patterns for explanations of clinical decision support systems. *International Journal of Human-Computer Studies*, 154, 102684.
53. Antoniadi, A. M., Du, Y., Guendouz, Y., Wei, L., Mazo, C., Becker, B. A., & Mooney, C. (2021). Current challenges and future opportunities for XAI in machine learning-based clinical decision support systems: A systematic review. *Applied Sciences*, 11(11), 5088.
54. Cirqueira, D., Nedbal, D., Helfert, M., & Bezbradica, M. (2020). Scenario-based requirements elicitation for user-centric explainable AI: A case in fraud detection. In *Machine Learning and Knowledge Extraction: 4th IFIP TC 5, TC 12, WG 8.4, WG 8.9, WG 12.9 International Cross-Domain Conference, CD-MAKE 2020*, Dublin, Ireland, August 25–28, 2020, Proceedings 4 (pp. 321–341). Springer International Publishing.
55. Roshan, K., & Zafar, A. (2021). Utilizing XAI technique to improve autoencoder based model for computer network anomaly detection with shapley additive explanation (SHAP). arXiv preprint arXiv:2112.08442.
56. Bianchi, M., & Briere, M. (2021). Robo-Advising: Less AI and More XAI? Available at SSRN 3825110.
57. van der Waa, J., Nieuwburg, E., Cremers, A., & Neerincx, M. (2021). Evaluating XAI: A comparison of rule-based and example-based explanations. *Artificial Intelligence*, 291, 103404.
58. Xu, F., Uszkoreit, H., Du, Y., Fan, W., Zhao, D., & Zhu, J. (2019). Explainable AI: A brief survey on history, research areas, approaches and challenges. In *Natural Language Processing and Chinese Computing: 8th CCF International Conference, NLPCC 2019*, Dunhuang, China, October 9–14, 2019, Proceedings, Part II 8 (pp. 563–574). Springer International Publishing.
59. Rai, A. (2020). Explainable AI: From black box to glass box. *Journal of the Academy of Marketing Science*, 48, 137–141.
60. Dwivedi, Y. K., Ismagilova, E., Hughes, D. L., Carlson, J., Filieri, R., Jacobson, J., ... & Wang, Y. (2021). Setting the future of digital and social media marketing research: Perspectives and research propositions. *International Journal of Information Management*, 59, 102168.
61. Atakishiyev, S., Salameh, M., Yao, H., & Goebel, R. (2021). Explainable artificial intelligence for autonomous driving: A comprehensive overview and field guide for future research directions. arXiv preprint arXiv:2112.11561.
62. Atakishiyev, S., Salameh, M., Yao, H., & Goebel, R. (2021). Towards safe, explainable, and regulated autonomous driving. arXiv preprint arXiv:2111.10518.
63. Hussain, F., Hussain, R., & Hossain, E. (2021). Explainable artificial intelligence (XAI): An engineering perspective. arXiv preprint arXiv:2101.03613.
64. Atakishiyev, S., Salameh, M., Yao, H., & Goebel, R. (2021). Towards safe, explainable, and regulated autonomous driving. arXiv preprint arXiv:2111.10518.
65. Rojat, T., Puget, R., Filliat, D., Del Ser, J., Gelin, R., & Díaz-Rodríguez, N. (2021). Explainable artificial intelligence (xai) on timeseries data: A survey. arXiv preprint arXiv:2104.00950.
66. Shukla, B., Fan, I. S., & Jennions, I. (2020, July). Opportunities for explainable artificial intelligence in aerospace predictive maintenance. In Fifth European Conference of the PHM Society (PHME20), Turin, Italy, 1–3 July, 2020 (Vol. 5, No. 1, pp. 11).
67. Theissler, A., Pérez-Velázquez, J., Kettelgerdes, M., & Elger, G. (2021). Predictive maintenance enabled by machine learning: Use cases and challenges in the automotive industry. *Reliability Engineering & System Safety*, 215, 107864.
68. Karlsson, N., & Bengtsson, M. (2022). Explainable AI For Predictive Maintenance, Independent thesis Advanced level (professional degree, Halmstad University, School of Information Technology, Halmstad, Sweden), Digitala Vetenskapliga Arkivet, DiVA, id: diva2:1746875.

69. Nwakanma, C. I., Ahakonye, L. A. C., Njoku, J. N., Odirichukwu, J. C., Okolie, S. A., Uzondu, C., … & Kim, D. S. (2023). Explainable Artificial Intelligence (XAI) for intrusion detection and mitigation in intelligent connected vehicles: A review. *Applied Sciences*, 13(3), 1252.
70. Das, D. B., & Birant, D. (2023). GASEL: Genetic algorithm-supported ensemble learning for fault detection in autonomous underwater vehicles. *Ocean Engineering*, 272, 113844.
71. Machlev, R., Heistrene, L., Perl, M., Levy, K. Y., Belikov, J., Mannor, S., & Levron, Y. (2022). Explainable Artificial Intelligence (XAI) techniques for energy and power systems: Review, challenges and opportunities. *Energy and AI*, 9, 100169.
72. Islam, M. R., Ahmed, M. U., Barua, S., & Begum, S. (2022). A systematic review of explainable artificial intelligence in terms of different application domains and tasks. *Applied Sciences*, 12(3), 1353.
73. Putera, N. S. F. M. S., Saripan, H., Bajury, M. S. A., & Ya'cob, S. N. (2022). Artificial intelligence-powered criminal sentencing in Malaysia: A conflict with the rule of law. *Environment-Behaviour Proceedings Journal*, 7(SI7), 441–448.
74. Oconitrillo, L. R. R., Vargas, J. J., Camacho, A., Burgos, A., & Corchado, J. M. (2021). RYEL system: A novel method for capturing and represent knowledge in a legal domain using Explainable Artificial Intelligence (XAI) and Granular Computing (GrC). In *Interpretable Artificial Intelligence: A Perspective of Granular Computing* (pp. 369–399). doi: 10.1007/978-3-030-64949-4_12.
75. Loyola-González, O. (2019). Understanding the criminal behavior in Mexico City through an explainable artificial intelligence model. In *Advances in Soft Computing: 18th Mexican International Conference on Artificial Intelligence, MICAI 2019*, Xalapa, Mexico, October 27–November 2, 2019, Proceedings 18 (pp. 136–149). Springer International Publishing.
76. Bhardwaj, A., & Kaushik, K. (2022). Investigate financial crime patterns using graph databases. *IT Professional*, 24(4), 27–36.
77. Loh, H. W., Ooi, C. P., Seoni, S., Barua, P. D., Molinari, F., & Acharya, U. R. (2022). Application of explainable artificial intelligence for healthcare: A systematic review of the last decade (2011–2022). *Computer Methods and Programs in Biomedicine*, 226, 107161.
78. Langer, M., Oster, D., Speith, T., Hermanns, H., Kästner, L., Schmidt, E., … & Baum, K. (2021). What do we want from Explainable Artificial Intelligence (XAI)?-A stakeholder perspective on XAI and a conceptual model guiding interdisciplinary XAI research. *Artificial Intelligence*, 296, 103473.
79. Górski, Ł., & Ramakrishna, S. (2021, June). Explainable artificial intelligence, lawyer's perspective. In *Proceedings of the Eighteenth International Conference on Artificial Intelligence and Law (ICAIL'21)*, São Paulo, Brazil, June 21–25, 2021 (pp. 60–68).
80. Landthaler, J., Glaser, I., & Matthes, F. (2018). Towards explainable semantic text matching. In *Legal Knowledge and Information Systems* (pp. 200–204). IOS Press. doi: 10.3233/978-1-61499-935-5-200.
81. Gawantka, F., Schulz, A., Lässig, J., & Just, F. (2022, December). SkillDB-An Evaluation on the stability of XAI algorithms for a HR decision support system and the legal context. In *2022 IEEE 21st International Conference on Cognitive Informatics & Cognitive Computing (ICCI* CC)* (pp. 183–190). IEEE.
82. Skeem, J., & Eno Louden, J. (2007). Assessment of evidence on the quality of the Correctional Offender Management Profiling for Alternative Sanctions (COMPAS). Unpublished report prepared for the California Department of Corrections and Rehabilitation. Available at: https://socialecology.uci.edu/faculty/skeem/.

2 Looking at Exploratory Paradigms of Explainability in Creative Computing

Parag Kulkarni and L. M. Patnaik

2.1 INTRODUCTION

Researchers are working to bring creativity and artificial intelligence (AI) onto the same platform to deliver value. In these efforts, subtle aspects of learning to differentiate or coming up with new solutions are at the center of attention. Creativity is defined as something with novelty, non-obviousness, and usefulness. Though this definition is widely accepted, it is still challenging to differentiate when it comes to determining whether machine actions are creative or not. Creativity is not continuous but rather contextual. Machines can learn from past creative samples, but producing similar artifacts may not necessarily be attributed to creativity. There could be multiple possibilities to produce new artifacts through exploration, combination, or expansion of existing artifacts. Machines can produce numerous such combinations, but the question remains unanswered: Can one of these artifacts qualify as a creative one? As it is a temporal concept, it depends on the moment when this artifact is produced. Sometimes it could be looked at from a hierarchical perspective or even some sort of incremental perspective. In all these scenarios, the previous creation is a reference for measurement along with historical creations. As Margaret Boden described in her book *The Creative Mind*, there is historical creativity and psychological creativity. Typically, psychological creativity has a reference of confined reference space, while historical creativity breaks the shackles to have a broader reference space like the complete country or the universe. The moment of creation also matters the most, while we could also think about the duration of creation. In all these scenarios, two important aspects are evaluation and impact.

2.2 EXPLAINABILITY AND INTELLIGENCE

Intelligence involves exhibiting behavior like humans, while creativity entails exhibiting behavior like humans where utility, non-obviousness, and novelty are at the center. While behavior improves based on a data bank, it cannot always contribute to creativity. Though on most occasions, the focus is on effectiveness and efficiency. In the case of typical AI systems, learning based on patterns works as a black

DOI: 10.1201/9781003442509-2

box – it could be an added advantage, but at times, it is crucial to have explainability. There could be two aspects to it – first, it can help towards the evolution of the systems, and second, making the system evolve with humans in a controlled manner. So, there is a need to have explainability attached to intelligent evolution – or some sort of why models are to be decoded to understand what is happening underneath. Accountability and explainability go hand in hand, especially when we are looking at creative activities; it is key to qualifying for a certain class.

In the case of deep learning and similar techniques of learning where transparency is complex and relationships are presented as black boxes, explainability concepts pose more challenges. It becomes astoundingly difficult as layers increase. Researchers discuss the importance of the explainability of machine learning models for different applications. It could be easier to visualize from an accountability and scrutiny perspective for financial and business decision-making applications. One approach could be the machine providing explanations through another model sitting on top of it. In this case, this model will learn for an explanation based on a huge data set. But who will provide the learning sets for explanation? Will it be humans? Before delving deep into this complex yet interesting set of questions, let us try to understand what explainability is in the case of AI and especially for creative AI.

Explainability is basically about logically putting forward a convincing explanation about decisions or actions that resulted from AI. In simple words, it could be an explanation about the working of a model and the underlying process of an AI system. When black box learning models attempt to learn directly from data without human intervention, this concept becomes of higher importance. Take for example, AI-generated Music [1], where tagging explainability could be extremely challenging. It is necessary and most relevant for human-machine co-creation [2]. In both scenarios, once the model is created, it is difficult to understand why the model suggests a particular path or creates certain combinations. My friend and head of AI at Tech Mahindra believes that all symptoms that are not explainable in healthcare are difficult to treat. In a very similar way, something that is not explainable in case of creative artifacts cannot be reproduced by another human-machine co-creation exercise. Furthermore, it cannot be used for subsequent enhancement of creativity for the next generation. In such a scenario, machine dominance in the outcome can spoil the possibility of uncovering and learning from certain facts. Co-creation of functionality through explainable models could be a way out, but it seems challenging with increasing complexity.

Another philosophy behind explainability is about explaining the process to technical and non-technical stakeholders. Transparency could increase reachability and contribute towards human participation in evolution. Human participation in evolution is extremely important for creativity. Basically, creativity here has three aspects: Making machines exhibit creative behavior like humans, enhancing human creativity with the aid of machines, and co-creating creative artifacts by humans and machines. Thus, it boils down to the development of additional supporting tools for explainability in all the three cases mentioned above. Thus, understanding the models can contribute to knowing how data is converted into intelligent actions. Explainability brings trust and when models become complex and cannot be decoded, there is always an outcome-based approach for evaluation, which in the long run raises many questions and could even prove to be dangerous. Take for example, an intelligent recruitment scenario where such models could not give insights into why a particular candidate is selected or why some other candidate is rejected. This may raise many questions regarding fair selection. Secondly, it will also become very difficult to scrutinize decisions. In long-term projects, it could be dangerous as the evaluation in a particular circumstance will be outcome-based, and the outcome is expected to come much later after the decision point. Explainability of AI models, thus, could be about not only understanding the process by non-technical members but also providing means for fair evaluation of the process while the decision is being taken.

Additionally, whichever model may be in focus, it may develop some sort of biases over time. These biases are overlooked in the absence of explainability. The ability to understand these biases can help in working on improved settings, leading to fair and better decision-making. This further contributes to trust building for models and the complete process.

2.3 TOOLS AND TECHNIQUES FOR EXPLAINABILITY

Researchers have worked on different aspects of explainability to develop different types of explainability techniques. The type of technique is specific to the model and the way the model is developed and used. One of the most important aspects considers the local dependencies and specific parts of the system. This could be named as local model explainability. In such an approach, the focus is on answering specific questions regarding answers or decisions provided by a specific model. Stakeholders or users of the models may be interested in knowing why a certain decision is made by the model. It could be for the selection of candidates or an audit of the process. Revisiting the recruitment and appraisal scenario – where a candidate may challenge a decision regarding promotion or layoff. Extending the same to the mortgage scenario, the customer may challenge the decision regarding the rejection of his/her loan application. The questions can lead to an audit of micro-level decisions regarding the amount sanctioned, interest rate decided, valuation of mortgage property, and so on. The local explainability is typically about a specific scenario that arises after the deployment of the model. Local explainability helps in understanding the scope of impact of decisions under focus. It helps in clarifying the business decisions for organizations and could be very crucial to winning the trust or rather confidence of stakeholders. In some specific scenarios leading to anomalies, local explainability tools could be used. It further helps in assessment, understanding of frequent issues, and undertaking ongoing improvements.

Second, cohort model explainability is more relevant during the production phase and typically applied to a subset of the data. This is useful in validating the model. It focuses on model generalization and its behavior in boundary conditions. This further is necessary to find out the behavior of the model in the case of unseen data. Cohort model explainability aids in answering questions regarding bias concerning a specific subset of the data. It helps perform a comparative analysis of behavior and accuracy for different data subsets.

On the other hand, global model explainability takes a holistic approach to focus on features that impact outcomes across the models. It helps answer questions regarding overall performance and decision-making of models. Typically, it addresses questions that are very general and are probably posed by stakeholders without proper data experience or knowledge of machine learning. Basically, it queries how the model functions and makes decisions. It can aid data scientists during the training phase to understand the model in greater detail. It can also help data scientists to identify the most relevant features impacting the performance and their relationships. Furthermore, it can help data scientists understand feature dependency issues and help deal with them gracefully.

Moving ahead with the discussion regarding explainability, one of the most evident hypotheses could be that explanation could add to social value and subsequent performances. This could be more evident in the case of creative artifacts. In the case of creative artifacts, explanations could be regarding the process of development and additionally, it could be regarding the intent of creation. These drivers are also relevant when we think regarding the evaluation of the creation. Percept and mental state of individuals are key factors in evaluation scoring.

Visualization is one of the key techniques for explainable AI. Visualizing data and its relationships, as well as changes in those relationships over time, gives insights into data dependencies and helps in explainability. In the case of medical applications, there could be predictions of medical conditions, leading to question about why a certain medical condition is predicted and what actions can be taken to avoid it? Logistic regression and decision tree models can help explain the AI outcomes and processes.

We have been working with an oil and gas company. The company wanted to detect the optimal time for maintenance based on the AI model. The purpose is to improve efficiency and safety and also save money by optimally charting out the maintenance schedule. Though the outcome of the whole process may be positive, it raises many questions. These questions are regarding adaptability and also regarding bringing this process into regular practice. Target users may need to know the underlying processes and the case studies support that this knowledge not only leads to an increase in adaptability but also improvement in performance and enhancements. Trust by target users is closely associated with explainability. Another aspect of explainability is participation. There are a few stakeholders, who want to participate in the overall process. In such a scenario, they may ignore small episodes of decisions that are logical and explainable. Even if there is a chain of such explainable events, stakeholders are fine with it. Thus, another way to resolve this problem is to present fully explainable episodes, even as part of black box. In short, explainability has three parts – reasoning, participation, and coherence. Some decisions may not seem very logical but could be explained. Interestingly, creative decisions or creative artifacts demand differentiation, novelty, while balancing it with utility. So before going for machine or AI explainability in terms of creative activities, let us try to look at the explainability of human creativity. Here the idea is to explain human actions or a set of actions. On the other hand, in the case of computational creativity, we expect two-way communication between systems and users.

In the case of computational creativity, co-creation is the key; there is a need for humans to understand the machine's perspectives and be able to decode them. On the other hand, it is also necessary for the machine to understand the human thought process to build on top of it. In co-creation, there is a need to understand creativity points where deviation from the routine thought process or path creates possibilities of creative outcomes. At any moment, one needs to understand the progression of these points.

2.3.1 Define Explainable AI

Explainable AI is a collection of tools or processes that help users to understand different stages of machine learning algorithms, the final outcomes, and convincing them of the way data is processed, thereby improving overall trust. This improvement contributes to trust between humans and machine learning products. Explainability expands horizons and creates opportunities for creativity. In the case of humans, the process of creativity is difficult to explain, and the same holds true for machines. There could be multiple approaches towards explainability in the case of AI.

a. Explainability by parts by dividing the process: In this case, the process of learning or the path from input to output is divided into multiple segments. These logical parts help us understand how the process unfolds. This approach could work well if the process could be divided into multiple parts.
b. Explainability by parts by dividing architecture: The first approach may not work in the case of black box-type processes where dividing the process is not possible. In this scenario, explainability is achieved through parts of the architecture. Take an example of a neural network where the functionality of each of the layers helps to explain the learning. The convolution layer, pooling layer, and flattening layers have their explanations for decoding the whole process of CNN.
c. Connecting evident parts: Some of the expected parts in the process relate to logical expressions.
d. Data and example-driven explainability: Here, the overall learning process is explained based on representative simple examples.

2.4 COMPUTATIONAL CREATIVITY

Computational creativity encompasses more than merely making machines creative [3]. There are three basic aspects of computational creativity:

a. Machine creating independently creative artifacts
b. Machines helping humans to enhance their creativity
c. Machine coworking with humans on creative assignments

These three aspects are interrelated. Interestingly, all these three aspects are important parts of computational creativity. When any of these three are satisfied, can we vouch to answer whether machines are creative? This definition is a bit inclined toward machines. It is accepted by some researchers that machines could pass basic creativity tests. But to be truly creative, all three needs to be met. The most important aspect of computational creativity is co-working with humans. The communication and interactions between the users and computationally creative systems are core to it. These interactions need to be simple, effortless, and knowledge-driven. It could be in the form of creative episodes where learning and creativity opportunities are created through interactions. Take the example of creative music systems or music improvisers like speak system [3,4] where we expect machines and humans to need continuous interactions. There is a need to have better interactions across the opportunities. It could be bidirectional and need-based communication with evaluation mechanisms at both ends. Such a new interaction mode could be the most important addition to existing systems.

Interaction, evaluation, and improvisation are three important aspects of computational creativity. If information about the operations and rationale behind it could be provided by machines, then human respondents will be equipped to converse and at times provide new inputs. These assimilated inputs are evaluated and incorporated during improvisation. The next interaction debate is not only on the quality and process but also on paths and new possibilities. Thus, during interactions, new possibilities are created and evaluated.

We are still on the hunt for definitions of creativity. Over the years, we have discussed that something new and useful is considered creative. The subjectivity in the definition of creativity makes evaluations and interactions even more challenging. Many times, what one person perceives as creative, others may not be ready to endorse. When two individuals debate any process, they may have different evaluation criteria. Not having standardization is necessary to nurture new possibilities leading to creativity. Co-creativity can further inspire individual creativity in episodes [5]. Visualizing hidden layers with examples [6,7] may help make this possible. In interactions, context is the key.

Creativity is not absolute and is driven by the context. Here, context refers to the situation, scenario, place, time, and participants. What was considered creative yesterday may not qualify as a creative attempt today. The explainability aspects of the creativity experience can be visualized as interaction design where interactions are designed to make the process evident. Figure 2.1 explains the CNN flow and steps.

2.5 PSYCHOLOGY, EXPLANATION, AND CREATIVITY

Causality refers to relationships between two events. Human tries to relate two events, and the occurrence of an event can be explained with causal relationships. In such tools, experiential analysis is at the center of explainability. Experiments are conducted on known data, and outcomes are mapped to behavior. The outcomes of experiments explain the learning process or system behavior. Experimental outputs could be mapped to different landmark decisions. The causation theory provides insights into system behavior. and also helps understand possible behavior in new data sets. In new scenarios, these insights may be revised with newer datasets and sometimes human inputs. Causal inference refers to behavioral estimation strategies and designs related to causal relationships. Causal inference and causal discovery are two parts of understanding causal behavior. While causal discovery is completely dependent on experiments and derives knowledge from observations,

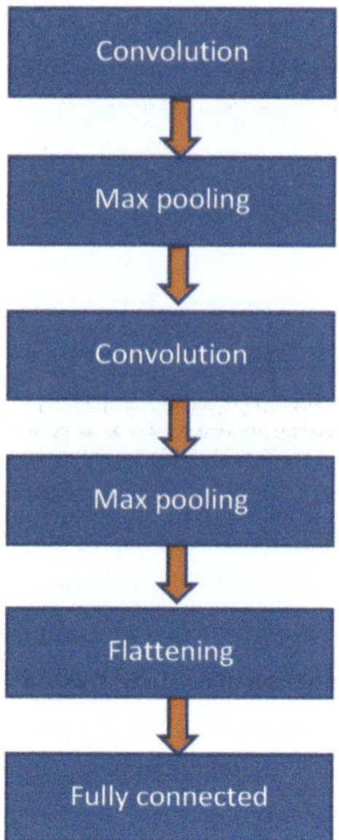

FIGURE 2.1 Convolutional neural network.

causal inference deals with variations in inputs or different parameters to understand behavior in different situations.

Causal discovery focuses on delving deep into data to analyze it and create models that derive different causal relationships based on experiments. On the other hand, causal inference is about understanding possible behavioral changes due to alterations and changes in data. Typically, causal models are mathematical representations of these behavioral changes for inference. Explainability is about fairness, accountability, and transparency [7], supported through probabilistic analysis and belief evidence. These mappings develop pointers that contribute to the explanation. Thus, implicit and explicit interactions need to be understood for interactive insights. Hence, design aspects related to interactions are equally important for co-creation. The human side of AI is the foundation for AI explainability in creativity. This requires capturing human reactions at every interval. Human inputs are not only necessary for learning but also important to understand how the outcome is perceived. This perception is crucial to determine whether the user is convinced and has faith in the system. Graphical interfaces, reasoning, and empirical evaluation help develop this required trust. Knowledge representation and contextual awareness are at the heart of meeting user expectations. Hence, interactive machine learning considers actions, intents, and perceived scenarios. The context is developed through a series of interactions, and creative paths are reinforced as the user and the system interact. Creativity evaluation could be based on voting or evolving evaluators. Such evaluators are not static and typically evolve with exposure to data. Though creativity is always defined with reference to humans, the idea of human creative actions being evaluated by machines and machine creative decisions being evaluated by humans as well as machines cannot be discarded. How about evaluation also taking place through interactions and cooperative learning between machines and humans to derive the creativity index. Figure 2.2 depicts the process of trust.

During interactions, information is shared and evolves. Generally, existing systems focus on the sharing of information. Sharing and analytical insights are a rudimentary part of intelligence. In the case of creativity, it is necessary to go beyond patterns to produce some surprisingly useful results. Creativity is generally associated with usefulness. However, it is debated whether all creative activities and outcomes are useful or if there is a need for a broader definition of usefulness. Or is there subjectivity associated with usefulness? There have been many attempts to define creativity, and let us take a deeper look at Creative AI systems. Creative artifacts are those that are new and interesting. The level of usefulness varies, and some outcomes may not meet conventional usefulness criteria.

Computer colleagues or companions are founded on real-time evolutionary learning and improving interactions along with generating creative contributions or actions. The creative contributions can be generated through different actions but evaluating and improvising them remains a bigger challenge. Additionally, explainability and creativity in such scenarios will be challenged. The viable alternative, or probably the most needed creativity path, is co-creation. Co-creativity has a self-evaluation mechanism as it encourages participants to improvise and evaluate creative outcomes based on the decisions of their peer humans or peer machines. The fusion of ideas can be a starting point to build new ideas. Thus, creative products emerge through cooperation and collaboration by interactions, evaluation, and negotiation by multiple participants ready to learn from the environment [8]. The creative agents are built on these principles where agents cooperate with other agents and look for possible pathways through fusion for creative ideas.

Mimicry, selective fusion, structured interactive improvisation, and shared mental models yield interesting results when considering creative co-creation. Synchronized musical improvisation through analysis is an example of this approach. In co-creation, agents use sensory inputs to get data, which is used to develop a learning

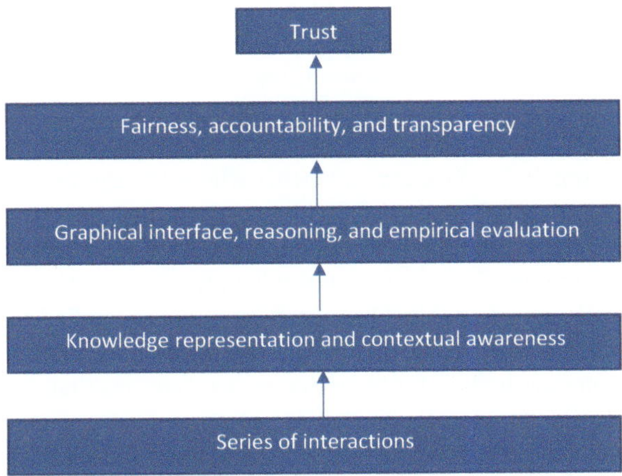

FIGURE 2.2 Process of trust.

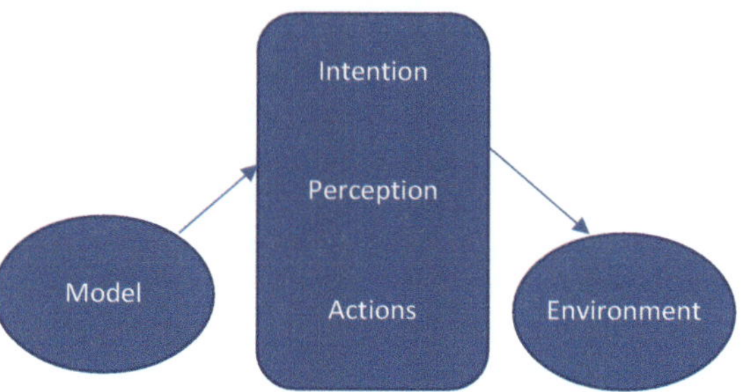

FIGURE 2.3 Learning and model building.

model. An agent has sensors and actuators. Sensors sense the environment to understand the world and the overall context at the time of selecting action. The interpretation of changes in data is used for behavioral association and interpretation for meaningful improvisation [9]. Figure 2.3 depicts learning and model building in an intelligent agent.

Continuous updating and reframing of sensemaking are necessary for critical and participative interactions, where there is a need for negotiating with changes in the environment. While negotiating with the environment, new possibilities may be explored. Enactive cognition is about how living organisms interact with the environment [10]. The creative domain often begins with single or multiple vague ideas, then under some guidelines for evaluation, re-evaluation leads to further exploration of creative possibilities. The creative exploration leading to ideas proves merit in some way, and the system is rewarded, and the exploration continues. Thus, adjustment and fusion of actions lead to new possibilities that are further evaluated for creativity.

2.6 EXPLORATION AND EXPLAINABILITY IN CREATIVE COMPUTING

In previous sections, we stressed on the importance of explainability and established its need with reference to different scenarios. When considering creative computing and explainability together – the obvious option is exploration. It involves delving into different possibilities, accumulating rewards and penalties along the way. While the system aims to maximize rewards, it determines the direction for exploration. The idea goes beyond learning direction to explainability where rewards are associated with the value function that is closely associated with explainability. Episodic reward calculations could come in handy in this scenario, understanding learning regarding different episodes and their relationships. When multiple options are explored, certain options are creative as well as explainable. One paradigm involves embedding explainability within the learning process itself, differing from developing tools for explainability for existing learning methods. This approach seeks solutions that are explainable and optimal, necessitating the development of a new exploratory paradigm for explainability in creative computing. While an intelligent agent senses the environment and builds the context – any recommended action changes the state and a new state may lead to certain rewards or penalties. These rewards and penalties help in deciding on the next action. In the domain of explainability, a rational action by an intelligent agent is obvious, logical action that maximizes rewards, and is also explainable [11]. One of the available options is chosen to maximize rewards [12]. Such a set of actions finally contributes to learning strategy. Finally, explanation is systemic, and hence relationships among different episodes and their systemic impact are the key to understanding and evaluation [13]. Learning based on patterns may not be advisable in this case, whereas learning based on outcomes can provide certain insights, referred to as reverse hypothesis machine learning [14].

2.7 CONCLUSION

Achieving explainability for AI and fostering explainability for creativity are both challenging endeavors. Accomplishing this task may not be possible with existing paradigms. Probably, there is a need to learn in an explainable way to achieve explainability in the case of creative computing. The episodic and exploratory paradigms of learning could come in handy if we intend to achieve this objective. The approach of explaining what is learned after learning is over probably brings many obstacles in the way. To avoid this, explainability needs to be made part of the learning process or even while developing AI systems. Whether it is the black box model or any other model, practicing explainability and introducing it as a part of learning could help in developing new creative AI models that would be explainable. This approach will not only contribute to increasing faith among users but also give new pathways for co-creation and can help to make it possible to use AI as a creative companion. The creative agents are expected to be exploratory and learn in association with the environment.

2.8 FUTURE SCOPE

There is a need to define and analyze evaluation functions from an explainability perspective. The fitness value is derived from the environment and represents the fitness at a particular time instant in the given context. In the future, temporal significance needs to be considered for explainability. The work on contextual explainability could find answers for these numerous creative possibilities. It could be the timing of a humorous quote or even, at times, an innovation when exploring the scientific domain. Many politically creative caricatures become simply insignificant with changes in political scenarios. The same holds true for many textual artifacts. There are many fleeting creative experiences along with some sustained creativity outcomes. These two scenarios need to be addressed by knowledge-driven approaches. Mapping this concept to a connectionist model is indeed an interesting idea that can be explored for the selective firing of multiple neurons in light of an event. Thus, as a future direction, explainability in the creative space could be looked at incrementally, with every small step explained, leading to creative actions with reference to the environment. Anyway, explainability in the case of creativity is always challenging, but evolutionary learning with temporal tracking could lead to some interesting future directions.

REFERENCES

1. Duncan, C., Imasato, N., & Nagai, T. (2022). Is Explainability a Prerequisite for Creativity?. In *人工知能学会全国大会論文集 第 36 回 (2022)* (pp. 2S5IS2c01–2S5IS2c01). 一般社団法人 人工知能学会.
2. Zhu, Jichen, Antonios Liapis, Sebastian Risi, Rafael Bidarra, and G. Michael Youngblood. "Explainable AI for designers: A human-centered perspective on mixed-initiative co-creation." In *2018 IEEE Conference on Computational Intelligence and Games (CIG)*, Maastricht, Netherlands, pp. 1–8, IEEE, 2018.
3. Llano, Maria Teresa, Mark d'Inverno, Matthew Yee-King, Jon McCormack, Alon Ilsar, Alison Pease, and Simon Colton. "Explainable computational creativity." arXiv preprint arXiv:2205.05682, 2022.
4. Yee-King, Matthew, and Mark d'Inverno. "Experience driven design of creative systems." In *Proceedings of the 7th Computational Creativity Conference (ICCC 2016)*, Paris, France, pp. 85–92, 2016.
5. Davis, Nicholas, Chih-Pin Hsiao, Yanna Popova, and Brian Magerko. "An enactive model of creativity for computational collaboration and co-creation." In *Creativity in the Digital Age*, pp. 109–133, 2015. Doi: 10.1007/978-1-4471-6681-8_7.
6. Zhu, Jichen, Antonios Liapis, Sebastian Risi, Rafael Bidarra, and G. Michael Youngblood. "Explainable AI for designers: A human-centered perspective on mixed-initiative co-creation." In *2018 IEEE Conference on Computational Intelligence and Games (CIG)*, Maastricht, Netherlands, pp. 1–8, IEEE, 2018.
7. Abdul, Ashraf, Jo Vermeulen, Danding Wang, Brian Y. Lim, and Mohan Kankanhalli. "Trends and trajectories for explainable, accountable and intelligible systems: An HCI research agenda." In *Proceedings of the 2018 CHI Conference on Human Factors in Computing Systems*, Montreal, Canada, pp. 1–18. 2018.
8. Mamykina, Lena, Linda Candy, and Ernest Edmonds. "Collaborative creativity." *Communications of the ACM* 45(10): 96–99, 2002.
9. Hoffman, Guy, and Gil Weinberg. "Shimon: An interactive improvisational robotic marimba player." In *CHI'10 Extended Abstracts on Human Factors in Computing Systems*, Atlanta, GA, USA, pp. 3097–3102. 2010.
10. Varela, F., Evan Thompson, and Eleanor Rosch. *The Embodied Mind: Cognitive Science and Human Experience*. MIT Press: Cambridge, MA, 1991.
11. Kulkarni, Parag, and Prachi Joshi. *Artificial Intelligence: Building Intelligent Systems*. PHI Learning Pvt. Ltd.: New Delhi, 2015.
12. Kulkarni, Parag. "Choice modelling: Where choosing meets computing." In *Choice Computing: Machine Learning and Systemic Economics for Choosing*, pp. 15–55. Springer Nature Singapore: Singapore, 2022. doi: 10.1007/978-981-19-4059-0_2.
13. Kulkarni, Parag. *Reinforcement and Systemic Machine Learning for Decision Making*, Vol. 1. John Wiley & Sons: Hoboken, NJ, 2012.
14. Kulkarni, Parag, and Parag Kulkarni. "Co-operative and collective learning for creative ML." In *Reverse Hypothesis Machine Learning: A Practitioner's Perspective*, pp. 119–124, 2017. doi: 10.1007/978-3-319-55312-2_6.

3 Applications of XAI in Modern Automotive, Financial, and Manufacturing Sectors

Kaushik K., Pavithra L. K., and Subbulakshmi P.

3.1 INTRODUCTION

Artificial Intelligence (AI) is the machine simulation of human intelligence with the capability to perform various tasks. It has brought rapid changes in modern world applications and has improved work efficiency. The purpose of AI is to provide appropriate solutions with the help of various Machine Learning (ML) and Deep Learning (DL) algorithms for a given set of problems. However, the main issue revolving around AI is trustworthiness and the tremendous growth of AI in recent years has put forth the need for an explanation of the proposed solution. Hence, Explainable Artificial Intelligence (XAI) comes into the picture.

XAI refers to the explanatory part of the solution provided by the machine to humans in an effective manner [1]. The transparency of the solution and the ability to answer "Why", "What", and similar types of questions related to the recommended solution can be achieved through XAI.

There is a huge need for XAI in modern days, and it will remain an important part of AI/ML/DL models. XAI will effectively build trust among end-users upon the predictions provided by the system, and this helps in increased productivity. Subsequently, this will boost the business revenue and help in the growth of the company. Also, XAI will play a crucial role in minimizing risks and ensuring effective decision-making for the companies from the predictions [2]. Figure 3.1 shows the flow diagram of XAI.

XAI majorly focuses on explaining the data, the algorithms, and the predictions of the model. It concentrates on providing a positive impact on the solution and makes the end-users understand the reasons for the appropriate recommendations. XAI has a wide range of features and its application boundary can be extended to various sectors. XAI lies between the boundary lines of AI and ML, which will soon become an integral part of every sector.

XAI is broadly classified into three categories based on the level of explainability: Neural Network Post Hoc Explanation, Model-Agnostic Post Hoc Explanation, and Interpretable ML [3]. Neural Network Post Hoc technique helps in interpreting the behavior of DL models with other neural networks. Model-Agnostic Post

DOI: 10.1201/9781003442509-3

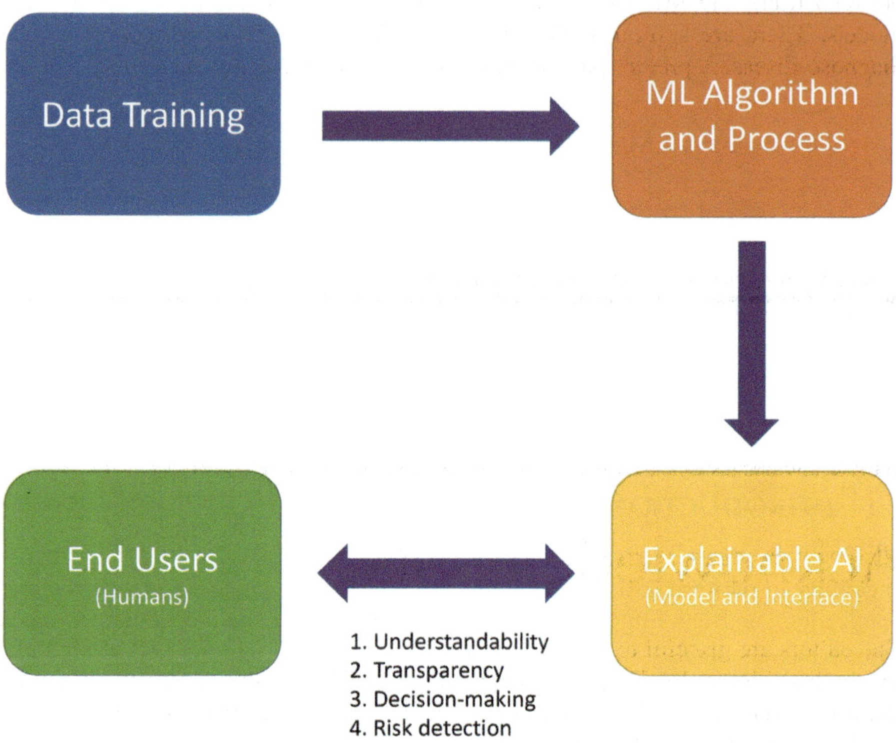

FIGURE 3.1 Basic flow diagram of Explainable Artificial Intelligence.

Hoc Explanation technique quantifies model characteristics and defines the problem. Interpretable ML focuses on improving the performance of ML models by interpreting the results provided. Some of the XAI applications are shown in Figure 3.2.

The current automotive industry is moving toward autonomous vehicles where AI is widely used for assisting the vehicle. Companies like Tesla have achieved autopilot mode, enabling the vehicle to automate on its own. However, there have been a lot of cases reported of vehicle crashes. XAI will play a crucial role in this and will try to solve the problem of vehicle accidents [4] by listing down the reasons and the situations that made previous accidents happen. It will enable the AI system learn and act accordingly in such future situations. Thus, the possibility of vehicle accidents in the future will be reduced, enhancing safe driving. XAI can be implemented in the banking and finance sector to elucidate the outcomes of the predictions made on the financial data [5]. It can be used to state the reasons for a loan being approved/denied/waived. It can also be employed for identifying fraudulent transactions as it explains why the transaction is marked as fraudulent [6]. Algorithmic trading is another financial application of XAI where the trading is done with pre-defined instructions and each trading instruction is explained properly. There are wide use cases for XAI in the insurance sector, including customer service, adjustment of insurance policies, increasing productivity, and detection of risks [7]. XAI plays a vital role in the manufacturing and design sector. Supply chain management can be effectively handled

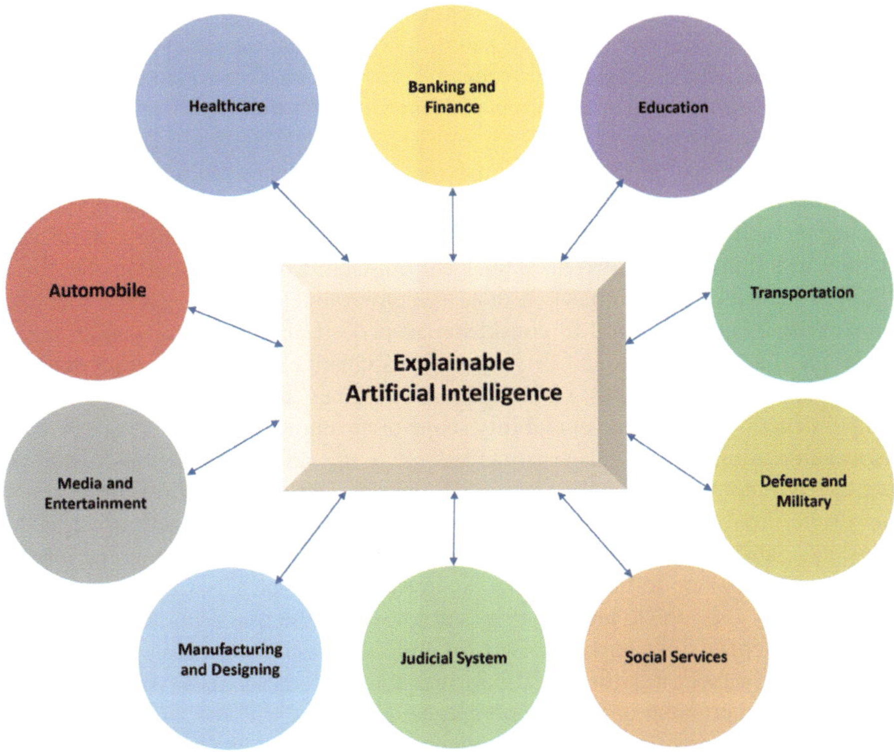

FIGURE 3.2 Applications of Explainable Artificial Intelligence.

with the use of XAI, which interprets the decisions taken by the system for every state. There is a huge scope for XAI in varied applications, and this chapter focuses on the applications of XAI in automotive, finance, and manufacturing sectors.

3.2 XAI IN AUTOMOTIVE SECTOR

3.2.1 Common Problems Associated with Present Automotive Sector

For the past several years, there has been an increase in technology-oriented advancements in terms of design, driving, user experience, and facilities in the automotive industry. Many improvements have been made, and AI-enabled machines have started replacing humans in vehicle simulation, which is so-called "Autonomous driving system". The main purpose of these driving systems is to sense the environment around them and operate the vehicle on its own without any involvement of human activity.

The emergence of these vehicles in the market initially gained many positive reviews from the public as it minimizes the human workload of driving vehicles. However, there is a huge cause of concern when it comes to the reliability of such autonomous vehicles. In the current automotive industry, almost all autonomous vehicles depend on black-box deep learning models for their functioning [8]. The major problem associated with such neural models is that they do not state the reasons for the activities they have performed. This creates a huge gap in the interaction between the driving system and humans. Hence, it paves the way for trustworthiness issues with the system.

The ultimate goal of autonomous driving systems is the safe transportability of mortals from one destination to another without any vehicle crashes. The operability of such a machine in a closed environment does not determine the glory of the technology, as it must be operated in an open environment which is a safety-critical zone [8]. The identification of objects around the environment is crucial for an autonomous vehicle, and it is vital to consider road pedestrians while the vehicle travels along the road path. Also, self-driving vehicles must be in the state of recognizing traffic signals and road signs; violating these rules may lead to accidents and significant destruction. Hence, the operability scope of an autonomous vehicle is extensive. Proper functioning of various sensors installed in the vehicle is essential, and building appropriate models for the system is important for the overall performance of the driving systems. The system's failure to achieve this can cause unimaginable accidents, leading to vehicle crashes, death, and destruction. Also, it is the responsibility of the system to inform all the actions it has performed. Certain operations performed by the vehicle are sometimes unknown to the users. For example, if the sensors recognize spike strips on the road and the vehicle stops, then it must notify the passengers about it. However, the current autonomous vehicles fail to do so, and this has been a problem in the automotive sector.

Though autonomous vehicles have been advantageous machines for humans, the system has also many malfunctioning problems associated with its operations. Global Positioning System (GPS)-based navigation models that are used in autonomous vehicles have faced routing issues and route planning, which is problematic for the driving systems [9]. Several cases are reported regarding the selection of the wrong path to reach the destination. Therefore, the pathway selection from the available choice list determines the significance of the autonomous driving system, and this is a major issue with the current existing models.

When the entire transport network is automated by autonomous systems without any involvement of human drivers, then the inter-sharing of vehicle dynamics information among them can help manage various tasks effectively, known as "vehicle platooning". Platooning technology optimizes modern transportation and can drastically reduce traffic jams on roads. This technology efficiently shares information and is most widely used in trucks. Since the technology operates on the sharing of information through packets, there are high chances of cyber-attacks. Packet falsification poses a risk and can lead to vehicle crashes [10]. Packet falsification can result in speed acceleration/deceleration, application of brakes in wrong circumstances, incorrect use of side indicators, incorrect choice of headlight selection, and honking. This, in turn, may cause road accidents. Hence, cybersecurity threats are a cause of concern in platooning.

In recent times, various cases have been reported regarding the loss of control over the steering wheel in autonomous vehicles. In such cases, it is very dangerous for individuals to travel inside the vehicle. Hence, there is a big necessity to manage the controllability of the steering wheel in autonomous vehicles.

Analyzing the various motions and predicting the behavior of other objects is crucial for a self-driving system while it functions. Failing to identify and analyze can cause major accidents, and it is necessary for a machine to have an eye on all actions and behaviors of the objects in and around the environment.

Several case studies have revealed that many passengers prefer human drivers over AI simulation for driving the vehicle [11]. This common mentality among the public has been an added cause of concern. It has reduced the reachability of the new technology to millions of people, making them feel less trustworthy upon the use of self-driving systems.

Henceforth, to overcome and address major problems in the automotive sector, it is necessary to adopt XAI in the system. Inclusion and effective implementation of XAI help solve various complications associated with the current automotive sector and can increase the performance of autonomous vehicles. It can bring drastic changes to present transportation models and can build well-innovated autonomous vehicles.

3.2.2 Role of XAI to Tackle the Problems Associated with Current Automotive Sector

XAI helps in identifying pertinent challenges and supports in providing accurate recommendations for various problems. The operations involved in autonomous vehicles include perception, localization, planning, control, and system management [12]. These operations can be improved when involving XAI in the autonomous system. XAI in autonomous vehicles promises to provide trust, comprehensibility, interpretability, transparency, understandability, and explainability [3,13].

The performance levels of ML/DL/CNN models implemented in autonomous vehicles can be refined with the use of XAI. Though XAI does not directly contribute to the increase in the accuracy levels of these models, it can guide humans to better understand the current working models and can help build better models or add additional features to existing works. XAI helps in identifying areas of improvement in the system and thus, contributes in increasing the performance level of autonomous vehicles indirectly. Model induction of XAI can help state the explanatory part of the solution provided by the black box models [9]. Implementing it on a large-scale aspect can level up the transparency of the system and boost trust between the autonomous vehicle and people. Additionally, XAI can be effective even with minimal available datasets. It helps improve debugging, feature selection, and model optimization. Combining ML/DL/CNN with XAI helps interpret the decisions of the models and improves the understandability of autonomous vehicles. Hence, integrating XAI with the existing models in self-driving systems can promise better transportability in modern society.

At present, the acceptance rate of self-driving systems is very low, and there is a necessity to increase these levels to effectively implement the technology in modern transportation systems. Concentrating on the user experience (UX) of the system can help achieve the target. A recent experiment has shown that there is an increase in the acceptance level after the travel experience of people in an XAI-enabled autonomous driving simulator set-up compared to the AI-enabled driving system [14]. Interviews and two major questionnaires were conducted with the people after their experience with the system. One questionnaire was the Autonomous Vehicle Acceptance Model (AVAM), which addresses the engagement level of users and the second questionnaire was the User Experience Questionnaire-Short (UEQ-S), which measures the UX. Both questionnaires have revealed that travel experience in an XAI-based autonomous vehicle helps turn a negative user experience of the ride into neutral because of an increased feeling of control over the machine. Thus, user experience will increase if the current autonomous systems are equipped with XAI models.

Explanations for every activity performed by the vehicle can guide the users, which in turn helps build a better user experience with the system. The implementation of XAI to provide text and voice assistance for the users can further boost the level of user experience with the system and enhance the quality metrics of autonomous vehicles.

XAI supports in creating a special tool with visualization to explain the decisions taken by AI. This helps the user to understand the reason behind every action of AI and can induce trust among the users regarding the further processes of the machine. This ensures transparency, explainability, and understandability of the autonomous system.

XAI is helpful in promoting efficient routing algorithms and transportation scheduling algorithms [9]. It can assist users by suggesting suitable routes to reach the destination, listing various ex-factors that directly or indirectly influence the ride. When XAI is involved, the recommendations from the system-end provide better reasoning for its suggestions and guide humans to understand the scenario in a better way.

XAI plays a major role in the understandability of models involving anomaly detection. Anomaly detection is crucial for a model to predict finer outcomes. Most present traditional models fail to identify the importance of outliers and provide bad forecasting of the data. Hence, consideration of anomalies is crucial and ignorance of outliers is not appreciable [9]. Anomaly detection helps in examining potential outliers and provides better performance of the model in predictions. XAI aims to provide explanations for such predictions of the model and states the processes involved in the identification of anomalies. XAI helps enforce interpretability and explainability, especially for unsupervised ML models of anomaly detection.

As discussed previously, vehicle platooning is prone to cyber-attacks, making it important to protect the system from packet falsification. The knowledge extraction deepens when humans enter the logical world of ML/DL/CNN, and this is possible through XAI. XAI helps engineers in designing countermeasures to packet falsification in platooning [10]. It provides insights into the problem of packet falsification and effectively assists cybersecurity analysts in building powerful models from the analysis. This ensures the safety as well as proper functioning of autonomous vehicles without any cyber-attacks. XAI can help the system to take control over Cyber Physical Systems (CPS) with inherent knowledge extraction.

XAI is useful for humans to understand the background approaches involved behind prediction algorithms. It can be implemented in vehicle power consumption, helping to understand the execution of algorithms in determining the working conditions of a battery in the driving system. The algorithms examine vehicle and battery information to provide predictions/suggestions for efficient energy management. Heating issues associated with the batteries, faster battery draining, and their analysis based on various other parameters will be enhanced and better reported when XAI is involved. It further helps in correlating vehicle battery performance with the choice of route for trip plans. Battery range anxiety is a concern in current autonomous models, causing anxiety among passengers about reaching the spot with the available battery without needing a recharge. Also, in case of long travels, it is necessary to charge the vehicle in between, and therefore, it is important to consider the availability of charging stations along the route path. The status of battery life after a particular location along the route, the availability of battery charging stations on a route, and the total time taken for each route considering charging time can be well-explained and recommended by XAI-enabled systems. This ensures an enhanced level of intelligent assistance in decision-making in route selection.

Vehicle predictive maintenance is another application of XAI in prediction algorithms [9]. There are many incidents reported in Western countries on accidents of autonomous vehicles. It is crucial to prevent such potential accidents committed by the system to build trust among the user-end, and the role of XAI can help accomplish this task. Any malfunctioning or breakdown of any parts in the vehicle can be deducted by the sensors, and it can be reported to the users in an explanative manner when XAI is enforced. It uses the vehicle dynamics and mechanical information for implementing the predictive maintenance algorithm. Also, XAI helps provide warnings along with explanations when the machine encounters any sort of problem. This can prevent vehicles from major accidents and can safeguard the passengers by intimating the problems in-prior. The system technology is designed to avoid crashes and restrict the level of accidents to negligible. But there are possibilities of accidents to happen with autonomous vehicles, and the system must also be built to handle such cases. It should be designed to report the reasons behind the crash if the vehicle meets with any accidents.

The use of eXplainable Convolutional Neural Networks (XCNN) is widely preferred for post-accident investigations [12,15]. XCNN, which is the combination of XAI and CNN, can help figure out the errors committed by the computer vision of the system. It can effectively explain the model's working principle and guide the developers to incorporate appropriate changes to the model. It helps the engineers in re-designing the machine to prevent the occurrence of similar accidents in the future. One major drawback of XCNN is the necessity of a large dataset for its operation. XCNN is powerful only when an adequate amount of dataset is available, and hence, the vehicle must provide all the required amount of data from its sensors to effectively investigate the accident. Overall, XCNN helps control accidents and helps investigate by reporting the reasons for any vehicle crashes.

The existing models used in self-driving systems show decent performance in the identification of objects, but they do not meet the standards of better operability. The level of trust that people have in autonomous vehicles has reduced drastically because of their wrong operations in the detection of objects in certain situations. As per various records, there are many incidents where autonomous vehicles have collided with the road pedestrians and have violated traffic laws [16]. This is due to a failure in the object detection system. Engineers try to rectify and re-propose the model. However, it is difficult for them to understand the model architecture. The use of XAI can help in sorting this problem. XAI assists the developers in understanding how the machine recognizes the objects around the environment, how the objects are segmented, what priorities the system sets in detecting the objects, at what level of sensing range the machine can operate, what the level of adaptability is, what the working conditions of the machine are, etc. [17]. The explanations provided to all such questions build transparency of the system and guide technicians to better analyze the model architecture for further upgradation of the system. In the case of users, it improves trust and user experience with autonomous vehicles. Also, the scope of XAI can be extended to steering wheel control. It can be used to provide reasoning behind the operability of steering wheel control of the automobile vehicle based on its object sensing. XAI can be developed based on rule-based models which can explicitly encode driving rules/regulations, including speed limits, traffic signals, road signs, and traffic laws. This can make the system follow the traffic guidelines while traveling and prevent violations of laws. Consequently, it safeguards the vehicle and the passengers from accidents. Figure 3.3 shows the overview of XAI advantages to system engineers and users.

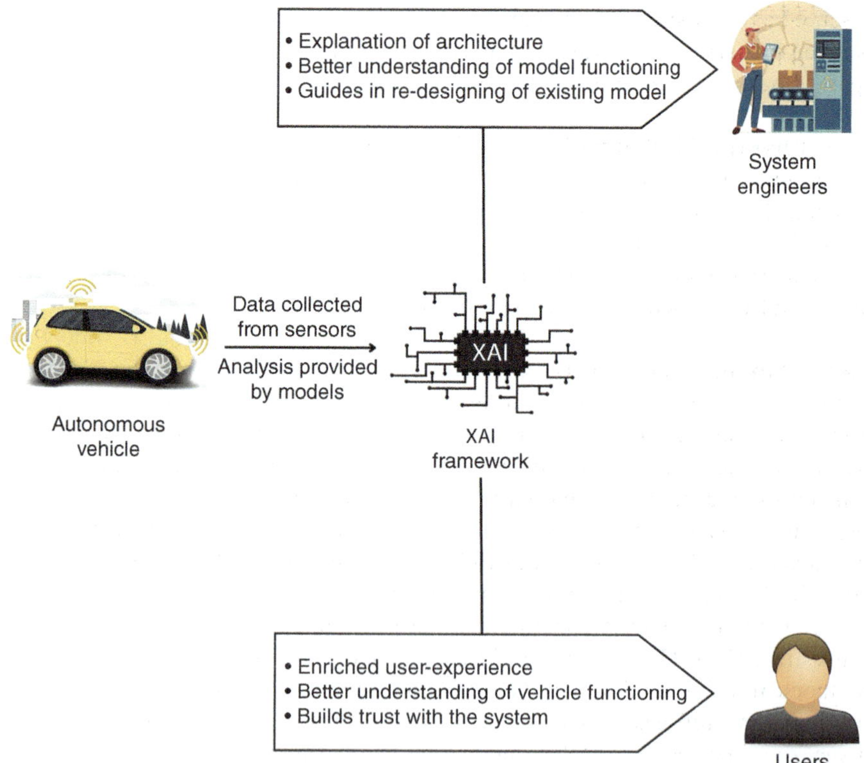

FIGURE 3.3 Overview of XAI advantages to system engineers and users.

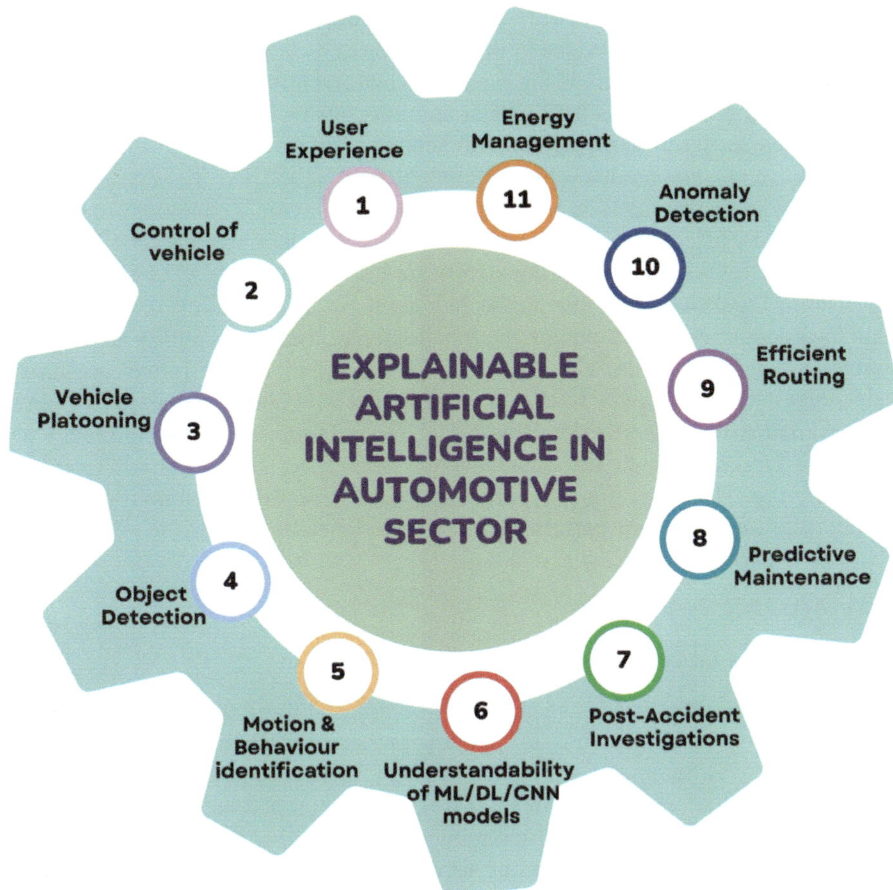

FIGURE 3.4 Role of XAI in automotive sector.

The emergence of XAI in autonomous vehicles at a larger scale can revolutionize the current transportation system. It promises to modernize the present autonomous vehicles with the use of advanced innovative technologies. The support of XAI helps engineers and technicians to analyze/identify new problems in the model and also sets a greater path for them to invent solutions to these newly identified distinct problems. The role of the automotive sector is shown in Figure 3.4.

3.2.3 Challenges in Adoptability of Autonomous Vehicles in Indian Roads and Role of XAI

The operability of autonomous vehicles in a country like India is critical [18]. The road conditions and infrastructure level in India are not favorable for self-driving vehicles to operate well. Potholes on roads can cause the systems to misjudge the scene and can mislead in functioning. There is no proper setup of sign boards on Indian roads, and this will make the vehicle not sense the environment precisely, which can lead to potential accidents. Improper functioning of traffic signals and road lights at night can cause trouble for the system in operating the vehicle and can result in crashes. The height of speed-breakers is not maintained at certain level which can cause collisions. People tend to cross the roads at anytime, posing a danger for the system to identify such scenarios and respond with actions within a fraction of time. The lack of knowledge on autonomous vehicles among the public restricts the accessibility and adoptability of such systems on Indian roads. Also, the weather conditions in India are unpredictable. Hence, there are a lot of challenges for adopting autonomous vehicles in India. The role of XAI can bring significant improvements. XAI can increase the transparency and trust level among the Indian citizens; enhance the user experience with the system; report and explain the functioning of the system during adverse conditions; guide the software architects to re-define the model for Indian conditions; help suggest efficient routes based on weather, highway tolls, availability of power stations, and road infrastructure. Road accidents in India are increasing rapidly and the major reasons behind such accidents are sleepiness/carelessness of drivers and drunk driving. Therefore, this paves the way for the necessity of autonomous vehicles in India and for better transportability in modern society.

3.3 XAI IN FINANCE

3.3.1 XAI in Loan Automation

In the financial sector, XAI is increasingly being used to provide transparency and accountability in automated decision-making processes, particularly in areas such as credit scoring [19], automated loan underwriting [20], fraud detection [21], and investment recommendations [22]. Financial institutions are often required to provide clear explanations of their decision-making processes, and XAI can help ensure that these requirements are met efficiently and effectively. It helps build trust and confidence in AI-based systems. By providing clear explanations of how the decisions are made, XAI can help reduce the risk of bias, errors, and other issues that could negatively impact customer's trust and confidence in financial institutions. In non-automated loans, decisions are taken by the human underwriter [20]. To taking such decisions, they need to analyze a huge amount of data like repaying history of the customer, their income level, and other loans. XAI is more helpful for non-standard loans such as peer-to-peer loans and payday loans wanted by the customer. The interest rates of these kinds of loans are higher than the other loans offered by the financial institutions. Nowadays, the AI/ML model actively adopts the standard rules while deciding on loan applications. This yields better results when it has the borrowing history of the customer in its training data. The accuracy level and the speed of AI/ML-based decision-making model are higher than the manual decision-making process. If any new rule is introduced in the automated loan processing system, the corresponding AI/ML model needs to be trained with the new rule to enhance the present level of accuracy. The complex ML model like an ensemble and deep learning classifier will give more efficient results, but it lacks in explaining the inner process involved inside the model. The ML model like logistic regression is easily explainable, but it fails to give more accurate results.

The need for a complex model explanation is high when the model declines the loan request even though it has all the documents required for applying loan. XAI gives more confidence about the model used, and it allows the model to overcome the weaknesses by giving the appropriate feedback. There are two different classes of XAI: model-specific and model-agnostic. The model-specific XAI only explains the working nature of the complex ML models whereas the model-agnostic XAI explains the local features considered in the decision-making process as well as their importance graphically. Moreover, it explains the sequence of steps taken toward the decision process. The Knowledge Acquisition and Documentation Structuring-based expert system (KADS) and the Expert Mortgage Arrears Threat Adviser (X-MATE) used the IF-THEN rules generated by expertise in the loan approval field. Automated decision processes on residential mortgage loans in the UK (the United Kingdom) also used the IF-THEN rule [23]. Here, the transparency of the final result of standard and non-standard loan applications is revealed. Decisions regarding loan approval or rejection for non-standard loan applications, involving non-standard income and less credit scores, requires more detailed investigation reports about the customer. This system has two different rule bases while making a decision: factual and heuristic rules. These two rules form a hierarchy in the knowledge base model. Factual rules are available at the lower level of the hierarchy and do not need any training. The factual rule base has counts for bankruptcy, payday loans, Individual Voluntary Arrangements (IVA), and County Court Judgments (CCJ) [24]. On the other hand, heuristic rules are derived from experience and map the set of input to the corresponding output with more valid information. The heuristic rule base does not consider loans that have already been declined in the factual rule base stage. It only considers the applications that have passed the low level and checks attributes such as secured/unsecured loans, credit score, property value, and loan criteria. Attributes of the model are directly taken from the electronic loan application, while other attributes are derived from the inputs and data collected from other resources. These derived attributes exhibit unique patterns and need to be transformed into proper input forms with the help of maximum likelihood evidence reasoning, processed by the knowledge base model [25]. This model provides the decision along with textual explanations as output, which is more helpful for customers as well as authorities to understand the reasons behind the decision.

3.3.2 XAI in Stock Prediction

Nowadays, the need for XAI extends to stock predictions and recommendations. In order to predict future stock values, it is necessary to identify patterns and trends by analyzing the historical market data. Machine learning models can be used to make predictions based on a various features, such as previous stock prices, news, and financial data. However, these models can be complex and difficult to understand, making it challenging for investors to trust their predictions and make decisions. However, XAI can offer insights into how the ML model arrives at its forecasts in the context of stock prediction.

For instance, an ML model might forecast that the price of a specific stock will rise tomorrow. Investors can use XAI approaches to pinpoint the characteristics, such as a strong earnings report or a bullish market sentiment that most strongly influenced this prediction. Investors can use this information to make more informed choices about whether to buy/sell a particular company's stock. Here, the features of each model play a major role [26]. Improving prediction performance involves separating significant feature variables from a collection of available features while disregarding irrelevant ones [24]. These significant features are used to identify pertinent characteristics for each stock separately, accounting for the behavior of various equities. The permutation importance feature selection approach assigns a score to each feature depending on how much the estimator's ability to predict the future is affected when the feature is replaced with another. This method's prediction performance is better than other feature selection methods such as Mean Decrease Impurity (MDI) in the random forest model and Local Interpretable Model-agnostic Explanations (LIME) [27].

Technical fundamentals and awareness of the market's hidden behavior form the basis of stock pricing. Each stock's sentiment is examined to forecast its price. For stock price analysis and explanation of price fluctuations, the Long Short-Term Memory block (LSTM) and XAI-LIME tool combination are employed [28]. Stock sentiments are collected from Yahoo Finance data and pre-processed. Then, the trend of each stock is extracted by the LSTM. Finally, the LIME tool is used for reading the model and understanding the biases present in the dataset. Therefore, using this XAI-LIME tool, novice investors can learn how and when stock prices rise or fall and how to base a decision on the same by making forecasts based on news.

Ensemble learning combines different prediction models to create a more accurate composite predictive model. Bagging, stacking, and boosting are the three kinds of ensemble methods. Bagging, also known as bootstrap aggregation refers to the variety of prediction models trained using various training subsets produced by random sampling. To create a final forecast value, the ensemble members' predictions are then subjected to a combination method (such as voting, averaging, or any other set of rules). To increase forecast accuracy, the stacking strategy uses the outputs of the number of different prediction models as inputs to another prediction model [22]. A newly developed two-stage stacking ensemble model based on ML, Empirical Mode Decomposition (EMD), and XAI is suggested for stock market direction forecasting [29]. Regression (R) and Classification (C) are carried out using an Artificial Neural Network (ANN) in the initial step. The random forest model forecasts the direction of the stock in the stage using the input from the ANNR and ANNC. Finally, to explain the logic behind the direction of stock movement, this random forest prediction model is integrated with the LIME-XAI tool.

3.3.3 XAI in Credit Risk Management

Credit risk refers to the possibility of loss to a bank when the borrower fails to repay the loan amount to them along with the interest, which fails in the borrower to meet the agreed terms. It is a measure that helps the bank to understand and operate the functioning of loan approvals to its customers. It is very important for the banks to identify the risk factors before lending the money and hence, credit risk management is crucial.

Credit risk management helps assess, analyze, identify, and mitigate the risks that are associated with lending money to customers. Effective credit risk management helps the bank to protect their investments, maintain financial stability, and set proper guidelines for lending practices [30]. The past decade has seen a huge growth in the use of modern technologies to effectively maintain and manage credit risks in the financial sector. The use of Artificial Intelligence, Machine Learning, and Deep Learning in credit risk management has significantly improved the working system of loan approvals in financial institutions. It has drastically reduced the human workforce, minimized the time taken for credit analysis of borrowers, and secured banks from providing loans to vulnerable customers. The existing algorithms for credit risk management are difficult to interpret and explain. This paves the way for concerns over fairness, trust, bias, and transparency over decisions made by the systems. Hence, implementing XAI in the models can help overcome such concerns. XAI helps understand and explain the algorithms implemented in credit risk assessment. It provides insights into the functioning of credit risk models and identifies the factors that influence in decision-making process. This can help the lenders to understand the model and suggest suitable changes to the system according to the financial conditions.

Credit Information Bureau (India) Limited (CIBIL) is the authorized credit agency of the Reserve Bank of India (RBI), and this bureau collects and manages credit information of a person, company, or other private and public organizations in India [31]. The CIBIL score which is so called "credit score" represents the creditworthiness of the borrower and provides information on how likely the borrower can repay the loan amount to the financial institutes. It helps the banks in the decision-making process in sanctioning loan applications. An individual is considered to have good credit when the credit score is high; a low credit score indicates bad credit history of the borrower. Loan history, amounts owed, length of credit history, new credit, and types of credit in use are some important factors that influence the credit score of the individual. These scores are tallied using advanced algorithms, and ML models support in making decisions for loan approvals. The decisions made by the system in such cases are unknown to the humans which makes it less transparent. Also, people are unaware of how their credit score is calculated, and the lenders too find it difficult to elaborate on the decisions made by the system to their customers. This causes a lack of trust, frustration, and confusion among the public. Thus, XAI can be an effective remedy for this problem. Using XAI techniques, the assessments that are carried out to calculate the credit score can be explained along with distinct factors that influence the overall score. It makes the system more transparent and explainable. By making credit scoring transparent, individuals can understand the calculations involved in the evaluation of their credit score, and this can help them plan actions to improve their credit score in the future. This benefits both the lenders and the borrowers.

XAI techniques that can be adopted in credit risk management include LIME, Anchors, BEEF (Balanced English Explanations of Forecasts), LORE (LOcal Rule-based Explanations), and SHAP (SHapley Additive exPlanations) [30]. The use of these tools in the system can guide the credit risk managers to find the areas of improvement in the decision-making process and also help them to redefine the system according to their bank policies/strategies. The explanations of the creditworthiness of the applicants can provide a detailed study to the lenders, enabling them to take effective actions in sanctioning loans. This prevents risky investments and safeguards the business from potential risks.

XAI can help in flagging unusual patterns in the credit history of the borrower, which is not easily interpreted by humans [32]. This helps the lenders to be cautious in approving the loans and enables the banks to rebuild the system model efficiently. Consequently, XAI improves reliability and accuracy levels of the system. XAI in credit risk management can assist financial institutes in setting proper guidelines, establishing appropriate credit policies, regulating credit limits, and defining commercial strategies. The use of XAI in credit risk management promises to provide trust, and transparency, and reliability among people in the decision flow on credit assessment.

3.3.4 XAI in Fraud Detection

Due to advancements in technology, financial fraud is increasing rapidly. It is important to detect such frauds to protect the customer's trust and money. Financial frauds happen in various ways across the globe, and it is crucial to build stronger technology to safeguard and identify such threats. Identity fraud, phishing, card fraud, skimming, counterfeit cards, advance fee scams, fund transfer scams, fake prizes, inheritance scams, international lottery frauds, Ponzi schemes, money laundering, insider trading, insurance frauds, bankruptcy frauds, embezzlements, accounting frauds, wills, and legacies are some of the major financial frauds that are happening worldwide [33,34]. XAI can be useful in preventing and detecting such financial frauds [35].

There has been an increase in online fraud transactions and hacking of personal accounts by scammers, especially during the lockdown period of the COVID-19 pandemic [21]. Though cyber-security engineers use advanced technologies to protect the system, hackers pass the gateway and gain access to vital information, enabling them to execute their frauds. In such cases, the use of XAI efficiently supports cyber engineers in distinguishing normal transactions from fraudulent ones over a period by detecting anomalies [34]. XAI can guide in identifying activity behavioral patterns and help in alerting financial institutes. This helps prevent such frauds in the future by remodeling the existing algorithms. Identity frauds can be minimized by XAI, which involves the process of user verification and enforcing modern biometric authentication. With the help of an XAI-based system, identifying discrepancies by comparing them with the past history of user's biometric records can help in effective detection of frauds. The detailed reports provided by XAI can further promise the bank officials to renovate the model to prevent such frauds in the near future.

Phishing, advance fee scams, fund transfer scams can be controlled extensively to prevent customers from such frauds with the help of XAI in the financial sector. XAI can be used in analyzing the content used in phishing emails and can determine if there are any try-on phishing/scam attempts. Mails with suspicious links, asking for account credentials, and improper messages can be better reported to the financial institutes and can notify the customers regarding the XAI-enabled system. Card fraud can be detected with the help of XAI by identifying the pattern change in money withdrawals. Anomalies such as sudden rise in transactions of money from user's account and frequent money deduction of the same amount at regular periods can be detected by XAI systems, which can alert the customer as well as the bank with detailed reports [36].

Skimming fraud refers to the stealing of card information during a legitimate transaction when the card is swiped through the skimming device. The device stores the information that is available on the magnetic strip of the credit/debit card which helps the fraudsters to be involved in financial crimes. Counterfeit cards refer to the cards that have been produced by fraudsters based on the retrieved information. These frauds can be prevented by implementing XAI in image processing technology. Effective detection of legitimate bank cards from counterfeit cards can help in reporting such fraud happenings to the banks which safeguard the money of customers [37,38].

Fake prizes, Ponzi schemes, inheritance scams, international lottery frauds, wills, and legacies are major attacks that focus on inducing a money mindset in the public, making it easier for fraudsters to proceed with their fraudulent acts [39]. In certain cases, people believe them to be legit and send all of their account details to them. This makes it easier to loot all the money from the account. These frauds can be controlled by effectively identifying such messages and emails by the XAI system. Reporting such messages and emails to banking officials helps identify the source of fraudsters and can help terminate the entire fraud community.

Insurance frauds and embezzlements can be better understood and explained to humans by XAI in real-time scenarios. The frauds that happen in the financial sector cannot be easily identified by experts, and fraudsters discover all possible ways to steal money from the public. Using XAI-enabled systems can help in pointing out changes in the normal pattern of fund transactions. This helps bank officials to detect such frauds and take effective actions on time. Accounting frauds and money laundering have been in existence for several years, making it difficult for frauds to be detected by current financial algorithms [40]. Frauds take place by using the loopholes in the financial system, and this can be prevented with the help of XAI. Predictive analytics, transaction monitoring, risk assessment, network analysis, anomaly detection, explainability, and real-time monitoring of data from XAI can help in reducing these types of frauds. XAI uses the history of financial records to understand the financial system and later, trains itself to help in detecting the fraud by reporting them to the banks and the customers.

Bankruptcy frauds can be controlled using XAI by analyzing financial ratios (debt-to-equity ratio, quick ratio, current ratio), analyzing the company's cash flow statements, analyzing the market environment, and predictive analytics. XAI helps in improving the performance level of detecting the bankruptcy frauds.

Insider trading/dealing is an illegal way of trading on stocks for personal gain using confidential details [41]. It is necessary to prevent such frauds, and XAI is again a useful tool to prevent them. The use of XAI techniques can help in analyzing industry data for detecting abnormal stock purchases. It can help in identifying suspicious patterns in trading practices by analyzing the order book data and transaction data. XAI can help notice the news and social media data for potential leaks of company information. It can also help discover the insider who is providing confidential information to external sources by examining the call history, email records, and other communication data for suspicious activities. Thus, XAI helps in spotting insider trading practices in modern stock exchanges.

3.4 XAI IN MANUFACTURING SECTOR

AI is implemented in almost every aspect of life, the reason being the availability of huge amounts of real-time data. One such aspect is manufacturing. It is no wonder that it has completely transformed the industrial sector by opening new opportunities for process monitoring, optimization, customization, and reconfiguration of machine parts. It assures cost savings and product quality enhancement by using the information provided to it.

However, the lack of transparency and interpretability in AI algorithms is an increasing concern as these algorithms get more complicated. It becomes difficult for the user to understand which of the past information played a major role in predicting the best action to be taken to reach an optimal outcome for the problem faced in the manufacturing industry. This is where XAI plays a crucial role in offering insights into how the AI models make decisions and what the important variables are that play a major role in prediction.

The implementation of XAI in a system follows four principles which are explanation, meaningfulness, explanation accuracy, and knowledge limits [42]. One of the best examples of these principles being implemented is the predictive maintenance system in the manufacturing sector which informs the user when there is a failure on any equipment, an error in the functioning of any equipment, maintenance to be done for an equipment, or a replacement of the equipment itself [43]. As maintenance and replacement are expensive processes, the customer of the system will likely query about the accuracy of the outcome generated by the system. Therefore, the explanation principle will focus on answering three main queries: what algorithm is being used in the system, how the model works with the algorithm in the system, and which input features or parameters of the data are playing a significant role in determining the output of the system [44]. By addressing these queries, the customer will get an idea about whether the system is performing in the right way or not so that the customers can make their decision on whether they will be utilizing the system or not in their workplace.

The next principle would be noting whether the explanation given by the system is meaningful or not [45]. The first and foremost priority would be whether the explanation given is understandable by the targeted user for it to be meaningful to them. To satisfy such requirements, different explanations need to be given for different kinds of targeted users like developers, maintenance personnel, end-users, etc., according to the knowledge they possess and the experience they have [46].

The next principle is justifying the explanation by providing explanation accuracy [47]. The system explains the process in which the Artificial Intelligence system is used to generate the output, and if the explanation given provides a wrong output, then XAI becomes redundant. Such an act will be an immediate red flag for the customer, affecting his trust factor in the XAI system, potentially leading to incorrect decisions and financial losses. Therefore, a correct explanation of the system is critically essential, with excellent accuracy and the required level of detail.

The final principle is knowledge limits, which restrict the system from providing unjust and false results. It ensures that the system accesses only relevant data to the model, therefore, convincing the user that system will never lead them astray.

3.4.1 XAI in Supply Chain Management

Supply chain management (SCM) is the process of managing the flow of products and services. It involves the movement and storage of raw materials, inventories for work-in-progress, and the transit of goods from the place of origin to the location of consumption. The optimization of the supply chain may result in significant energy savings and a commensurate reduction in carbon emissions.

In the manufacturing area, SCM focuses on production planning and control, production R&D, maintenance and diagnosis, and quality management. SCM uses a wide range of technologies, including sensors, barcodes, and IoT (Internet of Things), to integrate and coordinate each link in the chain by employing advanced techniques to extract useful knowledge and facilitate data-driven decision-making. Support Vector Machines (SVM), along with spatial/temporal-based visual analysis, are the key approaches in the development of optimization models. The K-means clustering algorithm is applied in clustering, classification, forecasting, and simulation models. The progressive move away from production support systems pose one of the biggest challenges for developing countries like India because several of them have unquestionably experienced significant income erosion in recent years.

The well-known traits of XAI are human-machine trust, AI model interpretability, and comprehensibility. XAI is about making AI models and systems visible, fair, dependable, and understandable. Aversion and decision-making overrides are caused by mistrust and a lack of transparency. The creation of a completely automated and self-adjusting decision-making system is the primary goal of integrating AI into the supply chain. Businesses can effectively forecast demand surges and modify the amounts and routes of material flows with AI-powered supply chain management. Intelligent automated processes can operate without error for an extended period, reducing the number of errors. The productivity of warehouse robots is higher, and they operate more quickly and accurately.

XAI can be used in validating the acquired detailed data (weather patterns, GPS data, and reroutes) that may affect delivery delays. The sales team can anticipate delivery timeframes more precisely from the gathered data. It immediately alerts consumers to in-flight inventory changes. As a result, companies may provide their present and potential clients with greater customer service. Implementing XAI in the manufacturing sector will help in timely delivery, wiser planning, effective inventory and warehouse management, eliminate the need for human labor, reduce operational expenses, and make the process safer and faster.

3.5 CONCLUSION

The use of XAI in various sectors improves the efficiency and performance of several tasks. The use of various XAI tools like LIME and SHAP estimates and states the reasons for the proposed decisions by the model. In the health sector, diagnosis explanation of the problem can be done with the help of XAI, and it can help doctors to conclude the diagnosis of the disease [48]. In the education sector, XAI can be deployed to develop efficient learning tools, tutoring systems, grade estimation systems, and admission decision-making models [49]. XAI in education also helps in human-computer interaction, learning analytics, and cognitive and learning sciences [50]. XAI provides better performance in the detection of fake news and increases the level of recommendations of appropriate news articles to end-users in the media sector [51]. The music, art, and movie recommendations for the users can be improved further when XAI is used in the entertainment sector [49]. XAI can provide better clarity in e-governance and judicial systems on decisions taken by the system [52,53]. Thus, enforcing XAI in various industries promises better understandability, interpretability, accountability, transparency, and reasoning.

3.6 FUTURE STUDIES

The need for XAI is gradually increasing and will soon play a key role in all sectors. However, in certain cases, the lack of a user-centric approach to XAI fails to provide a better explanation for stakeholders [54]. This can be improved by considering the respective human cognitive and thinking skills of a particular domain by the XAI system. This will make the XAI system adapt to their level of thinking ability and promise to provide effective explanations accordingly in terms of their respective domain. The user-centric perspective of XAI models will additionally increase the level of trust in the systems, paving the way for stakeholders to make effective strategies and decisions. As the use of AI becomes more prevalent in society, the need for a user-centric approach to XAI will increase. It is important to build and redefine existing XAI models to ensure that AI is working in a responsible and ethical manner.

REFERENCES

1. Doran, Derek, Sarah Schulz, and Tarek R. Besold. "What does explainable AI really mean? A new conceptualization of perspectives." *CEUR Workshop Proceedings arXiv: 1710.00794* (2018).
2. Gunning, David, Mark Stefik, Jaesik Choi, Timothy Miller, Simone Stumpf, and Guang-Zhong Yang. "XAI-Explainable Artificial Intelligence." *Science Robotics* 4, no. 37 (2019): eaay7120.
3. Arrieta, Alejandro Barredo, Natalia Díaz-Rodríguez, Javier Del Ser, Adrien Bennetot, Siham Tabik, Alberto Barbado, Salvador García, et al. "Explainable Artificial Intelligence (XAI): Concepts, taxonomies, opportunities and challenges toward responsible AI." *Information Fusion* 58 (2020): 82–115.

4. Gaur, Loveleen, and Biswa Mohan Sahoo. "Introduction to Explainable AI and Intelligent transportation." In *Explainable Artificial Intelligence for Intelligent Transportation Systems: Ethics and Applications*, pp. 1–25. Cham: Springer International Publishing, 2022. doi: 10.1007/978-3-031-09644-0_1.
5. Burgt, Joost van der. "Explainable AI in banking." *Journal of Digital Banking* 4, no. 4 (2020): 344–350.
6. Bussmann, Niklas, Paolo Giudici, Dimitri Marinelli, and Jochen Papenbrock. "Explainable AI in fintech risk management." *Frontiers in Artificial Intelligence* 3 (2020): 26.
7. Owens, Emer, Barry Sheehan, Martin Mullins, Martin Cunneen, Juliane Ressel, and German Castignani. "Explainable Artificial Intelligence (XAI) in insurance." *Risks* 10, no. 12 (2022): 230.
8. Tang, Chen. "Designing explainable autonomous driving system for trustworthy interaction." PhD diss., UC Berkeley, 2022.
9. Cupek, Rafał, Jerry Chun-Wei Lin, and J. H. Syu. "Automated guided vehicles challenges for artificial intelligence." In *2022 IEEE International Conference on Big Data (Big Data)*, Osaka, Japan, pp. 6281–6289. IEEE, 2022.
10. Mongelli, Maurizio. "Design of countermeasure to packet falsification in vehicle platooning by Explainable Artificial Intelligence." *Computer Communications* 179 (2021): 166–174.
11. Hartwich, Franziska, Cornelia Schmidt, Daniela Gräfing, and Josef F. Krems. "In the passenger seat: Differences in the perception of human vs. automated vehicle control and resulting HMI demands of users." In *HCI in Mobility, Transport, and Automotive Systems. Automated Driving and In-Vehicle Experience Design: Second International Conference, MobiTAS 2020, Held as Part of the 22nd HCI International Conference, HCII 2020*, Copenhagen, Denmark, July 19–24, 2020, Proceedings, Part I 22, pp. 31–45. Springer International Publishing, 2020.
12. Omeiza, Daniel, Helena Webb, Marina Jirotka, and Lars Kunze. "Explanations in autonomous driving: A survey." *IEEE Transactions on Intelligent Transportation Systems* 23, no. 8 (2021): 10142–10162.
13. Atakishiyev, Shahin, Mohammad Salameh, Hengshuai Yao, and Randy Goebel. "Explainable Artificial Intelligence for autonomous driving: a comprehensive overview and field guide for future research directions." *arXiv preprint arXiv:2112.11561* (2021).
14. Schneider, Tobias, Joana Hois, Alischa Rosenstein, Sabiha Ghellal, Dimitra Theofanou-Fülbier, and Ansgar R.S. Gerlicher. "Explain yourself! transparency for positive UX in autonomous driving." In *Proceedings of the 2021 CHI Conference on Human Factors in Computing Systems*, Yokohama, Japan, pp. 1–12, 2021.
15. Atakishiyev, Shahin, Mohammad Salameh, Hengshuai Yao, and Randy Goebel. "Towards safe, explainable, and regulated autonomous driving." *arXiv preprint arXiv:2111.10518* (2021).
16. Fleetwood, Janet. "Public health, ethics, and autonomous vehicles." *American Journal of Public Health* 107, no. 4 (2017): 532–537.
17. Kim, Hong-Sik, and Inwhee Joe. "An XAI method for convolutional neural networks in self-driving cars." *PLoS one* 17, no. 8 (2022): e0267282.
18. Kolekar, Suresh, Shilpa Gite, Biswajeet Pradhan, and Abdullah Alamri. "Explainable AI in scene understanding for autonomous vehicles in unstructured traffic environments on Indian roads using the inception U-Net Model with Grad-CAM visualization." *Sensors* 22, no. 24 (2022): 9677.
19. Moscato, Vincenzo, Antonio Picariello, and Giancarlo Sperlí. "A benchmark of machine learning approaches for credit score prediction." *Expert Systems with Applications* 165 (2021): 113986.

20. Sachan, Swati, Jian-Bo Yang, Dong-Ling Xu, David Eraso Benavides, and Yang Li. "An explainable AI decision-support-system to automate loan underwriting." *Expert Systems with Applications* 144 (2020): 113100.
21. Buil-Gil, David, Fernando Miró-Llinares, Asier Moneva, Steven Kemp, and Nacho Díaz-Castaño. "Cybercrime and shifts in opportunities during COVID-19: A preliminary analysis in the UK." *European Societies* 23, no. suppl 1 (2021): S47–S59.
22. Dash, Rajashree, Sidharth Samal, Rasmita Dash, and Rasmita Rautray. "An integrated TOPSIS crow search based classifier ensemble: In application to stock index price movement prediction." *Applied Soft Computing* 85 (2019): 105784.
23. Sun, Ron. "Robust reasoning: Integrating rule-based and similarity-based reasoning." *Artificial Intelligence* 75, no. 2 (1995): 241–295.
24. Carta, Salvatore, Alessandro Sebastian Podda, Diego Reforgiato Recupero, and Maria Madalina Stanciu. "Explainable AI for financial forecasting." In *Machine Learning, Optimization, and Data Science: 7th International Conference, LOD 2021*, Grasmere, UK, October 4–8, 2021, Revised Selected Papers, Part II, pp. 51–69. Cham: Springer International Publishing, 2022.
25. Li, Yanrui, Kaiyou Fu, Yuchen Zhao, and Chunjie Yang. "How to make machine select stocks like fund managers? Use scoring and screening model." *Expert Systems with Applications* 196 (2022): 116629.
26. Kumar, Satyam, Mendhikar Vishal, and Vadlamani Ravi. "Explainable Reinforcement Learning on Financial Stock Trading using SHAP." *arXiv preprint arXiv:2208.08790* (2022).
27. Gite, Shilpa, Hrituja Khatavkar, Ketan Kotecha, Shilpi Srivastava, Priyam Maheshwari, and Neerav Pandey. "Explainable stock prices prediction from financial news articles using sentiment analysis." *PeerJ Computer Science* 7 (2021): e340.
28. Aria, Massimo, Corrado Cuccurullo, and Agostino Gnasso. "A comparison among interpretative proposals for Random Forests." *Machine Learning with Applications* 6 (2021): 100094.
29. Çelik, Taha Buğra, Özgür İcan, and Elif Bulut. "Extending machine learning prediction capabilities by explainable AI in financial time series prediction." *Applied Soft Computing* 132 (2023): 109876.
30. Moscato, Vincenzo, Antonio Picariello, and Giancarlo Sperlí. "A benchmark of machine learning approaches for credit score prediction." *Expert Systems with Applications* 165 (2021): 113986.
31. Laxmanan, Saravanan. "Awareness of Credit Score Mechanism in India: A Study with Reference to Credit Information Bureau India Limited (CIBIL)." *TIJ's Research Journal of Commerce & Behavioural Science - RJCBS* (2017): 6.
32. Kłosok, Marta, and Marcin Chlebus. *Towards Better Understanding of Complex Machine Learning Models Using Explainable Artificial Intelligence (XAI): Case of Credit Scoring Modelling*. Warsaw: University of Warsaw, Faculty of Economic Sciences, 2020.
33. Reurink, Arjan. "Financial fraud: A literature review." *Contemporary Topics in Finance: A Collection of Literature Surveys* (2019): 79–115. doi: 10.1111/joes.12294.
34. Michalopoulos, Panagiotis. "Interpretable Unsupervised Fraud Detection in Financial Services." *Mathematics and Computer Science* (2020).
35. Psychoula, Ismini, Andreas Gutmann, Pradip Mainali, Sharon H. Lee, Paul Dunphy, and Fabien Petitcolas. "Explainable machine learning for fraud detection." *Computer* 54, no. 10 (2021): 49–59.
36. Wu, Tung-Yu, and You-Ting Wang. "Locally interpretable one-class anomaly detection for credit card fraud detection." In *2021 International Conference on Technologies and Applications of Artificial Intelligence (TAAI)*, Taichung, Taiwan, pp. 25–30. IEEE, 2021.

37. Vivek, Yelleti, Vadlamani Ravi, Abhay Anand Mane, and Laveti Ramesh Naidu. "Explainable Artificial Intelligence and Causal Inference based ATM Fraud Detection." *arXiv preprint arXiv:2211.10595* (2022).
38. Ji, Yingchao. "Explainable AI methods for credit card fraud detection: Evaluation of LIME and SHAP through a User Study" (2021), Dissertation, https://urn.kb.se/resolve?urn=urn:nbn:se:his:diva-20848.
39. Omolara, Abiodun Esther, Aman Jantan, Oludare Isaac Abiodun, Manmeet Mahinderjit Singh, Mohammed Anbar, and D. V. Kemi. "State-of-the-art in big data application techniques to financial crime: A survey." *International Journal of Computer Science and Network Security* 18, no. 7 (2018): 6–16.
40. Kute, Dattatray Vishnu, Biswajeet Pradhan, Nagesh Shukla, and Abdullah Alamri. "Deep learning and Explainable Artificial Intelligence techniques applied for detecting money laundering-a critical review." *IEEE Access* 9 (2021): 82300–82317.
41. Bishop, Matt, and Carrie Gates. "Defining the insider threat." In *Proceedings of the 4th Annual Workshop on Cyber Security and Information Intelligence Research: Developing Strategies to Meet the Cyber Security and Information Intelligence Challenges Ahead*, Oak Ridge, USA, pp. 1–3, 2008.
42. Phillips, P. Jonathon, Carina A. Hahn, Peter C. Fontana, David A. Broniatowski, and Mark A. Przybocki. "Four principles of Explainable Artificial Intelligence." *Gaithersburg, Maryland* (2020): 18. doi: 10.6028/NIST.IR.8312.
43. Kazim, Emre, and Adriano Koshiyama. "Explaining decisions made with AI: a review of the co-badged guidance by the ICO and the Turing Institute." Available at SSRN 3656269 (2020).
44. Bhatt, Umang, Alice Xiang, Shubham Sharma, Adrian Weller, Ankur Taly, Yunhan Jia, Joydeep Ghosh, Ruchir Puri, José MF Moura, and Peter Eckersley. "Explainable machine learning in deployment." In *Proceedings of the 2020 Conference on Fairness, Accountability, and Transparency*, Barcelona, Spain, pp. 648–657, 2020.
45. Lim, Brian Y., Anind K. Dey, and Daniel Avrahami. "Why and why not explanations improve the intelligibility of context-aware intelligent systems." In *Proceedings of the SIGCHI Conference on Human Factors in Computing Systems*, Boston, USA, pp. 2119–2128, 2009.
46. Hind, Michael. "Explaining explainable AI." *XRDS: Crossroads, The ACM Magazine for Students* 25, no. 3 (2019): 16–19.
47. Japkowicz, Nathalie, and Mohak Shah. *Evaluating Learning Algorithms: A Classification Perspective.* Cambridge: Cambridge University Press, 2011.
48. Pawar, Urja, Donna O'Shea, Susan Rea, and Ruairi O'Reilly. "Incorporating Explainable Artificial Intelligence (XAI) to aid the understanding of machine learning in the healthcare domain." In *AICS*, Dublin, Republic of Ireland, pp. 169–180. 2020.
49. Haque, A.K.M. Bahalul, A.K.M. Najmul Islam, and Patrick Mikalef. "Explainable Artificial Intelligence (XAI) from a user perspective: A synthesis of prior literature and problematizing avenues for future research." *Technological Forecasting and Social Change* 186 (2023): 122120.
50. Khosravi, Hassan, Simon Buckingham Shum, Guanliang Chen, Cristina Conati, Yi-Shan Tsai, Judy Kay, Simon Knight, Roberto Martinez-Maldonado, Shazia Sadiq, and Dragan Gašević. "Explainable Artificial Intelligence in education." *Computers and Education: Artificial Intelligence* 3 (2022): 100074.
51. Chien, Shih-Yi, Cheng-Jun Yang, and Fang Yu. "XFlag: Explainable fake news detection model on social media." *International Journal of Human-Computer Interaction* 38, no. 18–20 (2022): 1808–1827.

52. Carbonaro, Antonella. "Interpretability of AI systems in electronic governance." In *Electronic Governance with Emerging Technologies: First International Conference, EGETC 2022*, Tampico, Mexico, September 12–14, 2022, Revised Selected Papers, pp. 109–116. Cham: Springer Nature Switzerland, 2023.
53. Deeks, Ashley. "The judicial demand for Explainable Artificial Intelligence." *Columbia Law Review* 119, no. 7 (2019): 1829–1850.
54. Cirqueira, Douglas, Dietmar Nedbal, Markus Helfert, and Marija Bezbradica. "Scenario-based requirements elicitation for user-centric explainable AI: A case in fraud detection." In *Machine Learning and Knowledge Extraction: 4th IFIP TC 5, TC 12, WG 8.4, WG 8.9, WG 12.9 International Cross-Domain Conference, CD-MAKE 2020*, Dublin, Ireland, August 25–28, 2020, Proceedings 4, pp. 321–341. Springer International Publishing, 2020.

4 Explainable AI in Distributed Denial of Service Detection

Raj Kumar Batchu, Seshu Bhavani Mallampati, and Hari Seetha

4.1 INTRODUCTION

The Internet was introduced in the early 1970s as a result of ARPANET. The Internet's quick evolution has transformed it into a formidable platform for communication, business, entertainment, research, education, media, and other activities. However, data breaches, cyber-attacks, and security risks have risen with the growth in Internet connectivity. As a result, Internet-based security assaults have increased in frequency and complexity over the last few years. Attempts to protect networks from cyber-attacks have led to the creation of defensive software such as firewalls (intended to avoid intrusions) and intrusion detection systems (IDS) (to detect and prevent the illegal use of data). IDSs are meant to analyze network traffic against predefined parameters to determine possibly malicious network activities. An IDS is application software that monitors all network activity and looks for abnormal patterns that may signal a system or network attack by someone seeking to bypass the security procedures. IDS can be categorized in various ways, depending on the data collected, analyzed, and required actions.

Additionally, it can be classified into two types based on its installation location within the network, such as network-based IDS (NIDS) and host-based IDS (HIDS) [1]. A NIDS continuously monitors and analyzes the individual packets that traverse a network to detect malicious activity. On the other hand, a HIDS monitors actions such as process scheduling, system call tracing, and login attempts. An IDS is categorized into two kinds based on its detection mechanism: misuse detection and anomaly detection [2]. Misuse detection relies on signatures and is only effective against known attacks; it cannot detect novel attacks [3]. Anomaly detection is based on comparing an attacker's behavior to that of a regular user. Anomaly detection can detect unknown assaults but at a high false-positive rate [4]. The comparison of both of these methods is shown in Table 4.1.

In the recent past, among the various cyber-attacks generated, distributed denial of service (DDoS) attacks affect organizations in terms of revenue, loss of customers, and reputation, among others. According to Imperva, a security provider, reports indicate that the size, volume, and sophistication of these attacks have increased drastically. This makes defending against DDoS attacks a top priority for all companies.

DOI: 10.1201/9781003442509-4

TABLE 4.1
Comparison of Misuse and Anomaly-Based Detection

	Misuse Based Detection	Anomaly-Based Detection
Detection accuracy	High	Depends on the complexity of the model
Performance	Less false alarm rate (LFAR) High missed alarm rate (HMAR)	High false alarm rate (HFAR) Less missed alarm rate (LMAR)
Domain knowledge dependency	High	Low
Unknown attacks detection	No	Yes
Interpretation	Strong interpretative ability	Weak interpretative ability

To defend against DDoS assaults, it is necessary to comprehend how they work and to investigate some of the most popular techniques. A DDoS attack may appear complicated, but it is actually rather simple to comprehend. The first strategy is "Bombarding," where a target machine is bombarded with hundreds of communication requests from different workstations, and this is a frequent tactic. This causes the server to become completely overloaded and unable to respond to valid user requests. Another tactic is to disable all communication between users and the target server by limiting their access to the network. The attacking workstations are geographically dispersed and utilize numerous Internet connections, making it extremely challenging to govern the assaults. This can tremendously harm organizations, particularly those that rely substantially on their website, such as E-commerce, online reservations, banking and health care. In a communications network, the Open Systems Interconnection (OSI) model defines seven conceptual layers. DDoS assaults primarily target layers 3 (network), 4 (transport), and 7 (application).Layer 3/4 DDoS Assaults: The network and transport layers are the primary targets of DDoS attacks. These assaults take place when an overloaded network or server exhausts all of its resources due to the volume of data packets and other traffic.

Application-level DDoS attacks: A web application flaw or vulnerability. By taking advantage of it, offenders overwhelm the server or database that powers the web application, rendering it unusable. Such assaults are more difficult to spot since they imitate actual user traffic. To identify these attacks, various IDS approaches are designed using multiple machine learning (ML) techniques.

An IDS comprises several stages, including data collection, pre-processing, feature selection (FS), and classification [5]. Out of these stages, FS is challenging, as the IDS must deal with a large volume and variety of network traffic. FS is the process of selecting significant features for categorization, i.e., only a subset of the total features is considered. The fundamental difficulty with IDS is the computational overhead and classification performance, as IDS identifies a wide variety of intrusions in real time. In the past few years, FS approaches have gained considerable popularity for their ability to choose the most influential attributes. Because classification is dependent on the class label, it is critical to eliminate redundant and irrelevant information [6]. The flow of the intrusion detection model is shown in Figure 4.1.

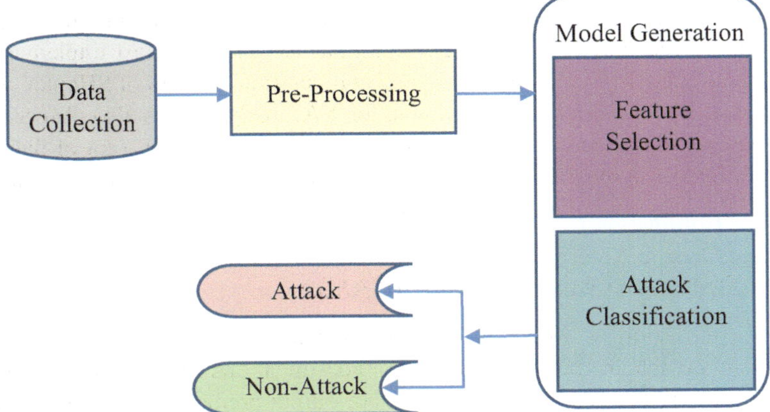

FIGURE 4.1 Flow of intrusion detection system.

A feature is a unique and significant characteristic of the analyzed input data. The cardinality of features in a dataset is called its dimensionality. The majority of real-world applications do not require all of the attributes due to the possibility of redundant, irrelevant, or noisy data. Dimensionality reduction significantly reduces the number of features by removing irrelevant and redundant data, increasing the model's efficiency and performance. According to Bellman, the number of samples to consider when training a model increases exponentially as the dimensions of the input data increase. This condition is called the "Curse of Dimensionality". The majority of datasets with the "big p, small n" problem (where p indicates the number of attributes, and n signifies the instances) have a greater probability of overfitting [7]. An overfitted model makes the training model complex and results in poor performance. Dimensionality reduction approaches are introduced as a pre-processing phase in visualization, data analysis, and modeling to solve the curse of dimensionality. The two most common dimensionality reduction techniques are feature selection and feature extraction [8].

This chapter elaborates on various approaches that provide optimal features for intrusion detection.

4.2 BACKGROUND

Some of the popular methods for feature selection applied by various researchers include filter, wrapper, and hybrid methods [9]. The filter technique evaluates features based on their correlations with the dependent variable, as determined by various statistical tests. The filter approach works without any predictive model (an ML model to forecast the outcomes based on historical and current data). When the number of characteristics is large, this is the quickest and usually the best option. It prevents overfitting yet may occasionally miss the essential attributes. The filtering techniques are classified into Univariate and Multivariate strategies, based on the features involved during the selection process.

The wrapper is a widely used method developed by Ron Kohavi and John in 1997 [10]. Initially, this method uses an iterative search that repeatedly supplies feature subsets to the learning algorithm to identify the essential attributes and thus generates appropriate feature subsets [11]. The final subset is chosen based on the learning algorithm's performance and the error rate. Though the approach is computationally expensive and slower than filter methods, it gives better performance. Some of the wrapper methods include Recursive Feature Elimination (RFE), Sequential Backward Elimination (SBE), Simulated Annealing, Sequential Forward Selection (SFS), and Genetic Algorithms [12,13].

Embedded or hybrid methods incorporate the feature selection inside the learning model. They combine the benefits of both filter and wrapper methods, using feature selection techniques to achieve feature selection and classification or regression simultaneously. There are three sorts of embedded approaches: algorithm-based, regularization, and pruning methods.

The algorithmic approach can be implemented using a tree-based algorithm such as Random Forest (RF), Decision Tree (DT), XGBoost, Extra Tree (ET), etc. Tree-based models select an attribute and partition the sample set into smaller subsets at each iterative stage of the tree development process. The relevance of an attribute is determined by the degree to which the descendant nodes of a subset belong to the same class.

Next, regularization approaches optimize the objective function of the learning algorithm to minimize the error in fitting the model. This includes Ridge (with the L2 penalty) and LASSO (with the L1 penalty) for creating a linear model. Both of these techniques can reduce features to zero or nearly zero [14]. The Elastic Net approach combines both L1 and L2 penalties. These models are faster, akin to filter techniques, and generally more accurate than filter techniques [15].

Finally, pruning methods train the learning algorithm with all the features and iteratively remove each feature based on its accuracy. Some of the models include Random Forest-Recursive Feature Elimination, Support Vector Machine-Genetic Algorithm, Support Vector Machine-Recursive Feature Elimination, etc. These approaches are computationally more efficient than earlier methods and are less prone to overfitting.

Using multiple ML approaches, diverse assaults have been identified. However, the majority of these methods were unable to describe the most influential attributes in forecasting the target. Knowing these contributed features and a specific instance of traffic is the first step in analyzing the model. Compared to conventional methods, Explainable Artificial Intelligence (XAI) facilitates the understanding of the importance of features. "XAI" is a collection of methods and approaches that enable human users to understand and trust the output and outcomes produced by ML algorithms. The concept of "explainable AI" refers to an AI model's potential impacts as well as any potential biases. It aids in determining model accuracy, fairness, transparency, and outcomes in decision-making supported by AI. Before employing AI models, an organization must build trust and confidence by using explainable AI. AI explainability also facilitates the adoption of a responsible approach to AI development within an organization.

4.3 THE NEED FOR EXPLAINABILITY

Recently, the data communication rate in networks has increased tremendously. Consequently, Distributed Denial of Service (DDoS) assaults continue to pose a severe threat to the availability of online services. Although various ML and deep neural network models have been developed to identify them, there needs to be more transparency as the reasons behind their decisions are unknown. Due to this lack of transparency, it is becoming challenging for experts to trust and deploy the model in production. Additionally, experts are aware that the traffic patterns of DDoS attacks are continuously evolving, causing existing models to fail at times. With this XAI, experts can conclude or understand the most influential characteristics during changes in the behavior of the attack traffic. Moreover, they can focus more on such characteristics while designing a model.

4.4 EXPLAINABLE AI TAXONYM

XAI research aims to improve the interpretation of AI systems. Van Lent et al. [16] initially used the term XAI in 2004 to express how well their system could explain the actions of AI-controlled characters in simulation games. Figure 4.2 depicts the XAI taxonomy.

The literature typically categorizes two main categories: model reliance and scope of explainability [17]. Model reliance is further divided into model-specific and model-agnostic, while the scope of explanation is categorized into local and global, as described below.

Model specific: Model-specific techniques are dependent on the specific structures of the machine learning or deep learning model being used. These strategies are specifically applied to a certain model architecture.

Model agnostic: Model agnostic methods are those that are not dependent on a particular class of ML technique. They function with black box models and can be applied to any machine learning technique. They acquire explanations

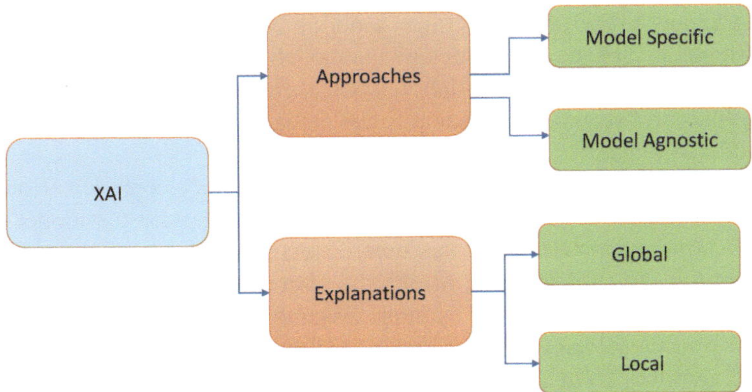

FIGURE 4.2 XAI taxonomic approach.

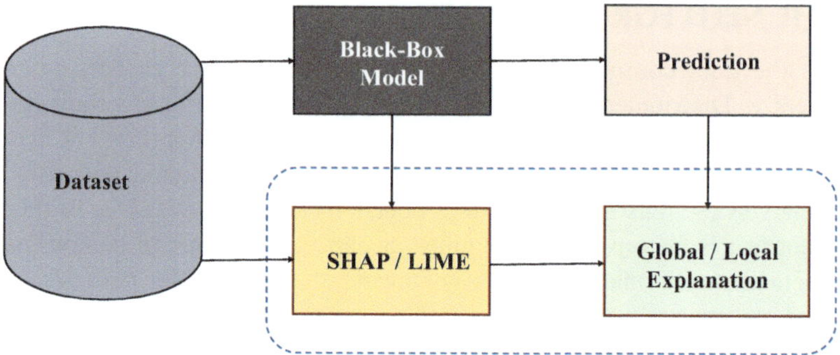

FIGURE 4.3 Explainable AI methods to interpret the model's predictions.

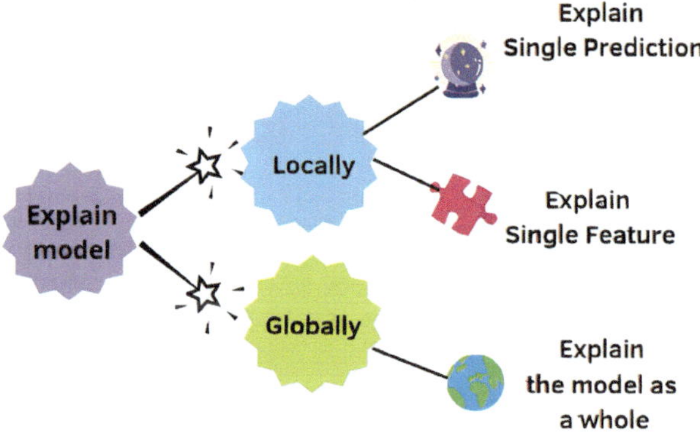

FIGURE 4.4 Explanations provided with explainable approaches.

by perturbing and modifying the input data and measuring the sensitivity of the performance of these changes in comparison to the performance of the original data. Figure 4.3. depicts the usage of model-agnostic methods SHAP and LIME to interpret the model's predictions. Figure 4.4 shows the different explanations provided by the model-agnostic XAI approaches.

Globally explaining the model: Global explainability enables the model designer to determine the extent to which each attribute contributes to the model's prediction across all the data.

Locally explaining the model: It is a self-explanatory approach. However, local explainability helps answer the question, "why did the model make this particular conclusion in this specific instance?". It means local explainability is essential for determining the core cause of a particular issue in DDoS attack detection. For example, assume that you have just discovered that your model has denied an instance of the network traffic, and you wish to know why? This local explainability would assist you in determining which factors (attributes) had the most significant impact on detecting the class of the instance.

4.5 EXPLAINABLE AI ARCHITECTURE

The architecture is determined by the specific methodologies and methods to enable interpretability and transparency in ML models. In general, the explainable AI architecture can be viewed as a mix of the following three components:

i. **The ML model:** It is the central component of explainable AI, representing the underlying methods used to draw predictions and conclusions from data. This component can be based on a variety of ML approaches, such as supervised, unsupervised, or reinforcement learning, and applies to a variety of applications, including cybersecurity, IoT security, healthcare (medical imaging), computer vision, manufacturing, insurance, autonomous vehicles, and so on.
ii. **Explanation algorithm:** The explanation algorithm is the component of explainable AI that provides insights and information about the model's most significant and relevant aspects. This component can be built on several explainable AI methodologies, such as feature importance, attribution, and visualization, and can provide valuable insights into the machine learning model's operation.
iii. **Interface:** The interface is the component of explainable AI used to display to humans the insights and data provided by the explanation algorithm. This component can be based on various technologies and platforms, such as web applications, mobile applications, and visualizations, and provide a user-friendly and intuitive interface for accessing and interacting with the data and insights obtained by the explainable AI system.

The architecture of explainable AI can be viewed as a mix of these three fundamental components that work together to enable transparency and interpretability in ML models. This architecture can make machine learning models more transparent, interpretable, trustworthy, and fair, providing valuable insights and advantages in various domains and applications.

4.6 EXPLAINABLE AI APPROACHES

In recent years, the model's complexity has increased due to the variety and huge volume of data. Assessing whether the selected features contribute to target identification is becoming difficult. Explainable AI techniques are gaining traction to address such issues. These techniques provide explanations for each feature contributing towards target prediction. Thus, the complexity of the model is reduced, and performance may be increased. The following are various approaches that provide explainability to the model.

4.6.1 SHAPLEY-ADDITIVE-EXPLANATIONS (SHAP)

This is an explainability technique that operates at the level of a single instance prediction. The method draws inspiration from the term "Shapley values," which is typically used in game theory to determine the pay-out for individual players within a unified coalition. In this context, an individual player's pay-out is determined by the magnitude or importance of the SHAP value corresponding to each player, and the player's contribution to the coalition's pay-out defines the magnitude. Lundberg [18] adapted the algorithm to calculate the Shapley value for each feature of an individual instance. The Shapley value now represents the feature's contribution to the final prediction. Consequently, each time the SHAP algorithm is applied to a specific instance, a set of Shapley values is produced, each of which is associated with a feature of the instance. Depending on the data, the contributions of each feature may vary.

Implementations of various Shapley value-based explanations are available in the SHAP library. They consist of the Tree-Explainer (which is quick and optimized for tree-based models), Gradient and Deep-Explainer (for neural networks), and Kernel-Explainer, which does not make any assumptions about the underlying model to be explained (similar to LIME).

4.6.2 Local Interpretable Model-Agnostic Explanations (LIME)

Local Interpretable Model-agnostic Explanations, commonly known as LIME, were introduced by Marco Tulio Ribeiro in 2016 to comprehend the complex reasoning behind any model's predictions. As its name suggests, this method uses local models to explain the predictions of any ML-based models. The authors claim that as LIME is independent of the classifier itself, the model may be used to explain any classifier, regardless of the prediction algorithm applied. Finally, LIME operates locally, meaning that it is specific to an instance, and, similar to SHAP, it will provide explanations for the prediction related to each instance. LIME attempts to fit a local model utilizing sample data points compared to the instance being explained. The local model could be included in the list of interpretable models, which also includes decision trees, linear models, and other models.

To understand how SHAP and LIME can be applied to various real-time problems, we analyze the work presented by Batchu and Seetha [19]. This work demonstrates clearly how SHAP and LIME can be utilized to interpret the predictions of any ML-based model. Their model was analyzed using the CIC-DDoS-2019 dataset.

By creating summary charts, this explanation aims to identify which attributes contribute more to the attack prediction. Each point (red and blue dot) on this plot represents a Shapley value and a feature ranging from low to high. Figures 4.5–4.9 display the global illustrations of the most contributing attributes collected using SHAP in conjunction with a wrapper-based recursive feature selection methodology. Summary plots are simple presentations that condense the entire dataset into a single graph. On the y-axis, each feature is ranked according to its importance. The feature at the top contributes the most to the forecasts, whereas the feature at the bottom contributes less or nothing. The x-axis is comprised of SHAP values. Contributions rise when the SHAP value moves away from zero, and zero implies that no contribution has been made at all. Whereas the direction of a value—positive or negative—indicates the class (attack or benign) that the feature is pushing towards the prediction. From the explanations obtained from these plots, it is noticed

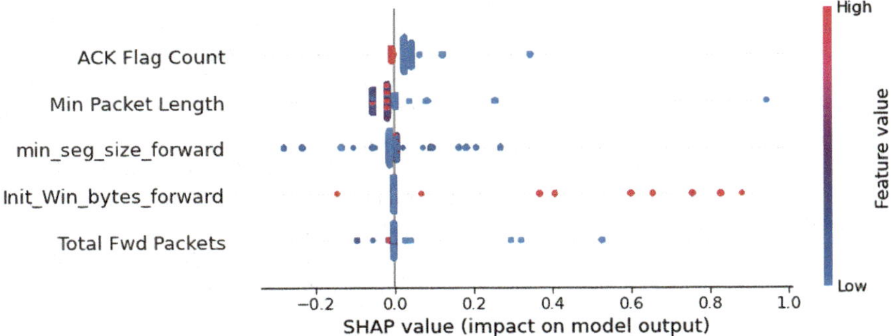

FIGURE 4.5 SHAP explanations with decision tree model.

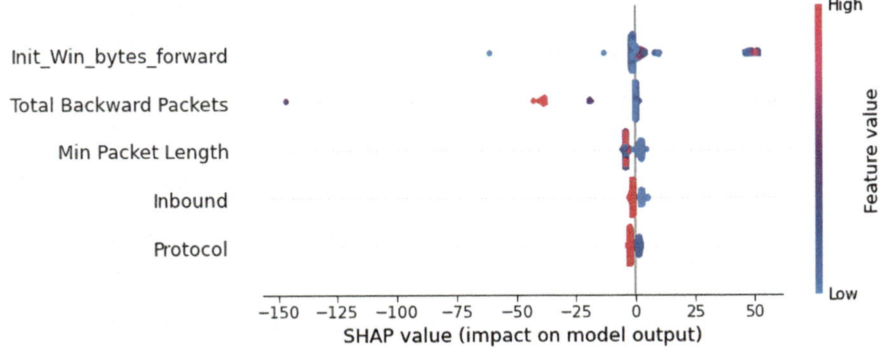

FIGURE 4.6 SHAP explanations with gradient boost model.

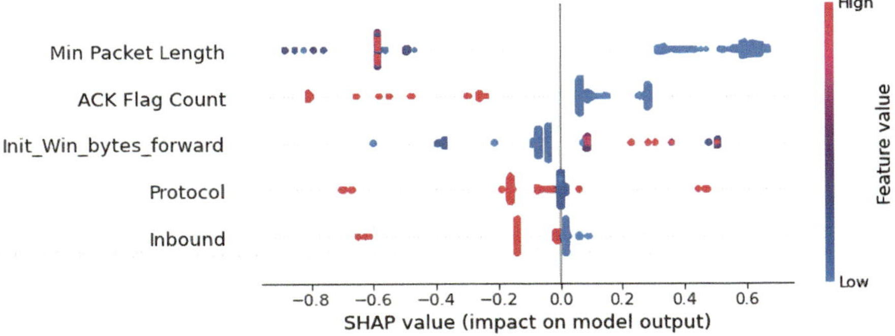

FIGURE 4.7 SHAP explanations with light gradient boost model.

that the extracted characteristics in their model contribute extensively to the model's prediction, with low values (blue dots) indicating an attack and high values (red dots) showing a benign packet. These explanations help the experts understand the model, which further increases trust [19].

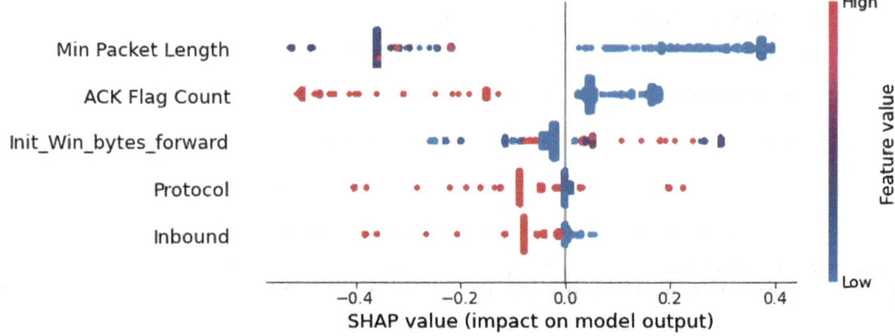

FIGURE 4.8 SHAP explanations with random forest model.

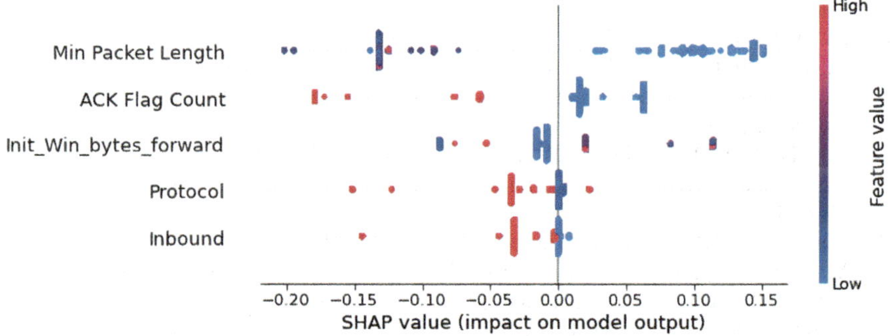

FIGURE 4.9 SHAP explanations with extreme gradient boost model

In addition to global explanations, local explanations are also provided for a specific instance utilizing the SHAP and LIME approaches. Figures 4.10 and 4.11 show how LIME generates local explanations. On a global scale, the decision function of a model may be highly complex, but it is simple and direct to approximate when examining an individual instance. LIME determines the characteristics that are most important for target class prediction. As illustrated in Figures 4.10 and 4.11 [19], the blue and orange colors reflect the contributing features and their prediction probabilities in estimating the target category. The prediction probabilities 0 and 1 represent the feature's probability score towards the attack and the benign case, respectively. The blue color features in Figure 4.10 for a sample are more likely to characterize traffic as an assault. For a sample in Figure 4.11, the orange color attributes were more likely to characterize the traffic as benign. These reasons differ depending on the traffic samples used. It has been discovered that attributes with a high likelihood score contribute more to determining whether a sample is benign or malicious.

In addition, the SHAP local explanations prediction process starts with the baseline value, which is the average of all forecasts. Then, the Shapley values of features are displayed as "forces". The Shapley value is represented in the force plot by an

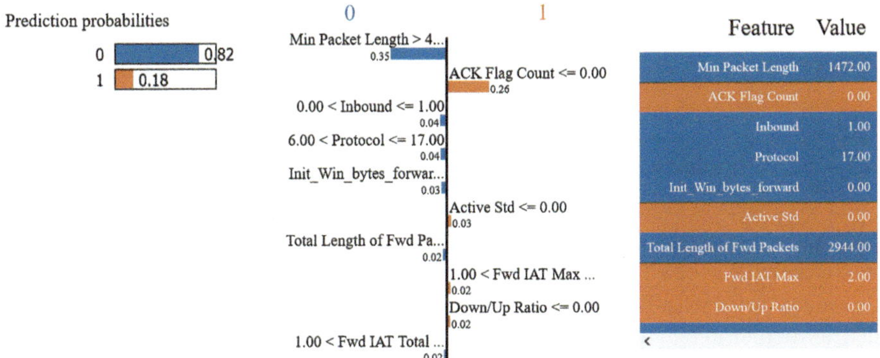

FIGURE 4.10 Lime local explanations for attack instance.

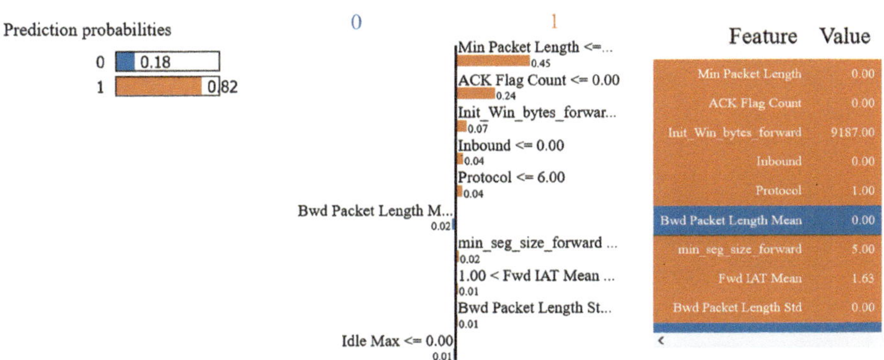

FIGURE 4.11 Lime local explanations for benign instance.

arrow that pushes the prediction to a significant positive or negative value, indicating that the data point strongly contributes to identifying the attack or benign class. For example, Figure 4.12 shows that the blue color features contribute to an attack instance with a baseline value of −13.70, and Figure 4.13 [19] shows that the red color attributes contribute to the benign sample. Thus, the local and global explanations help human experts identify the important attributes for the forecast of the target. Finally, these explanations can easily interpret any complex model and its decisions.

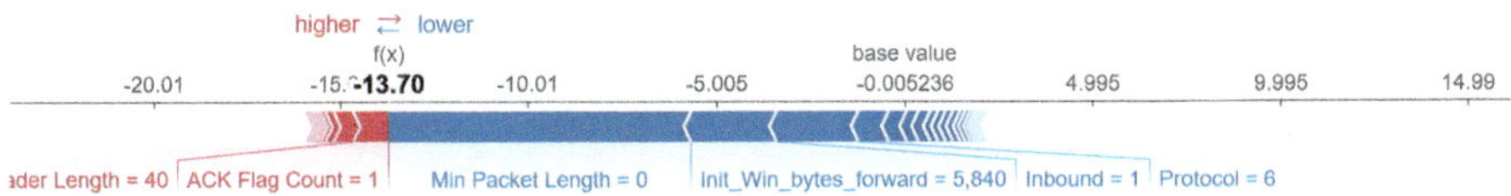

FIGURE 4.12 SHAP local explanation for attack instance.

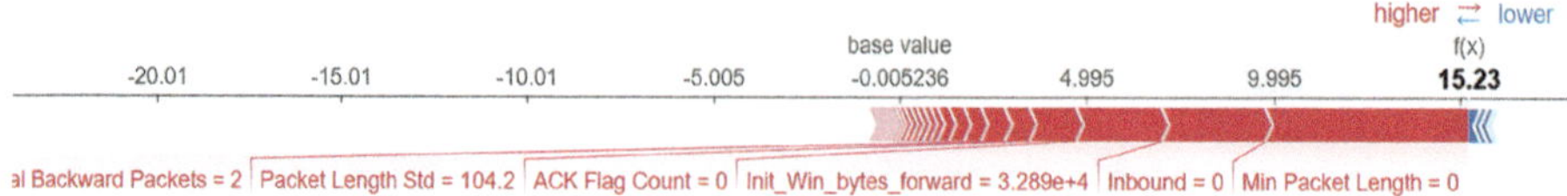

FIGURE 4.13 SHAP local explanation for benign instance.

4.6.3 ELI5

This technique is used to explain the predictions of a model, and several AI algorithms can work with ELI5. Moreover, it may be applied to models without features, such as depicting feature importance, which is worthwhile. Furthermore, it may determine Permutation Importance for specific prediction outcomes. As a result, it can be used as a tool to select local explanations. With the aid of ELI5, we can see how the most contributed features that the AI model utilizes to create classifications, and we can then produce both global and local explanations for the classification and the regression technique. Through this investigation, we were able to visualize the significance of the feature by ELI5 analysis partially. For analyzing feature importance by ELI5, we used a decision tree classifier.

Table 4.2 displays the importance of each independent attribute in the model. Green color indicates the most contributed features and red indicates less contributed

TABLE 4.2
Feature Score on Full Data Set

Weight	Feature
0.7600	Inbound
0.2346	URG Flag Count
0.0040	Active Mean
0.0007	Fwd Packet Length Max
0.0003	CWE Flag Count
0.0002	Init_Win_bytes_forward
0.0001	Flow IAT Mean
0.0001	Bwd Packets
0.0001	Flow Packets
0.0000	Flow IAT Mean
0.0000	SYN Flag Count
0.0000	Total Length of Fwd Packets
0.0000	Flow Duration
0.0000	min_seg_size_forward
0.0000	Fwd PSH Flags
0.0000	Down/Up Ratio
0.0000	Total Fwd Packets
0.0000	Max Packets Length
0	Active Std
	…24 more…

features. It is observed that features like Inbound and URG Flag Count are the top contributed features with the weight of 0.7600 and 0.2346. At the same time, Active Std is the less contributed feature.

Table 4.3 shows that the actual label of the 67th instance is 1, and the model predicted it as 1. It is indicated that features like Inbound, URG Flag Count, Active Mean, Max Packet Length, and Init_bytes_forward are the most contributed features to detect the instance as a DDoS attack. Table 4.4 shows that the actual label of the 756th instance is 0, and the model predicted it as 0. The features Inbound and URG Flag Count are identified as the most influential features for predicting the instance as a DDoS attack.

4.6.4 Boruta-SHAP

It is a wrapper feature selection model that combines the Boruta feature selection technique with Shapley values. This combination outperforms the original Permutation Importance technique in terms of both quickness and the integrity of the feature subset it produces. In addition to generating a high-quality subset of features,

TABLE 4.3
Feature Contribution on a Single Instance

	y = 1 (probability 1.00) Top Features	
Contribution?	**Feature**	**Value**
+0.500	<BIAS>	1.000
+0.381	Inbound	1.000
+0.118	URG Flag Count	0.000
+0.002	Active Mean	0.000
+0.000	Max Packet Length	0.149
_0.000	Init_Win_bytes_forward	0.000

TABLE 4.4
Feature Contribution on Single Instance

	y = 1 (probability 1.00) Top Features	
Contribution?	**Feature**	**Value**
+0.879	URG Flag Count	1.000
+0.500	<BIAS>	1.000
+0.001	Flow IAT Mean	0.000
+0.001	Init_Win_bytes_forward	0.000
+0.000	Flow IAT Min	0.000
+0.000	Flow Packets	0.013
+0.000	Bwd IAT Min	0.000
−0.381	Inbound	1.000

this strategy can also globally rank features in terms of their accuracy and consistency, making it suitable for model inference. Although BorutaShap offers speed improvements, the SHAP TreeExplainer scales linearly with the number of observations, making its usage in large datasets challenging. This issue is addressed by the sampling method of BorutaShap, which employs the smallest subsample of the data available at each iteration of the algorithm.

The initial step of the Boruta algorithm is to determine the feature importance. Typically, tree-based methods are used to accomplish this. However, in Boruta, the features compete with "shadow features," which are randomized replicas of their original versions.

To understand how the model generates feature importance, we begin by adjusting and fitting the LGBM model. To overcome the unpredictability in data selection, we repeat the process numerous times with different splitting seeds. Here is a list of the averaged feature importance scores shown in Figure 4.14. By observing Figure 4.14, the shadow random features are selected as important.

However, when the model is combined with Boruta+SHAP, it provides the most important features as shown in Figure 4.15. We integrated the feature selection technique with parameter tuning. To reduce unpredictability in data selection, we repeat the entire process for various splitting seeds. We store the chosen features for each attempt while taking into account both the SHAP and common tree-based feature importance. This allows us to visualize the frequency with which a feature is chosen at the conclusion of a trial. By observing Figure 4.15, it is noted that Boruta+SHAP did not consider the random features. It has determined the most contributed features of the CIC-IDS 2017 dataset for detecting DDoS attacks.

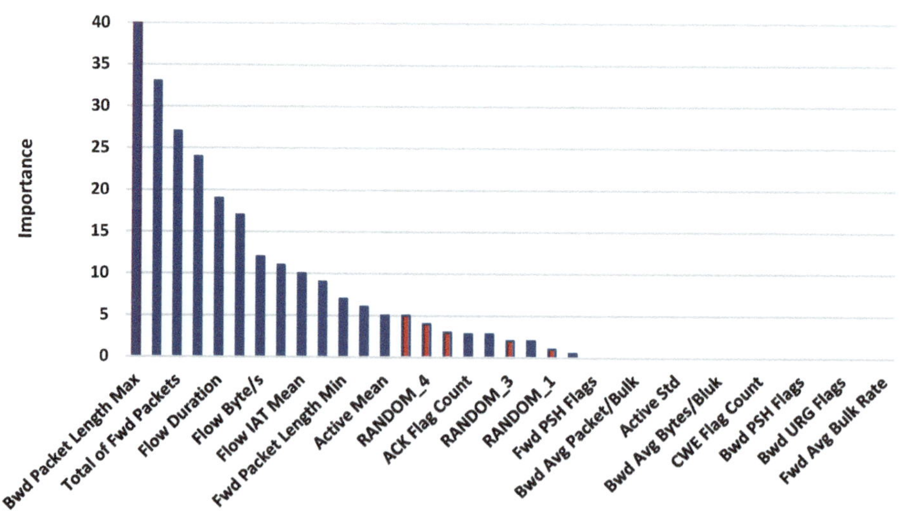

FIGURE 4.14 Averaged feature importance.

FIGURE 4.15 Final feature importance with Boruta+SHAP.

4.6.5 Partial Dependence Plot (PDP)

The partial dependence plot (PDP) can be used to highlight how one or two features may only slightly affect the outcome predicted by a machine learning model. It can show if the relationship between a target and a feature is linear, monotonic, or more complex. Visualization techniques can help identify an attack without depending on the classifier being employed. PDPs provide information regarding the way in which the desired response and an input feature of interest interact with one another. Therefore, we have provided the PDP plots for various features of the NSL-KDD dataset as shown in Figures 4.16 and 4.17 with the XGBM classifier. Figure 4.16 shows that as the feature diff_srv_rate value increases, the prediction value raises, which means the diff_srv_rate is one of the contributing features. By observing Figure 4.17, one would anticipate that this attribute would have no impact on the classification result as it is linear. Therefore, PDP plots can be effectively used for identifying feature importance. The primary motivation is that a flat PDP implies that the feature is unimportant, whereas the greater the variation in the PDP, the more essential the feature is.

Marino et al. [20] generated explanations for the incorrect classification of a set of samples made by IDS. The adversarial approach was created to identify the minimal changes needed to classify the misclassified occurrences appropriately. The relevant features responsible for the misclassification are then found and visualized using

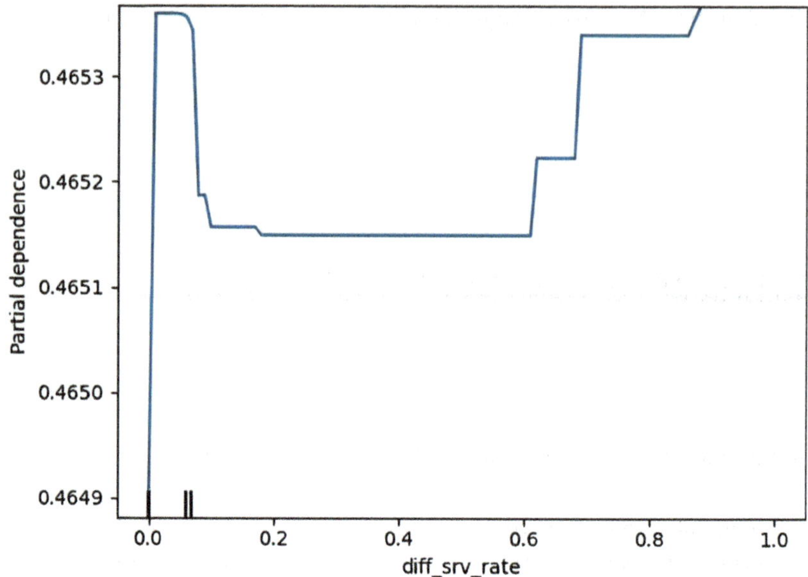

FIGURE 4.16 PDP plot for feature diff_srv_rate of NSL-KDD dataset.

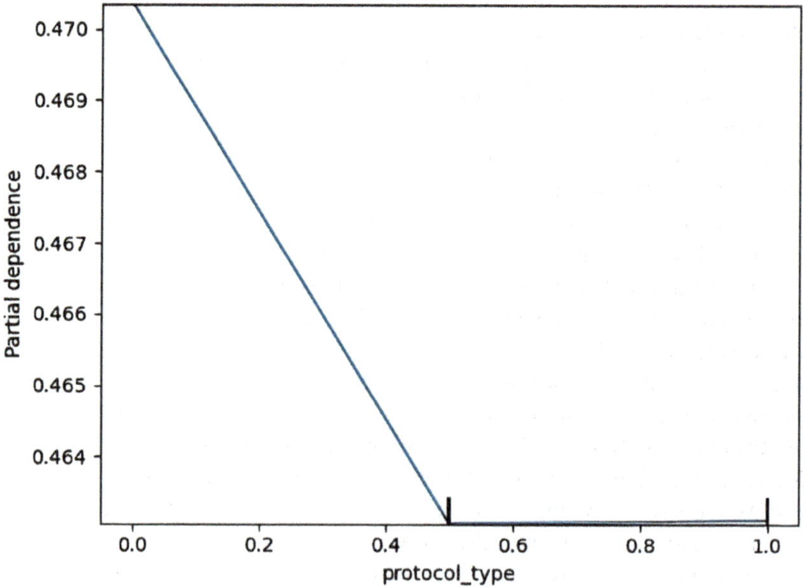

FIGURE 4.17 PDP plot for protocol type of NSL-KDD dataset.

these changes. The experimental evaluation of this approach was conducted using the NSL-KDD traffic dataset, employing multi-layer perceptron and linear classifiers.

To identify the reasons behind decisions made by the ML model, Wang et al. [21] developed a novel framework that generates local and global explanations. The global explanations are provided using the SHAP technique, which includes the key attributes, the relationship between attributes, and several assault types. In contrast, LIME is used to provide local explanations for a traffic instance. This framework was evaluated on the NSL-KDD traffic dataset, and the experimental evaluation shows consistent results with the attributes obtained for distinct attacks. To allow dynamic network access control (DNAC) by producing flow rules, Guo et al. [22] and Li et al. [23] followed an approach called LEMNA. This approach aims to explain the results of anomaly-based intrusion detection systems using an interpretable model and to produce access control policies based on the explanations. LEMNA seeks to deliver reliable results and contributes to developing standards for DNAC. However, this method must account for the interdependence between many records to increase explanation accurateness.

Therefore, XAI can assist cybersecurity professionals in understanding the predictions of IDS, enabling them to make more accurate decisions. Additionally, it helps security experts gain a better understanding of real-world cyber-attacks.

4.7 CONCLUSION

With the advent of modern technologies, network traffic volumes and complexity have increased drastically. Consequently, cyber-attacks such as DDoS attacks are growing enormously. To mitigate such attacks, various ML approaches have been implemented. However, due to the complexity and multiple attributes within the traffic, the ML model itself cannot perform well. Additionally, to enhance the model's performance, feature selection approaches play a vital role. In this work, various types of feature selection techniques like filter, wrapper, and embedded methods are discussed. Moreover, to build a trustworthy ML model, the importance of explainable approaches like SHAP and LIME, ELI5, PDP plots, and Boruta+SHAP are discussed concerning a use case. Evaluating novel XAI algorithms for detecting attacks in IoT and mobile networks will continue to be an essential topic for future research.

REFERENCES

1. Z. Q. Wang, D. K. Zhang, A. Khraisat, I. Gondal, P. Vamplew, and J. Kamruzzaman, "HIDS and NIDS hybrid intrusion detection system model design," *Adv. Eng. Forum*, vol. 2, no. 1, pp. 991–994, 2019, doi: 10.4028/www.scientific.net/aef.6-7.991.
2. R. K. Batchu and H. Seetha, "A generalized machine learning model for DDoS attacks detection using hybrid feature selection and hyperparameter tuning," *Comput. Netw.*, vol. 200, p. 108498, 2021, doi: 10.1016/j.comnet.2021.108498.
3. S. B. Mallampati and H. Seetha, "A review on recent approaches of machine learning, deep learning, and explainable artificial intelligence in intrusion detection systems," *Majlesi J. Electr. Eng.*, vol. 17, no. 1, pp. 29–54, 2023, doi: 10.30486/mjee.2023.1976657.1046.

4. P. García-Teodoro, J. Díaz-Verdejo, G. Maciá-Fernández, and E. Vázquez, "Anomaly-based network intrusion detection: Techniques, systems and challenges," *Comput. Secur.*, vol. 28, no. 1–2, pp. 18–28, 2009, doi: 10.1016/j.cose.2008.08.003.
5. R. K. Batchu and H. Seetha, "On improving the performance of DDoS attack detection system," *Microprocess. Microsyst.*, vol. 93, p. 104571, 2022, doi: 10.1016/j.micpro.2022.104571.
6. S. Han, H. Huang, and H. Qin, "Automatically redundant features removal for unsupervised feature selection via sparse feature graph," pp. 1–21, 2017, doi: 10.48550/arXiv.1705.04804.
7. Z. M. Hira and D. F. Gillies, "A review of feature selection and feature extraction methods applied on microarray data," *Hindawi Publ. Corp. Adv. Bioinforma.*, 2015, doi: 10.1155/2015/198363.
8. R. M. Terol, A. R. Reina, S. Ziaei, and D. Gil, "A machine learning approach to reduce dimensional space in large datasets," *IEEE Access*, vol. 8, pp. 148181–148192, 2020, doi: 10.1109/ACCESS.2020.3012836.
9. X. Li, P. Yi, W. Wei, Y. Jiang, and L. Tian, "LNNLS-KH: A feature selection method for network intrusion detection," *Secur. Commun. Netw.*, vol. 2021, 2021, doi: 10.1155/2021/8830431.
10. G. H. John and R. Kohavi, "Wrappers for feature subset selection," *Artif. Intell.*, vol. 97, pp. 273–324, 1997.
11. F. H. Almasoudy, W. L. Al-Yaseen, and A. K. Idrees, "Differential evolution wrapper feature selection for intrusion detection system," *Procedia Comput. Sci.*, vol. 167, pp. 1230–1239, 2020, doi: 10.1016/j.procs.2020.03.438.
12. Keerthipriya, N., Hemalatha, K., Prasad, H. and Arunkumar, T., "Wrapper based feature selection for disease diagnosis using optimization algorithms," *Int. J. Eng. Res. Technol.*, vol. 6, no. 04, pp. 1–10, 2018.
13. R. Panthong and A. Srivihok, "Wrapper feature subset selection for dimension reduction based on ensemble learning algorithm," *Procedia Comput. Sci.*, vol. 72, pp. 162–169, 2015, doi: 10.1016/j.procs.2015.12.117.
14. D. Sudaroli Vijayakumar and S. Ganapathy, "Feature reduction using Lasso hybrid algorithm in wireless intrusion detection system," *Int. J. Innov. Technol. Explor. Eng.*, vol. 8, no. 11, pp. 1476–1483, 2019, doi: 10.35940/ijitee.J9810.0981119.
15. R. K. Batchu and H. Seetha, "A hybrid detection system for DDoS attacks based on deep sparse autoencoder and light gradient boost machine," *J. Inf. Knowl. Manag.*, vol. 22, no. 1, 2022, doi: 10.1142/S021964922250071X.
16. M. Van Lent, W. Fisher, and M. Mancuso, "An explainable artificial intelligence system for small-unit tactical behavior," *AAAI Conference on Artificial Intelligence*, 2004.
17. S. Neupane, et al., "Explainable intrusion detection systems (X-IDS): A survey of current methods, challenges, and opportunities," pp. 1–25, 2022, [Online]. Available: https://arxiv.org/abs/2207.06236.
18. S. M. Lundberg and S. I. Lee, "A unified approach to interpreting model predictions," *Adv. Neural Inf. Process. Syst.*, vol. 2017, no. Section 2, pp. 4766–4775, 2017.
19. R. K. Batchu and H. Seetha, "An integrated approach explaining the detection of distributed denial of service attacks," *Comput. Netw.*, p. 109269, 2022, doi: 10.1016/j.comnet.2022.109269.
20. D. L. Marino, C. S. Wickramasinghe, and M. Manic, "An adversarial approach for explainable AI in intrusion detection systems," *Proc. IECON 2018-44th Annu. Conf. IEEE Ind. Electron. Soc.*, pp. 3237–3243, 2018, doi: 10.1109/IECON.2018.8591457.
21. M. Wang, K. Zheng, Y. Yang, and X. Wang, "An explainable machine learning framework for intrusion detection systems," *IEEE Access*, vol. 8, pp. 73127–73141, 2020, doi: 10.1109/ACCESS.2020.2988359.

22. W. Guo, D. Mu, J. Xu, P. Su, G. Wang, and X. Xing, Lemna: Explaining deep learning based security applications," *Proc. ACM Conf. Comput. Commun. Secur.*, pp. 364–379, 2018, doi: 10.1145/3243734.3243792.
23. H. Li, F. Wei, and H. Hu, "Enabling dynamic network access control with anomaly-based IDS and SDN," *SDN-NFV 2019- Proc. ACM Int. Work. Secur. Softw. Defin. Networks Netw. Funct. Virtualization, co-located with CODASPY 2019*, pp. 13–16, 2019, doi: 10.1145/3309194.3309199.

5 Adaptation of XAI for Smart Agriculture Systems

A. Anitha

5.1 INTRODUCTION

The concept of soft computing begins with an approximation of the values and aims to provide better solutions for approximation models rather than precision models. One of the best techniques used to solve approximation is the neural network. The neural network operates by updating weights and reducing error rates to achieve provided targets. Due to its internal processing being beyond human understanding and its complex computational process, the neural network is often referred to as a black box. Therefore, it is imperative to understand a model's design and to explain the process to enhance the quality of the working model.

5.1.1 Interpretability and Explainability

A neural network is a method in the domain of artificial intelligence that enables computers to process data in a manner that mimics the human brain. Thus, humans aim to make machines learn as humans do, neural network is a type of machine learning process. Concerning machine learning methods, the neural network employs a supervised learning process. However, the network structure and computation process reduces the transparency of connections and weight estimations in each layer of the neural network. The result produced by the neural network depends on various hyperparameters such as weight, bias, input vector values, activation functions, momentum, and vigilance parameters to obtain the desired output. Therefore, it is necessary to minimize hidden computations and ensure a transparent computation process to facilitate the decision-making process, build trust and adoption, mitigate regulatory and other risks, and in turn increase productivity. Despite AI models having better accuracy, fairness, transparency, and AI-powered decision-making processes, implementing them in real-time fosters trust and confidence to transition AI models into production. One of the major concerns with AI design is its black-box nature. It is challenging to estimate the associations between input characteristics and the reasons behind the results produced based on computations. The level of model interpretability is thus divided into post-hoc and intrinsic categories, as depicted in Figure 5.1.

DOI: 10.1201/9781003442509-5

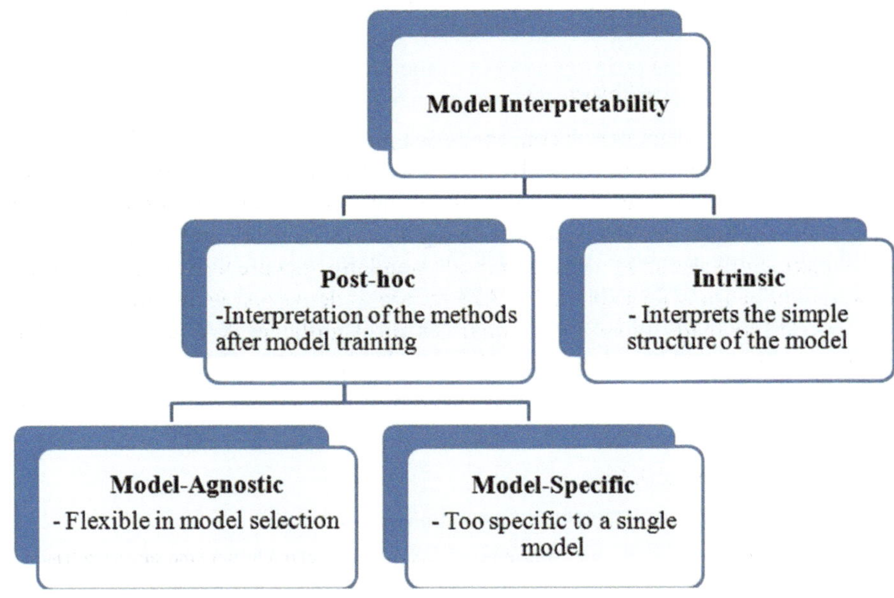

FIGURE 5.1 The taxonomy of the interpretability.

Machine learning techniques incorporate the explanation of AI models as part of their prediction process. For a better understanding of a model's behaviour, Explainable AI (XAI) provides explanations that are understandable by humans after the model makes predictions. A taxonomy of explainability is depicted in Figure 5.2.

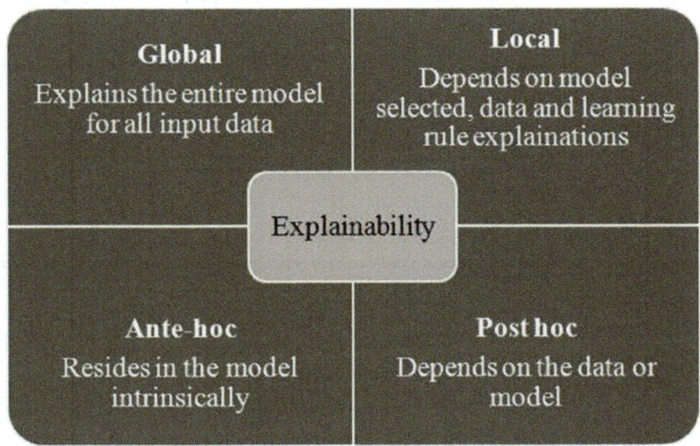

FIGURE 5.2 The aspects of the explainability.

5.1.2 Global Explainability

The explanations of the training process of the models fall into two categories. The first is global explanations, which focus on understanding the decision-making process of the models based on a comprehensive view of the features, including learned components such as weights, bias, error rate, and other parameters. One of the best models for global explainability is the global surrogate model. This interpretable model helps understand the predicted approximations made by the machine learning techniques, acting as a black box model. Surrogate models are constructed based on three components: (i) Selection of sample objects, (ii) Training the model parameters, and applying optimization techniques for (iii) Evaluating metrics used to obtain accuracy. The advantage of surrogate models help address issues in problem formulation and optimize results by improving understanding of the model through interactions. Some of the steps involved in building global surrogate models are as follows:

- Select a dataset for training the surrogate model.
- Obtain predictions using the black box model.
- Select a global interpretable model, such as a classifier.
- Choose an interpretable model suitable for the selected dataset and train it to generate the resulting prediction set.
- Measure the results given by the surrogate model with explanations of the results from the black box model.
- Interpret the surrogate model.

5.1.3 Local Explainability

This section focuses on the decisions made by individual models. It operates on the basic idea of transitioning from linear models to non-linear models. Since decisions are based on approximation, local explanations create an input space that approximates the neighbourhood values around the input. Local Interpretable Model-agnostic Explanations (LIME) serve as a good example of local explainability, explaining machine learning classifiers for a single instance with suitable local explanations. These explanations remain faithful to the observation of surrounding input values. Typically, regression models are applied to the obtained features, where the predicted output will have a similar magnitude as the original prediction model.

To understand the XAI model interpretation, this chapter aims to adapt agricultural crop suitability classification in various locations within the Vellore district, Tamil Nadu. Agriculture remains one of the most important sectors where automation is still pending, particularly concerning crop production. Since crop production depends on various parameters such as soil and water resources, rainfall, quality of seeds, and climatic conditions, the chapter aims to identify the important features that recommend a suitable location to cultivate the crops.

The chapter is structured with an introduction outlined in Section 5.1, followed by a review of related works in the agriculture domain under the purview of AI and XAI. XAI applications in the agriculture domain, including feature reduction, dimensionality reduction, and surrogate models, are discussed in Section 5.2. The proposed research model is explained in Section 5.3, with model interpretation using Eli 5. Section 5.4 demonstrates the experimental analysis, supported by graphs and tables, whereas Section 5.6 provides the conclusion of the chapter.

5.2 RELATED WORK

This section discusses the related research performed by researchers on AI, XAI, and surrogate mechanisms in agriculture.

5.2.1 AI in the Agriculture Domain

The optimized application of AI for automatic irrigation and weeding systems using robotics solves the problem of crop monitoring and pesticide spraying (Talaviya et al., 2020). The reduction of crop nitrogen inputs by integrating multi-source data provides accurate guidance for a better decision-making process (Yang et al., 2020). A rough set of fuzzy approximation spaces can be used for feature reduction, and the consistent data applied to neural networks to predict crop suitability in some regions of India provides better accuracy in dealing with uncertainty (Anitha and Acharjya, 2018). Detection of vegetable diseases using k-means clustering and support vector machines on features such as colour, shape, size, and texture for classification and detection process (Rahamathunnisa et al., 2020). AI is used in automating weed control by utilizing plant signalling for real-time identification using smart techniques (Su et al., 2020).

5.2.2 Explainable AI in Agriculture

Despite the existence of deep learning-based AI techniques, a new era of smart agriculture emerges with the potential for problem-solving using meaningful features. While deep learning offers greater accuracy than other machine learning models and deals with huge data, expressing the interpretations obtained and establishing trust in the decisions made remains challenging. As a result, several studies have been devoted to exploring explainability in the agriculture sector.

5.2.2.1 XAI via Dimension Reduction

Dimension reduction can be defined as the process of converting high-dimensional data into a smaller dimension while retaining the same information. Factor Analysis and Principal Component Analysis (PCA) are dimension-reduction techniques associated with AI methods. PCA dimension reduction is employed to provide an effective solution for early diagnostics of plant stress based on XAI using hyperspectral and thermal infrared sensor data, achieving significantly reduced computation time for local linear predictions (LySov et al., 2023). The interpretability and accessibility of machine learning are reviewed in terms of scientific, statistical, and semantic interpretations in agriculture, food processing, and health domains (Ranasinghe et al., 2022). A variational autoencoder with XAI as representation learning is utilized to provide optimal N-rate-reduced data recommendations for predicting nitrogen requirements in rice production (Iatrou et al., 2022). Crop suitability data are reduced using a fuzzy approximation space with a neural network (Anitha and Acharjya, 2019). A framework called DDS-XAI is employed to extract relevant and meaningful rules from a high-dimensional topographic map by applying project-based dimension reduction (Thrun et al., 2021). A rough set of fuzzy approximation spaces is utilized for dimension reduction in precipitation prediction using deep learning techniques (Manna and Anitha, 2023a).

5.2.2.2 XAI via Feature Extraction

For AI models to be interpretable, significant features should be explained based on their significance and correlations. The most commonly used techniques for feature extraction are autoencoders, CNNs, and wavelet transformations. This study compares nine statistical models to 15 conditional features and produces a saliency map that can be interpreted based on individual scale layers (Mundhenk et al., 2019). A ResNet model was used for feature extraction, while VGG and GoogleNet were used for interpretability. Maintaining the integrity of the ResNet-CBAM model is crucial for detecting plant leaf diseases (Wei et al., 2022). This paper employed CNNs for feature extraction during the cultural diagnosis of pest control and DNNs for classification (Couliably et al., 2022). The development of Botanic-AI aims to identify tomato leaf disease at an early stage using XAI, aiding farmers in overcoming the damage to the epidermis caused by the disease (Bhandari et al., 2023). An interactive unit for the analysis and prediction of stress for a certain period was developed by Maximova et al. (2021). Feature reduction using variance inflation factor and randomized low-rank approximation was integrated with a deep ensemble model for groundwater prediction (Manna and Anitha, 2023b).

5.2.2.3 XAI via Surrogate Models

An effective approach of XAI in the agriculture domain involves recognizing factors that enhance crop yield prediction using LIME, Explain Like I'm 5 (ELI5), or SHapley Additive explanations (SHAP). These models offer explanations for any classifier with surrogate interpretable and faithful representations.

LIME was utilized to help interpret the local-scale behaviour of reproducible DSM analysis with R for species distribution models (Ryo et al., 2021). XAI models ELI5, PDPbox, Skater, and PDPboxes were employed for data models to train agricultural workers with Python in identifying agricultural worker nature (Kawakura et al., 2022). Crop yield analysis with R language using LIME to identify the variables necessary for prediction by regression. It also determines the variables relevant for prediction, their importance to the response variable, and whether various machine learning approaches provide the same results (Sauder et al., 2020). Fuzzy logic is employed to decide on various data such as the number of sensors, type of crop, and adaptable soil and weather conditions, modifying the intelligent inference system to improve the interpretability of the proposed model (Sabrina et al., 2022). The survival analysis of bladder cancer patients using DeepSurv and Cox models was proposed by (Supriya and Anitha, 2024). The fuzzy logic model is used to estimate the crop's water requirements, and the results are compared to the actual water requirements. The findings indicate that the fuzzy logic model accurately estimates the crop's water requirements, effectively optimizing water usage in agricultural settings. The model is accurate and reliable and can be used to enhance water management in agricultural systems. A saliency-based processing pipeline for microscopic images of grape powdery mildew genetic analyses for disease resistance breeding is developed. The authors used interpretable methods to explain the need for saliency maps for disease-resistant breeding (Qiu et al., 2022). Forecasting the air quality index using stacked LSTM to identify better attributes and also by utilizing combinational LSTM (Manna and Anitha, 2022; Manna and Anitha, 2023c).

Limitations of the existing techniques

- High execution time
- High risk of overfitting and error loss
- Unidentified model interpretation and weights of each feature with the decision class
- Limited visualization of the model realization

Contribution to the chapter:

- Model interpretation using XAI methods
- Feature reduction using machine learning with better visualization of the weights of each feature upon decision class
- Obtained better accuracies with reduced error rate due to proper tuning of hyperparameter
- Representation of the contribution of weight factors to the network for the decision class using ELI5

5.3 PROPOSED RESEARCH METHOD

A detailed description of the proposed model of the desired problem is presented in this section.

5.3.1 Formulation of the Problem for the Proposed Model

In this section, an analysis of crops in the Vellore district is conducted to determine crop suitability. The objective of the research model is to predict the suitable place of cultivation for a particular crop over a specific period. The primary advantage is to assist laymen in identifying the crop to be cultivated based on soil richness, water characteristics, and rainfall. To analyze the proposed model, past and historical data were collected from the Krishi Vigyan Kendra of Vellore district. The dataset contains 4977 objects with 26 attributes. These attributes are classified into soil and water characteristics, with major components such as Nitrogen, Phosphorus, and Potassium (NPK). The soil characteristics include Soil pH, soil texture, soil type, moisture content, organic matter, nitrogen level, phosphorus level, potassium level, and micronutrients such as copper, zinc, manganese, and iron. The water characteristics include water pH and water-soluble components such as calcium, nitrate, magnesium, nickel, vanadium, silicon, and chloride. Superfluous attributes are identified

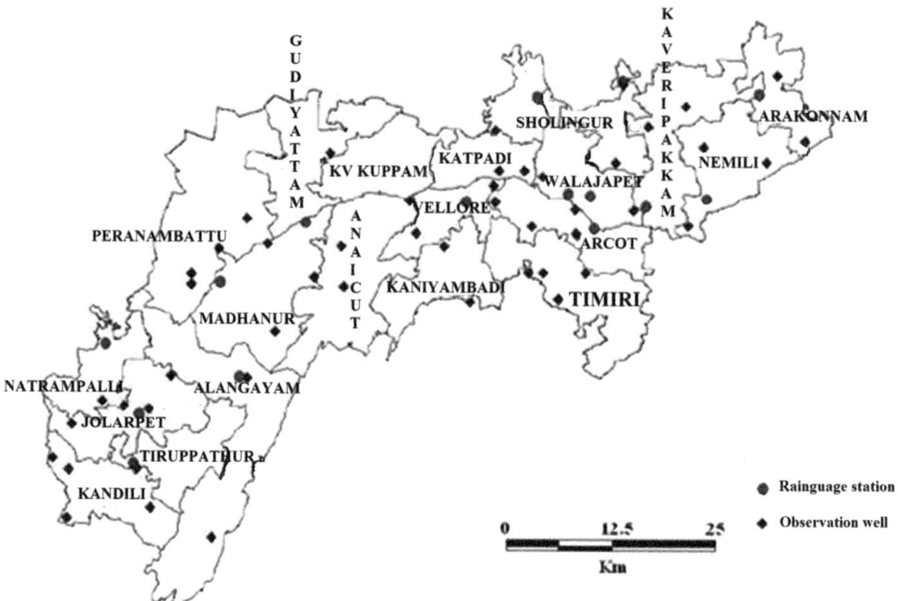

FIGURE 5.3 The study area of the proposed model.

using permutation importance with the SHAP package by mapping the data points to a high-dimensional data space and finding the hyperplane to divide the classes. This provides a way of determining feature importance for any black-box estimators based on the calculated weight to estimate which features contribute to the decision-making process. Thus, the superfluous attributes are removed, resulting in 15 important attributes for each crop and its suitable cultivation place. After pre-processing steps such as handling missing data, removing data duplication, and resolving conflicting objects, the dataset is reduced to 3987 objects for further analysis.

The crop production in the Vellore district includes soil and water components, as well as rainfall during the northeast and southwest monsoons. Organic matter and soil moisture are crucial contributors to crop productivity. Domain experts recommend checking the NPK levels in the soil at various stages before cultivation to minimize or avoid the need for inorganic chemical fertilizers. Moreover, the Vellore district is officially divided into 7 taluks and 20 blocks. The study area of the proposed model is illustrated in Figure 5.3, with notations represented in Table 5.1 for the attributes considered, their possible values, maximum range, and description. Table 5.2 presents a sample dataset considered for the proposed model.

5.3.2 XAI Feature Reduction Using SHAP

SHAP is an individual model-agnostic explainer used to access input data. It was introduced by Lloyd Shapley in 1951 based on set characteristic functions of permutations. When each feature contributes to decisions, their contributions are calculated as weights, either positive or negative, through many permutations for each combination called a coalition. SHAP possesses properties of efficiency, symmetry, dummy,

TABLE 5.1
Notation Representation Table for Crop Production

Agriculture Contingency Factors	Abbreviations	Notations	Possible Values	Max Value	Description
Soil pH	SPH	a_1	[5.4–8.5]	8.5	The pH level of the soil
Moisture	MOI	a_2	[5–12.2]	12.2	The Moisture content of the soil
Organic Matter	OM	a_3	[0.65–1.98]	1.98	Organic matter present in the soil
Availability of Nitrogen	N	a_4	[200–800]	800	Level of nitrogen
Availability of Phosphorus	P	a_5	[40–533]	533	Level of phosphorus
Availability of Potassium	K	a_6	[115–1045]	1045	Level of potassium
Copper	Cu	a_7	[0.05–2]	2	The copper content in soil
Zinc	Zn	a_8	[0.01–2]	2	The zinc content in soil
Manganese	Mn	a_9	[0.7–4.6]	4.6	The manganese content in soil
Iron	Fe	a_{10}	[1.98–99.6]	99.6	The iron content in soil
Water pH	WPH	a_{11}	[6.2–8.5]	8.5	The pH level of the water
Calcium	Ca	a_{12}	[11–420]	420	Calcium content in the water
Nitrate	No_3	a_{13}	[16–140]	140	Nitrate content in water
Magnesium	Mg	a_{14}	[21–280]	280	Magnesium content in water
Rainfall	R	a_{15}	[773.4–1111.2]	1111.2	Average rainfall of north-east and south-west monsoon
Places	PL	d	-	-	Place of crop suitable for cultivation

and additivity. The class Explainer() in the SHAP library utilizes the training dataset using the method shap_values() to calculate SHAP values. The explainer class takes 100 rows of the training dataset and calculates the SHAP values, providing feature vectors to be explained. The contribution of each feature, provided with SHAP values, indicates the importance of the feature for the decision-making process. However, as the number of features increases, the total number of possible permutations also increases, potentially reducing model performance. The time complexity is calculated based on the features (f) as given in Equation (5.1).

$$O\left(f2^{f-1}\right) = O\left(f^{2f}\right) \tag{5.1}$$

Thus, the time complexity for the f features with k rows can be $O\left(kf^{2f}\right)$ becomes computationally intractable. This issue can be overcome by using Kernel SHAP, an approximation method that uses weighted linear regression to compute the significance of the features. The Kernel SHAP values are given by Equation (5.2)

TABLE 5.2
Sample Dataset of Suitable Places for Crop Production

Objects	a_1	a_2	a_3	a_4	a_5	a_6	a_7	a_8	a_9	a_{10}	a_{11}	a_{12}	a_{13}	a_{14}	a_{15}	d
x_1	7.3	9	0.96	350	200	160	1.2	1.8	3.2	61.2	7.3	20	16	226	787.9	Alangayam
x_2	7.2	11.7	0.99	450	130	115	1.2	1.09	1.09	75	8.5	21	17	25	1045.4	Anaicut
x_3	7.21	11.5	0.91	360	200	645	0.5	1.5	0.8	69	7.1	72.3	45	77.3	1111.2	Arakonam
x_4	7.35	9.5	0.78	432	40	150	1.6	3.3	1.7	61	7.36	23	63	280	1052.2	Arcot
x_5	7.5	7	0.78	200	44	162	0.05	1.2	2.49	57	6.3	39	78	259	995	Gudiyatham
x_6	5.4	6.1	1.23	560	476	486	0.5	3.5	4.6	47	6.2	40.8	84	166	890	Jolarpet
x_7	7.47	8	1.32	475	120	310	0.45	1.1	1.2	1.98	7.43	11	78	26	999.3	KV Kuppam
x_8	6.2	6.7	0.98	345	527	1045	0.9	4.7	2.7	2.2	6.35	80	56	250	894.3	Kandeli
x_9	6.3	7	1.2	401	222	672	0.05	0.01	0.7	8.4	6.35	53.8	45	21	1037.5	Kaniyambadi
x_{10}	7.1	5	1.32	400	160	160	1.9	2.1	2.4	51.1	8.3	148	25	176	1004.4	Katpadi
x_{11}	7.45	8	1.67	540	242	370	1.5	5	3.3	8.8	7.41	420	56	110	998.7	Kaveripakkam
x_{12}	7.2	11.9	1.53	200	160	220	1.8	1.2	3.4	61.4	7.4	20	140	211	773.4	Madhanur
x_{13}	8.5	10	1.52	800	190	340	1.2	2.4	1.8	41.2	8.31	70	130	120	999	Natrampalli
x_{14}	7.32	12.1	1.98	645	140	120	1.4	2.2	3.5	64	7.42	12	18	27	1008.6	Nemili
x_{15}	7.4	9	1.32	450	160	325	1.6	1.6	1.8	62.1	7.2	45	41	23	885.2	Pernambattu
x_{16}	8.47	8	0.65	340	533	477	0.51	2.5	1.68	7.57	7	138	45	71	891.2	Sholingur
x_{17}	7.1	11.8	0.98	340	349	476	0.5	3.6	0.8	4.5	7.2	128	23	211	1012.6	Thimiri
x_{18}	5.5	10	0.88	650	170	150	1.1	1.9	4.6	51.5	7.28	60	126	130	880.5	Tirupathur
x_{19}	7.2	8	0.92	460	120	140	1.1	1.2	3.2	60.2	8.11	118	24	69.5	1032.2	Vellore
x_{20}	7.21	12.2	1.68	340	480	240	2	3.8	4.2	99.6	7.21	51	57	206	1008.1	Walajahpet

$$\mu_x(y_{in}) = \frac{(f-1)}{\left(f_{\text{choose}}|y_{\text{in}}|\right)|y_{\text{in}}|\left(f-|y_{\text{in}}|\right)^t} \quad (5.2)$$

where f = features, $|y_{\text{in}}|$ is the number of non-zero features in the simplified input $|y_{\text{in}}|^t$.

5.3.2.1 Kernel Explainer

Kernel Explainer is a function used to calculate SHAP values for the considered dataset using machine learning. The argument model is employed to provide the sample input as the matrix to the classification model. To train the surrogate model, a random forest is utilized in this chapter to generate the perturbed dataset. The missing values are imputed using the background dataset. For larger problems, either a single reference value is used or the k-means function is employed to find the missing values. L1 regularization is utilized for selecting features to explain their significance based on the values of the weights. The type of dataset is addressed as int64 or float [21].

5.3.2.2 Classification Using XAI Using ELI5 Model Interpretation

ELI5 is an XAI package that helps understand the machine learning classifiers and is also used to debug algorithms such as XGBoost, CatBoost, and Keras. ELI5 performs better with preprocessed data. Once the important features are identified using SHAP, the next step is to check for non-zeros and partition the dataset into training and testing using label encoding for categorical values. For training the model, both the decision tree classifier and XGBoost classifier are applied. The decision tree classifier employs a criterion of the "Gini" index, a minimum sample split of 2, and deprecated presorting with the splitter set to best. Out of 3987 objects, 2990 objects (75%) are used for training, while 997 objects (25%) are used for testing the model. The model is evaluated using the decision tree classifier, resulting in a confusion matrix with evaluation metrics including precision and recall, achieving an accuracy of 82%. Similarly, the preprocessed dataset is applied to the XGBoost classifier with parameters such as a base score of 0.5, booster set to gbtree, importance type as gain, learning rate of 0.3000012, maximum depth of 6, and random state set to zero. LabelEncoder() is used for data transformation, and a classification report is obtained with evaluation metrics such as precision, recall, $f1$-score, and support. The model is explained with the weight of each feature using the Gini index, sample percentage, and values in the tree structure. Furthermore, the top features with the decision attribute for each decision class and the contribution level of the features are indicated using a contribution chart, representing positive and negative weights.

5.4 EXPERIMENTAL ANALYSIS AND RESULT DISCUSSIONS

This section focuses on the experimental analysis carried out to explain the model of the random forest for feature reduction, decision tree, and XGBoost as a classifier for the considered agriculture dataset. The experimental analysis was performed using a Windows Operating system with 16 GB RAM, Intel core with 64-bit OS. The implementation code was executed using Google Colab with 1.45 GB RAM.

i. Feature extraction using SHAP XAI analysis and discussion

The SHAP values are calculated based on the tree explainer as discussed in Section 5.3.2 using the (5.1) and found that the feature extraction process was time-consuming as the dataset size increased. The same dataset is fed into the Kernel SHAP using random forest feature extraction, and the results produced to identify the base value for each feature are depicted in Figure 5.4. The force stop plot is utilized to plot the features for the SHAP values of the desired decision classes.

From Figures 5.4 and 5.5, it is observed that for the considered dataset, both the tree explainer and the Kernel SHAP explainer produce a base value of 0.08. The attribute copper (a_7) contributes the highest weight, whereas attributes a_{12}, a_{10}, a_{11}, a_{13}, a_6, and a_{14} have lesser contributions to the decision value as "Alangayam". Specifically, attributes calcium, iron, water pH, nitrate, potassium level, and magnesium have significant contributions to determining the suitability of the crop to be cultivated in the place "Alangayam".

Similarly, the features of the place "KV Kuppam" are provided in Figure 5.5. Therefore, for the place "K V Kuppam", the attributes phosphorus, zinc, copper, and water pH are the salient features. Thus, the weight of the important features is summed up to identify the significant features that contribute to the decision-making process, as illustrated in Figure 5.6.

ii. Classification using XAI ELI5 using decision tree and XGBoost classifier

The feature-extracted consolidated dataset is now ready to be fed into the XAI ELI5 package for classifying the objects based on the decision classes. As discussed, the Vellore district is divided into 7 agriculture divisions, with 20 blocks categorized into these 7 agriculture divisions as represented

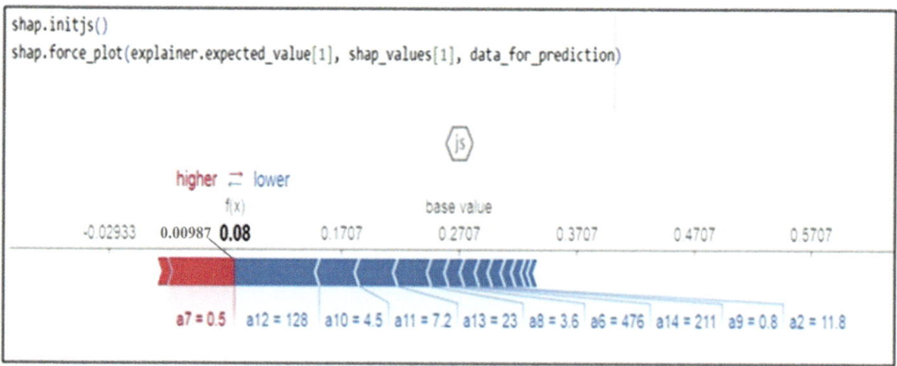

FIGURE 5.4 The tree explainer() – Alangayam.

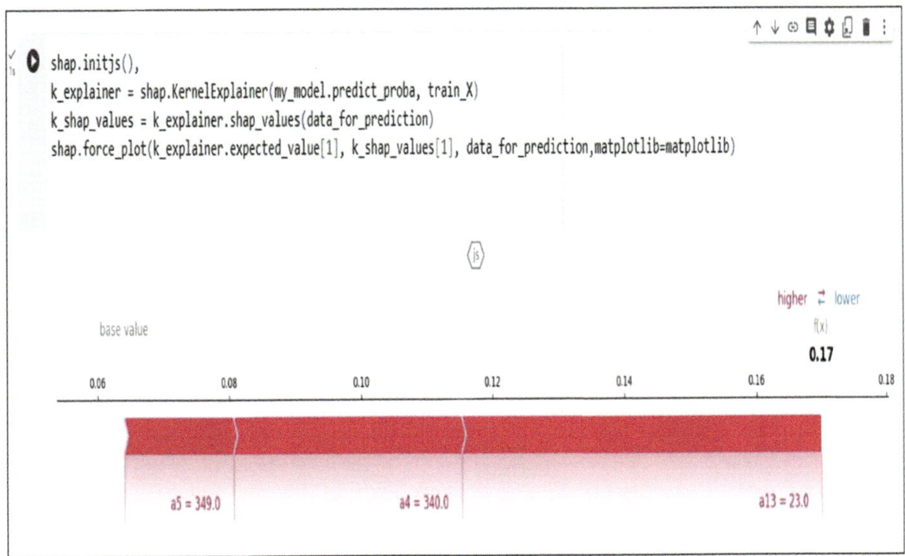

FIGURE 5.5 The Kernel SHAP explainer() – Alangayam.

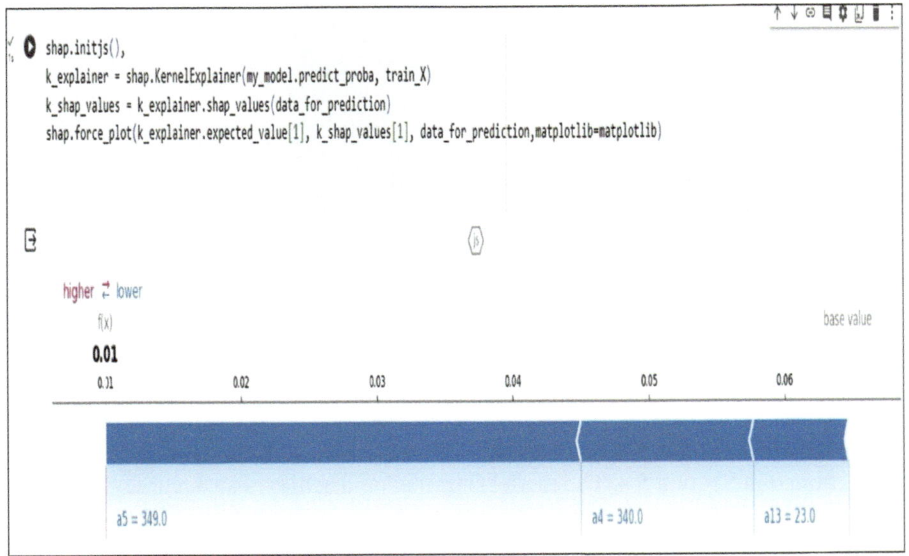

FIGURE 5.6 The SHAP value for K V Kuppam.

in Table 5.3. Thus, the decision class ranges from 1 to 7, and the same is represented in the sample dataset for the decision class. The pre-processed dataset is not fed into the ELI5 package, and the histogram of the 7 agriculture divisions is illustrated in Figure 5.7.

TABLE 5.3
Agriculture Divisions and Blocks Categorization

Sl.no	Agriculture Division	Blocks
1.	Vellore (1)	Vellore, Kaniyambadi, Anaicut
2.	Gudiyatham (2)	Gudiyatham, K.V. Kuppam and Katpadi
3.	Vaniyambadi (3)	Alangayam, Madhanur and Pernambut
4.	Thirupathur (4)	Thirupathur, Kandhili, Natrampalli and Jolarpet
5.	Walajah (5)	Walajah and Sholingur
6.	Arcot (6)	Arcot and Timiri
7.	Arakonam (7)	Arakonam, Nemili, Kaveripakkam

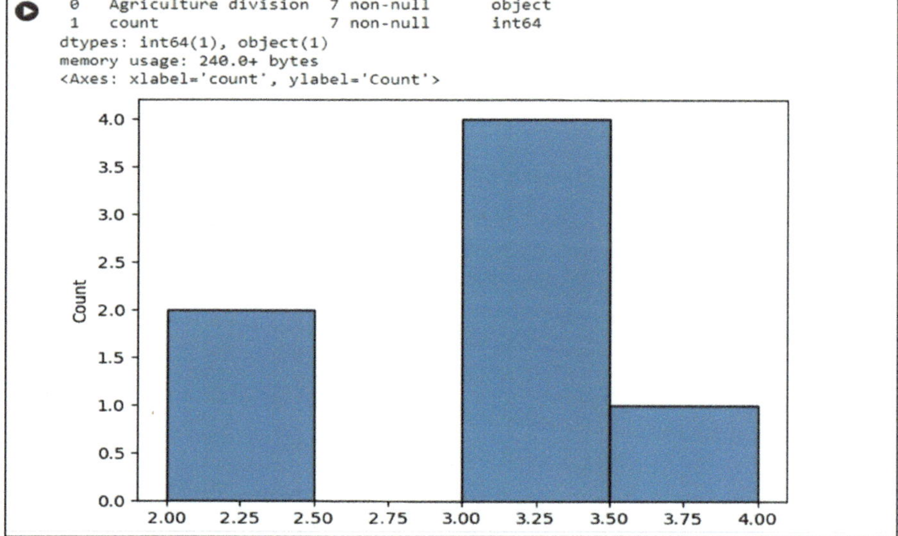

FIGURE 5.7 The histogram of the pre-processed data.

The dataset is fed into the XGBoost classifier with the constraints mentioned in Section 5.4. The model is trained and achieves an accuracy of 93% for the XGBoost classifier, whereas the decision tree classifier produces an accuracy of 82.3% accuracy. The evaluated and obtained results are represented using a confusion matrix and a tree structure for XGBoost and decision tree classifier, respectively, in Figures 5.8 and 5.9.

The weight contribution of the decision tree classifier and the XGBoost classifier is shown in Figure 5.10 and Figure 5.11, respectively. From Figure 5.10, it is observed that decision classes 1, 3, 5, 6, and 7 have similar contributions, whereas decision classes 2 and 4 exhibit variations. Thus, the obtained accuracies primarily depend on decision classes 2 and 4. The evaluation metrics of both classifiers are depicted in Table 5.4.

```
from sklearn.metrics import confusion_matrix
cm=confusion_matrix(y_test, y_pred)
sn.heatmap(cm,annot=True)
```

```
<AxesSubplot:>
```

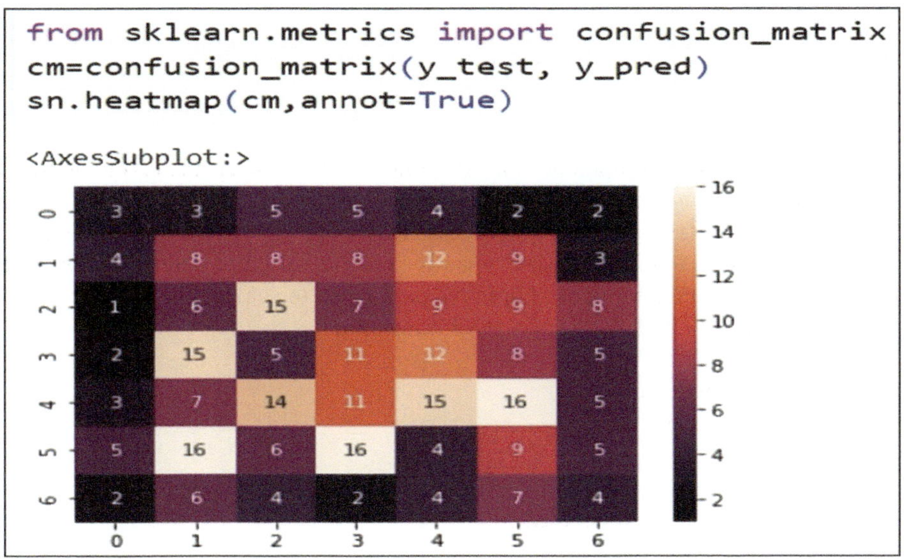

FIGURE 5.8 Confusion matrix as heat map notation.

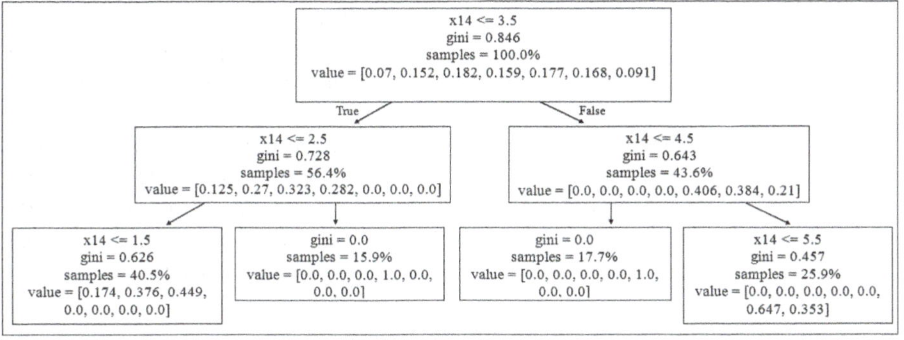

FIGURE 5.9 The decision tree structure.

```
eli5.show_prediction(classifier_dtc, X_test.iloc[1],
                     feature_names=list(X.columns),
                     show_feature_values=True)
```

y=1 (probability **0.000**) top features			**y=2** (probability **0.000**) top features			**y=3** (probability **0.000**) top features		
Contribution?	**Feature**	**Value**	**Contribution?**	**Feature**	**Value**	**Contribution?**	**Feature**	**Value**
+0.070	<BIAS>	1.000	+0.152	<BIAS>	1.000	+0.182	<BIAS>	1.000
-0.070	d	3.000	-0.152	d	3.000	-0.182	d	3.000

FIGURE 5.10 Weight of the decision tree classifier.

```
eli5.explain_prediction_xgboost(model,x_test.iloc[0])
```

y=1 (probability **0.981**, score **3.921**) top features

Contribution?	Feature
+1.231	mean concave points
+0.946	worst concave points
+0.936	<BIAS>
+0.480	worst texture
+0.429	compactness error
+0.354	worst perimeter
+0.343	worst area
+0.178	mean texture
+0.126	area error
+0.058	worst radius
+0.044	mean fractal dimension
+0.040	radius error
+0.032	perimeter error
+0.023	mean concavity
+0.019	mean area
+0.012	symmetry error
+0.006	texture error
-0.033	concave points error
-0.033	smoothness error
-0.117	worst symmetry
-0.140	mean smoothness
-0.382	worst smoothness

FIGURE 5.11 Weights of the XGBoost classifier.

TABLE 5.4
Evaluation Metrics of the Classifier

Evaluation Metrics	Precision	Recall	F1-score	Support	Accuracy (%)
XGBoost	89.45	84.78	89.02	91	93.2
Decision tree	86.78	74.3	81.04	84	83

5.5 CONCLUSION AND FUTURE ENHANCEMENT

Countries like India have agriculture as their main occupation. Advancements in the AI domain are making a huge difference in the classification and prediction processes. The chapter discusses crop suitability prediction in some places of the Vellore district of Tamil Nadu, India, using machine learning and deep learning techniques. Feature reduction was carried out using the XAI - SHAP technique by identifying the salient features based on the weights, and the classification process was explained using local interpretability models such as ELI5. The XAI models help to understand the internal processing of the attributes and the input data of the random forest feature reduction along with the classification process using XGBoost and decision tree classifier. Thus, the contribution of the work is to provide a new dimension in the growth of AI to give a clear picture to make the explanation very simple and easily understandable to tune the hyperparameters to achieve better results with less error rate. The future scope of the work is:

- To develop a cost-effective model in precision agriculture with greater ethical responsibility.
- To provide an automated decision-making system to improve the yield of the crops.

REFERENCES

Anitha, A., and Acharjya, D. P. (2018). "Crop suitability prediction in Vellore District using rough set on fuzzy approximation space and neural network". *Neural Computing and Applications*, 30, 3633–3650. https://doi.org/10.1007/s00521-017-2948-1.

Anitha, A., and Acharjya, D. P. (2019). "Agriculture crop suitability prediction using rough set on intuitionistic fuzzy approximation space and neural network". *Fuzzy Information and Engineering*, 11(1), 64–85. https://doi.org/ 10.1080/16168658.2021.1886813.

Bhandari, M., Shahi, T. B., Neupane, A., and Walsh, K. B. (2023). "BotanicX-AI: Identification of tomato leaf diseases using an explanation-driven deep-learning model". *Journal of Imaging*, 9(2), 53. https://doi.org/10.3390/jimaging9020053.

Couliably, S., Kamsu-Foguem, B., Kamissoko, D., and Traore, D. (2022). "Explainable deep convolutional neural networks for insect pest recognition". *Journal of Cleaner Production*, 133638. https://doi.org/10.1016/j.jclepro.2022.133638.

Iatrou, M., Karydas, C., Tseni, X., and Mourelatos, S. (2022). "Representation learning with a variational autoencoder for predicting nitrogen requirement in rice". *Remote Sensing*, 14(23), 5978. https://doi.org/10.3390/rs14235978.

Kawakura, S., Hirafuji, M., Ninomiya, S., and Shibasaki, R. (2022). "Adaptations of explainable artificial intelligence (XAI) to agricultural data models with ELI5, PDPbox, and skater using diverse agricultural worker data". *European Journal of Artificial Intelligence and Machine Learning*, 1(3), 27–34. https://doi.org/10.24018/ejai.2022.1.3.14.

Lysov, M., Pukhkiy, K., Vasiliev, E., Getmanskaya, A., and Turlapov, V. (2023). "Ensuring explainability and dimensionality reduction in a multidimensional HSI world for early XAI-diagnostics of plant stress". *Entropy*, 25(5), 801. https://doi.org/10.3390/e25050801.

Manna, T., and Anitha, A. (2022, December). "Forecasting Air Quality Index based on Stacked LSTM". In *2022 IEEE 7th International Conference on Recent Advances and Innovations in Engineering (ICRAIE)* (Vol. 7, pp. 326–330). IEEE. https://doi.org/10.1109/ICRAIE56454.2022.10054260.

Manna, T., and Anitha, A. (2023a). "Deep ensemble-based approach using randomized low-rank approximation for sustainable groundwater level prediction". *Applied Sciences*, 13(5), 3210. https://doi.org/10.3390/app13053210.

Manna, T., and Anitha, A. (2023b). "Precipitation prediction by integrating rough set on fuzzy approximation space with deep learning techniques". *Applied Soft Computing*, 139, 110253. https://doi.org/10.1016/j.asoc.2023.110253.

Manna, T., & Anitha, A. (2023c). Hybridization of rough set–wrapper method with regularized combinational LSTM for seasonal air quality index prediction. *Neural Computing and Applications*, 1–20. DOI: 10.1007/s00521-023-09220-6

Maximova, I., Vasiliev, E., Getmanskaya, A., Kior, D., Sukhov, V., Vodeneev, V., and Turlapov, V. (2021, July). "Study of XAI-capabilities for early diagnosis of plant drought". In *2021 International Joint Conference on Neural Networks (IJCNN)* (pp. 1–8). IEEE. https://doi.org/10.3390/e24111597.

Mundhenk, T. N., Chen, B. Y., and Friedland, G. (2019). "Efficient saliency maps for explainable AI". arXiv preprint arXiv:1911.11293. https://doi.org/10.48550/arXiv.1911.11293.

Qiu, T., Underhill, A., Sapkota, S., Cadle-Davidson, L., and Jiang, Y. (2022). "High throughput saliency-based quantification of grape powdery mildew at the microscopic level for disease resistance breeding". *Horticulture Research*, 9. https://doi.org/10.1093/hr/uhac187.

Rahamathunnisa, U., Nallakaruppan, M. K., Anitha, A., and Sendhil Kumar, K. S. (2020, March). "Vegetable disease detection using k-means clustering and SVM". In *2020 6th International Conference on Advanced Computing and Communication Systems (ICACCS)* (pp. 1308–1311). IEEE. https://doi.org/10.1109/ICACCS48705.2020.9074434.

Ranasinghe, N., Ramanan, A., Fernando, S., Hameed, P. N., Herath, D., Malepathirana, T., and Halgamuge, S. (2022). "Interpretability and accessibility of machine learning in selected food processing, agriculture and health applications". arXiv preprint arXiv:2211.16699. https://doi.org/10.48550/arXiv.2211.16699.

Ryo, M., Angelov, B., Mammola, S., Kass, J. M., Benito, B. M., and Hartig, F. (2021). Explainable artificial intelligence enhances the ecological interpretability of black-box species distribution models. *Ecography*, 44(2), 199–205. https://doi.org/10.1111/ecog.05360.

Sabrina, F., Sohail, S., Farid, F., Jahan, S., Ahamed, F., and Gordon, S. (2022). "An interpretable artificial intelligence based smart agriculture system". *Computers, Materials and Continua*, 3777–3797. https://doi.org/10.32604/cmc.2022.026363.

Sauder, D., Muhlbauer, C., and Koch, J. (2020). U.S. Patent No. 10,791,666. Washington, DC: U.S. Patent and Trademark Office.

Su, W. H. (2020). "Crop plant signalling for real-time plant identification in the smart farm: A systematic review and new concept in artificial intelligence for automated weed control". *Artificial Intelligence in Agriculture*, 4, 262–271. https://doi.org/10.1016/j.aiia.2020.11.001.

Supriya, K. and Anitha, A., "Survival Analysis of Superficial Bladder Cancer patients using DeepSurv and Cox models," 2024 Second International Conference on Emerging Trends in Information Technology and Engineering (ICETITE), Vellore, India, 2024, pp. 1–7, doi: 10.1109/ic-ETITE58242.2024.10493319

Talaviya, T., Shah, D., Patel, N., Yagnik, H., and Shah, M. (2020). "Implementation of artificial intelligence in agriculture for optimisation of irrigation and application of pesticides and herbicides". *Artificial Intelligence in Agriculture*, 4, 58–73. https://doi.org/10.1016/j.aiia.2020.04.002.

Thrun, M. C., Ultsch, A., and Breuer, L. (2021). "Explainable AI framework for multivariate hydrochemical time series". *Machine Learning and Knowledge Extraction*, 3(1), 170–204. https://doi.org/10.3390/make3010009.

Wei, K., Chen, B., Zhang, J., Fan, S., Wu, K., Liu, G., and Chen, D. (2022). "Explainable deep learning study for leaf disease classification". *Agronomy*, 12(5), 1035. https://doi.org/10.3390/agronomy12051035.

Yang, G., Huang, Y., and Zhao, C. (2020). "Agri-BIGDATA: A smart pathway for crop nitrogen inputs". *Artificial Intelligence in Agriculture*, 4, 150–152. https://doi.org/10.1016/j.aiia.2020.08.001.

6 Explainable Artificial Intelligence for Healthcare Applications Using Random Forest Classifier with LIME and SHAP

Mrutyunjaya Panda and Soumya Ranjan Mahanta

6.1 INTRODUCTION

In the recent past, applications of artificial intelligence techniques have seen exponential growth in every sphere of life, including computer vision, natural language processing, precision medicine, smart agriculture, and autonomous driving, among others, despite their poor transparency and interpretability. The emerging deep learning architectures are posing even more complexity in interpreting and explaining the inner details of the black box approaches they adopt. A diagrammatic representation of developed AI models based on complexity is shown in Figure 6.1. From Figure 6.1, it is quite evident that the most popular and widely used deep learning models are not only complex in design but also provide fewer explanations about their functionality compared to other existing AI approaches. As per European Union regulation 679 [1], the user has every right to not only understand the usability of its data by AI models but also challenge their predictions before accepting the solutions for necessary applications such as precision healthcare and autonomous driving. This way, it is envisioned that XAI might be a proper solution where AI with explanations may enhance trust in adherence to several regulatory provisions and generate profits for organizations. However, the challenge lies in making XAI interpretable, transparent, trustworthy, and complete [2]. Completeness comes from how accurately a system's inner details are being explained.

6.1.1 MOTIVATION

Even though traditional AI methods have been implemented to solve real-life situations, including healthcare informatics, self-driving cars, self-driving drones, chatbots, 6-G communications, and industrial IoT, for several decades, the "black box"

DOI: 10.1201/9781003442509-6

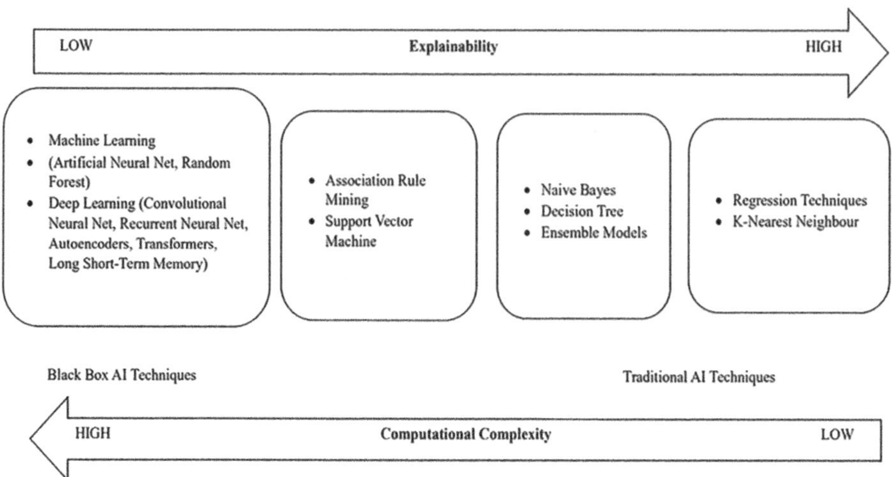

FIGURE 6.1 Classification of AI models in terms of complexity and explainability.

view of these methods, with very poor interpretability, explainability about the operations, and low transparency, poses a lack of trust in the decision-making process. Motivated by this, this chapter is intended to provide a study of several possible XAI frameworks for dealing with complex decision-making by means of the user's trust, using a medical dataset.

Looking into the recent developments in XAI, this chapter primarily focuses on the following:

- Which explainable AI methods are available?
- Performance measures to evaluate the XAI methods
- Types of explanations with fairness and confidence
- How to choose among different explainable AI methods?
- Trustworthiness of XAI

The rest of the chapter is organized as follows: Section 6.2 presents the review of related literature, stating the current application scenario of XAI. Section 6.3 presents XAI and its framework, followed by trust in XAI in Section 6.4. Section 6.5 discusses several XAI methods available in the literature. The experimental framework and results with discussions are presented in Section 6.6. Finally, conclusions and future work are highlighted in Section 6.7.

6.2 RELATED WORK

Mahbooba et al. [3] proposed using the concept of XAI to improve trust management issues in intrusion detection systems using a decision tree model. The authors in [4] proposed a fine-grained mechanism based on segmentation and RoI pooling applied to the Berkeley DeepDrive eXplanation (BDD-X) dataset and concluded that their approach outperforms earlier methods with more interpretable visual explanations in autonomous driving systems. It is discussed by the authors [5] that today's advanced bio-inspired algorithms, which are used to tackle real-life situations, are sometimes considered as weak AI or narrow AI, as they are useful for addressing specific applications of interest, whereas strong AI can be used for universal applications. Meske et al. [6] provided an investigative report on how explainability has been successfully applied in information systems to date and highlighted some quality criteria to judge the importance of XAI, along with future directions of research and pros and cons. Mane and Rao [7] applied deep learning methods to the NSL-KDD (network security lab-knowledge discovery in databases) dataset for network intrusion detection systems, providing more explanations and elaborating on how explanations generated from contrastive explanations help influence the degree of attack prediction. In [8], the authors discussed customer acceptance in the deployment of autonomous vehicles, highlighted current regulatory provisions, and detailed different autonomous driving operations such as perception, localization, planning, control, and system management. Kartikeya [9] presented two experimental frameworks to assess whether transparency enhances the model's trust or not. One framework used evaluation metrics, while the other relied on subjective questionnaire trust. After conducting statistical significance tests, the author found that both frameworks provided results contrary to each other in measuring the model's influence on trust with transparency. Markus et al. [10] pointed out that XAI can provide trustworthy AI in biomedical informatics; however, further investigation is needed to conclude its role in healthcare through the verification and validation of data quality and external regulations. Hussain et al. [11] discussed the usefulness of XAI in autonomous cars from an engineering perspective and its role in object detection, control strategies, and decision-making. The authors in [12] illustrated the applicability of XAI in several Internet of Things (IoT) enhancements, including Internet of Medical Things (IoMT), Industrial IoT (IIoT), and Internet of City Things (IoCT), with added security and support from sixth-generation (6G) communication services. Mankodia et al. [13] applied various XAI methods for semantic object detection, where deep learning models were used to segment and detect the road while the autonomous car is moving, achieving maximum Intersection over Union (IoU) scores of 94.59% and 96.21%, and accuracies of 97.61% and 97.86% for the training and testing dataset with ResNet-18. Javed et al. [14] presented a survey report on the usefulness of XAI in developing smart cities, considered to be a noble idea, and highlighted its research directions on system architectures. Renda et al. [15] propose the concept of federated learning of XAI in 6G communications for autonomous driving in intelligent transportation systems, with benefits of trust, quality of experience, and privacy-preserving management of the system design. Kim and Joe [16] applied XAI in autonomous driving using a convolutional neural network (CNN) by explaining the differences in output values obtained in the final hidden layer of CNN using image sensitivity analysis on image data and finally concluded that XAI could categorize the images with high accuracy. Madanu et al. [17] discuss the role of AI in pain modelling using machine learning and deep learning techniques and advocate that XAI models might add more detailed diagnostic analysis with incorporations of explainability to certain illnesses with root cause analysis.

For this purpose, pre-existing models were applied to the Google Jigsaw dataset to achieve the best prediction accuracy. Explainable methods such as LIME (Local Interpretable Model-Agnostic Explanations) were then applied to the HateXplain dataset [18]. Variants of the BERT (Bidirectional Encoder Representations from Transformers) model with neural network combinations were created to achieve good performance in terms of explainability using the ERASER (Evaluating Rationales and Simple English Reasoning) benchmark [19].

6.3 EXPLAINABLE AI AND ITS FRAMEWORKS

Several XAI frameworks are discussed in this section to address the following:

- Explainability
- Interpretability
- Transparency
- Trustworthiness
- Evaluation criteria

6.3.1 Explainability

Transparency aims at getting the inner workings details of the AI model being developed so that a sense of human understanding is achieved, and mutual trust can be ensured while using the model. In this aspect, it is observed that sometimes researchers find it hard to clearly define the explainability and finally trapped in the definitions of interpretability and/or into transparency [20].

6.3.2 Transparency

There are three dimensions of transparency for human understanding regarding the inherent details of the AI model being considered for applications under study. They are:

- **Simulation ability:** This is the first level of transparency where the AI model's ability to be simple and compact allows a human to simulate the model with ease, without compromising efficient decision-making. Artificial neural networks with no hidden layers might fall into this category.
- **Decomposition of the model:** This is the second level of transparency where the AI model can be divided into several parts in terms of input, mathematical computations, and setting of the model parameters, and then a detailed explanation of each individual part is obtained for better understanding of the human. At this level, there are only a few AI models available that could satisfy these criteria.
- **Transparency in AI model procedure:** In this third transparency, one can envisage a better understanding of the procedures being undertaken to develop the AI model and the way the output is obtained. One such example is the k-Means clustering algorithm, where the distance-based similarity criteria are well-mentioned so that samples with high similarity can be

placed in the same cluster. In contrast, when we examine deep neural network models, the loss function used seems to be elusive, and the objective of training the neural network model is accomplished through approximate reasoning, resulting in a lack of transparency in understanding the inner details while reaching the output solution. Hence, an AI model with mathematical analysis falls into this transparency level [21].

Looking at all these discussions, AI models are classified as either transparent, black box, or opaque models. Examples such as Decision Trees, Linear Regression, K-Nearest Neighbor (KNN), and fuzzy classifiers fall under transparent models, while all variations of neural networks except for single perceptron neural networks, Random Forest, Support Vector Machine (SVM), and Microsoft's recent fiasco with the experimental Tay Twitter Chatbot fall into opaque/non-transparent models. It is also worth noting that opaque AI models are more effective than their counterparts due to no reasoning constraints but with higher risk. So, there is a trade-off between these choices for companies or individuals considering their specific requirements.

6.3.3 Quality Criteria

The evaluation of the AI model is done through the quality of information it generates for an effective and efficient decision-making process. There are two aspects of these quality criteria for model evaluations: Explanation aspects and model aspects. In explanation aspects, the following quality criteria are often used to make the AI model explainable:

- **Comprehensibility:** Comprehensibility refers to the ability of an AI model to represent its learning outcomes in a human-understandable manner [22].
- **Accessibility:** Accessibility intends to involve the end-users in the improvement of the AI model without having any deep understanding of the AI programming [23].
- **Fidelity:** Fidelity discusses how accurately the explanation method justifies the underlying model from which the difference between the users could be able to measure the difference in its descriptive model accuracy with that of system-generated ones [24].
- **Identity, separability and novelty:** Where the explanations between different instances are compared with identical instances, they should have identical explanations. Non-identical ones should have separate explanations, and the instance should not have come from a region in instance space too far from the training data.

Apart from these quality criteria, the following ones are used for obtaining explanations from the model aspects. The output from the users' aspect or mental model is fed back to the system via the criteria appropriate trust and reliance [25].

- Accuracy is one of the model evaluation criteria through which one can evaluate the level of model's correctness in comparison to the actual target.

- Fairness, which is also understood as opposite to model biasness, presents the degree of error patterns present in the AI model.
- Reliability presents soundness of the model to the user for a specific application.

6.3.4 Types of Explainability Techniques

The explainability techniques in XAI are basically of two types: Global and Local. In global explainability, model explanations are made in general with their generic operations, whereas in local ones, the explanation comes for every single data with model reasoning and appropriate rules through which a decision is obtained.

- **Types of explanations:** In general, there are two categories of explanations: intrinsic and post hoc explanations (or interpretability). In the case of intrinsic explanations, simple AI models that are interpretable due to their structural simplicity, such as sparse linear models and decision trees are considered, whereas interpretations are obtained after the model is trained in the case of post hoc explanations. In this chapter, our focus is on post hoc explanations. Examples of post hoc explanation methods for image and time series datasets are factual, semi-factual, and counterfactual methods.
- **Post-hoc explanations:** Post-hoc explanations are traditionally viewed as explanations or justifications of the decision-making process by example, where some factual (or nearest neighbor) instance is created to justify some target query [26]. However, there are several other explanatory possibilities available based on the type of application using the explanations. Let us consider an opaque classifier with a student dataset with student placement attributes available in a tabular format to obtain the decision of whether a student will get placement or not. Now, supposing that you did not get selected in campus placement and for which you asked for explanations about the non-selection. Then, the AI model could return an answer to your query explanation through a factual example-based explanation such as "you are not selected because your interview was similar to another student who also did not get placement". Alternatively, the AI model could give you some counterfactual explanations like "if you had scored better in the interview like Student-X who got selected for placement". Lastly, there is also a chance of getting semi-factual explanations with a reply like "even if you performed well in the interview, you would still not have the positive attitude of Student-X who got selected". It is customary to say that factual post-hoc example-based explanations are most widely used by researchers, as in case-based reasoning classifiers, and these alternate explanations are considered to be non-trivial, but it is observed that there is a growing demand for using counterfactual [27] and semi-factual explanation methods [28].

6.4 TRUST IN XAI

Trust can be considered as an abductive speculation which seems to be the "best hypothesis" to measure the trustworthiness in AI. As this consideration is purely based on the intelligibility and anticipated predictability of the AI model, hence is undoubtedly fallible [29]. To this, one can envisage that what a computer scientist understands about the implications of computer assertions, others may not, and thereby ask for justification to believe that the computer assertions are valid and trustworthy. Looking into this, trust in AI is defined in several ways:

- Absolute Trusting, where the novice user considers the computer assertions to be completely valid and trustworthy in all conditions.
- Contingent Trusting is one where the customer accepts the computer assertions as valid and trustworthy under some specific conditions.
- Progressive Trusting, on the other hand, takes users' experience on given computer assertions over time to decide whether it is valid and trustworthy or not.
- Digressive Trusting by the user finally considers some part of the computer assertions to conclude whether they are valid and trustworthy based on the users' experiences over time.

The trust in XAI through clarification and explanation with fairness in the AI model development process can be seen in Figure 6.2.

In problem identification, the aim is to determine whether the algorithm under study presents ethical solutions to the problem at hand. Accordingly, the dataset is created by checking the representations from several groups and identifying any bias present in the data. If bias is detected in either the class label or features of the dataset used, steps are taken to remove or minimize the bias for better model predictability. Then, in the development of AI algorithms, fairness is included as a constraint to the objective function so that the transparency of selecting the algorithm is well understood by the user. Next, the dataset is split into training and testing sets. During training, the model is properly trained and then checked with a possible fairness matrix during the testing phase. In the deployment phase, the model is verified for its intended uses and its effectiveness in diverse applications. Finally, the model is monitored for any unfair means adopted during the whole process with obtained feedback, and the process continues for further improvement.

6.5 XAI METHODS

XAI methods or interpretation tools are broadly categorized into two types: model-agnostic and model-specific, depending on whether the explanations are required for all types of models or tailored for certain model structures, respectively. In model-agnostic XAI methods, even though the inner details and structure of the models cannot be accessed, they provide an understanding of any AI models being used. Conversely, in model-specific methods, the interpretation depends on the working capabilities of the specific model being considered. Furthermore, some

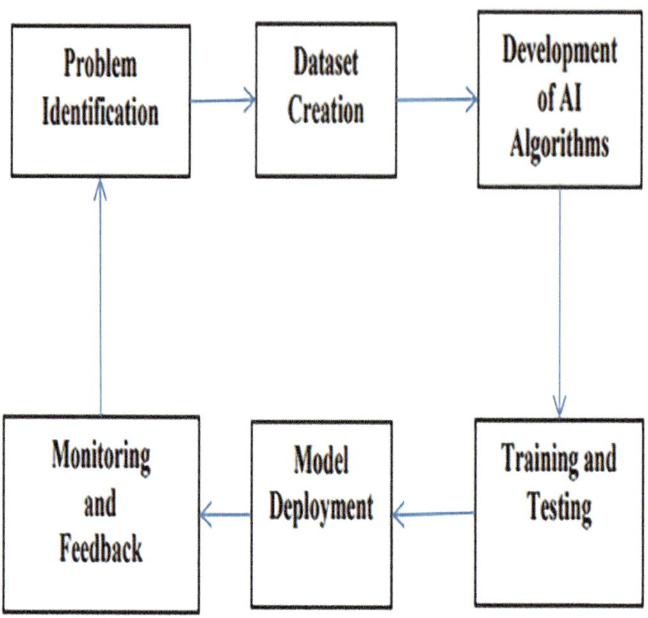

FIGURE 6.2 Explainability and fairness for measuring trust in XAI.

XAI methods are considered to be global, where the model tries to approximate the function underlying the model in the whole space. In contrast, in local methods, a small representative of the whole space is chosen, and then an approximation of that specific part is chosen for explanations. It is also observed that XAI methods provide interpretations based on feature importance. Accordingly, a rightful way is envisaged through correlation to find the important variables, which are then used to provide explanations. Global explanations provide interpretations to non-data scientists about what the model uses to make predictions. For example, in customer churn prediction by a telecom company, the marketing team shall explore several important features that contribute to predicting customer churn. On the other hand, local explanations identify the most important dimension of a single input to predict its output, as can be seen in medical diagnosis using deep neural networks.

Based on the above discussions, the following are some local and global model-agnostic XAI methods available in the literature to obtain explanations: Partial Dependence Plot (PDP), Accumulated Local Effects (ALE), Individual Conditional Expectation (ICE), Local Interpretable Model-agnostic Explanations (LIME), SHapley Additive exPlanations (SHAP), Generalized additive model (GAM), moDel Agnostic Language for Exploration and eXplanation (Dalex), Contextual Importance and Utility (CIU), etc. Out of all these, PDP, ALE, CIU, and ICE are considered to be global interpretable model-agnostic explanations, whereas LIME, SHAP, Dalex, and GAM are the local ones. Brief discussions about these methods are provided below.

6.5.1 Partial Dependence Plot (PDP)

The Partial Dependence Plot (PDP) is a model-agnostic and global interpretable method in XAI that provides panoramic explanations about the influence of features on the target variable within the frame of the whole dataset. It is observed that PDP has a very nominal effect on the complement features (i.e., other than the intended ones) when making predictions using AI models [30]. Furthermore, with PDP, one can determine whether the relationship between the target variable and an input feature is linear, monotonic, or more complex. The formal definition of a PDP function for regression operations can be represented as presented in Equation (1).

$$f_s'(x_s) = E_{X_c}\left[f_s'(x_s, X_c)\right] = \int f_s'(x_s, X_c)\,dP(X_c) \qquad (6.1)$$

In Equation (6.1), x_s represents the features for which the PDP plot is to be obtained and *Xc* represents the complement features present in the AI model, while *f's* are random variables. In this equation, set *S* contains one or two features that are used to understand the prediction outcome. The full feature space can be formed by combining the feature vectors *xs* and *Xc*. PDP works by computing the output of the black box model over the feature distribution in set *C*, through which the PDP function can show the association between features of interest from set *S* and the predicted output. On the other hand, if we disparage the complement features, then the obtained PDP function hang on between features in set *S* and the other features from set *C*, not in set *S* are included. The PDP function *f's* can be calculated by a popular method known as the Monte Carlo method by contriving means in the training dataset. Even though PDP is straightforward in displaying the linkage between a feature and the target, the independence assumption between the features of interest and the other features has become a major issue. Without meeting this assumption, interpretable and reliable PDP may not be feasible. At the same time, the drawback of PDP is that it can work well up to two features to have a comprehensible plot. In Python, the Sklearn Inspection module may be used for getting PDP display for better interpretation of the AI model. Another plotting method similar to PDP is ICE (Individual Conditional Expectation) to envision and inspect the interaction between the set of interest features with the target output.

6.5.2 Accumulated Local Effects (ALE)

Accumulated Local Effect (ALE) overcomes the drawbacks of PDP and ICE plots in estimating feature effects with much accuracy. However, approximation through ALE is not far from limitations. The drawbacks of ALE lie in its proneness to out-of-distribution (OOD) sampling for small training datasets and its inability to scale well for high dimensional input datasets [31].

6.5.3 Local Interpretable Model-Agnostic Explanations (LIME)

LIME is the short form for Local Interpretable Model-Agnostic Explanations. Each part of the name reflects something that we desire in explanations. Local refers to local fidelity – i.e., we want the explanation to reflect the behavior of the classifier "around" the instance being predicted.

6.5.4 SHapley Additive exPlanations (SHAP)

SHapley Additive exPlanations (SHAP) dependence plot is a local, model-agnostic XAI framework that is used for image, tabular, and text datasets for interpretability with Shaply values, a concept inspired by cooperative game theory. The prime objective of the SHAP is to provide explanations in AI model prediction by calculating the contribution of each individual feature to the output prediction. In this, feature values of a data sample are denoted as players in coalition, and the SHapley values are calculated as the mean marginal contributions of a feature value across all possible coalitions [32] [SHAP1]. The features having large SHapley values are considered to be more important, and the plotting of features is done based on their decreasing order of importance. SHAP dependence plot has advantage of fast implementation in tree-based models by removing the slow computation barriers of SHapley values. In Python, the SHAP package is available for XAI analysis. It is also observed that SHAP has some proneness while computing SHapley values for a lot of data samples. Additionally, it may introduce biases through some malicious data scientists by intentionally designing misleading explanations.

6.5.5 Generalized Additive Model (GAM)

A Generalised Additive Model (GAM), an extension of the multiple linear models, is extremely flexible in choosing any kind of non-linear and linear regression models that are more appropriate to different types of output predictions. On the one hand, GAM uses the additivity property for explaining the contribution of each individual predictor while fixing other predictors, on the other hand, this also poses some constraints in correlating the predictors' non-linear explanations automatically [33].

6.5.6 Contextual Importance and Utility (CIU)

Contextual Importance and Utility (CIU) arithmetic uses the concepts of Multi-Attribute Utility Theory, with the novel concept of contextual influence makes it possible to compare CIU directly with so-called additive feature attribution (AFA) methods for model-agnostic outcome explanation. It is to be noted here that the 'influence' concept used by AFA methods is deficient for outcome explanation even for simple models. Further, CIU generates faithful explanations using contextual importance (CI) and contextual utility (CU) for outcome predictions where influence-based methods fail [34].

6.6 EXPERIMENTS AND RESULTS

In this section, the experimental framework with results obtained is discussed in detail.

6.6.1 Experimental Framework

The experimental framework of the proposed approach is shown in Figure 6.3.

From Figure 6.3, the top part illustrates the black box model of AI, where the user has many questions in mind to understand the inner details of the process through which a decision has been reached about experiments. However, in the XAI framework, as shown in the bottom part of Figure 6.3, a happy user is depicted who has understood the details about how and why the decision is made with satisfaction. In the XAI framework depicted in Figure 6.3, the process begins with the collection of the input dataset, followed by all necessary preprocessing steps to ensure data quality. Then, the modified data is applied to a new machine learning model such as

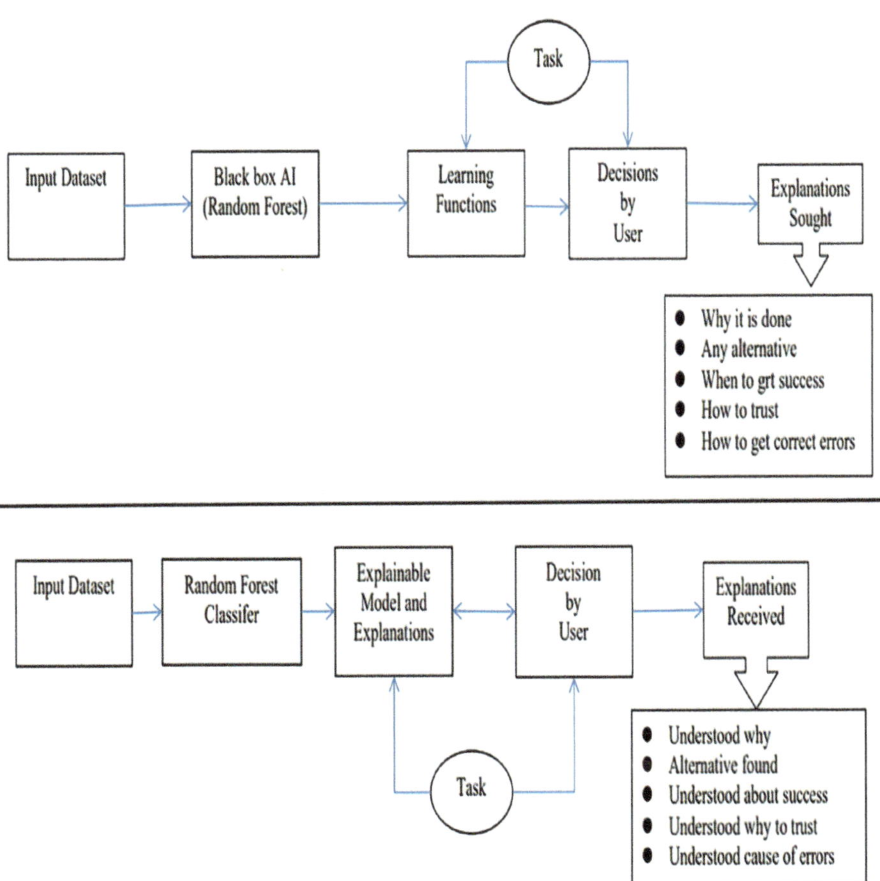

FIGURE 6.3 The proposed XAI framework.

a Deep Neural Network, and predictions are obtained. The predicted outcome from the deep neural network is passed through any XAI model (such as LIME, SHAP, etc.) to obtain detailed explanations about the decisions being taken in the process. Finally, an explanation interface such as temeons XAI, a platform-agnostic free-standing upshot, is used to integrate the XAI model with any arbitrator or user. In this way, XAI pursues a paradigm shift towards a more user- and society-friendly AI categorization without compromising efficacy.

6.6.2 Experimental Results and Discussion

All the experiments are conducted in an HP Pavilion Intel Core i5 PC with 1 TB HDD and 8GB RAM using Python in Jupyter Notebook. The publicly available Diabetes dataset collected from Kaggle and Random Forest as a black box AI method is used for the development of the prediction of diabetic disease. Finally, two surrogate XAI models such as LIMA and SHAP are used for explanations of the inner details of the AI model. At first, the black box model performance by the random forest algorithm in terms of precision, recall, and f1 score is presented in Figure 6.4. Here, 0 indicates the person having no diabetes and 1 for with diabetes.

Now, to add explanations to the model developed, the global interpretation of the feature importance in the dataset is presented in Figure 6.5 using the SHAP explainer.

From Figure 6.5, it can be observed that all the features equally contribute to both classes (Class 0 without diabetes and Class 1 with diabetes), as red and blue are evenly present in all the features. Furthermore, Glucose is shown to have the highest significance in predicting the disease, followed by age and BMI index. Another way to represent the feature importance map is shown in Figure 6.6 using the SHAP explainer, with its SHAP values. Here, the red dot indicates more significance compared to the blue dot. For Glucose, more red dots in the positive x-axis imply a positive impact on the model predictions, whereas the blue dots in the positive x-axis indicate a low probability of being detected as diabetes.

Finally, the SHAP dependency plot for age is used for more detailed explanations, as age provides confusion about the feature significance due to its equal distribution

	precision	recall	f1-score	support
0	0.79	0.85	0.82	150
1	0.68	0.58	0.63	81
accuracy			0.76	231
macro avg	0.74	0.72	0.72	231
weighted avg	0.75	0.76	0.75	231

FIGURE 6.4 Prediction performance of the random forest algorithm.

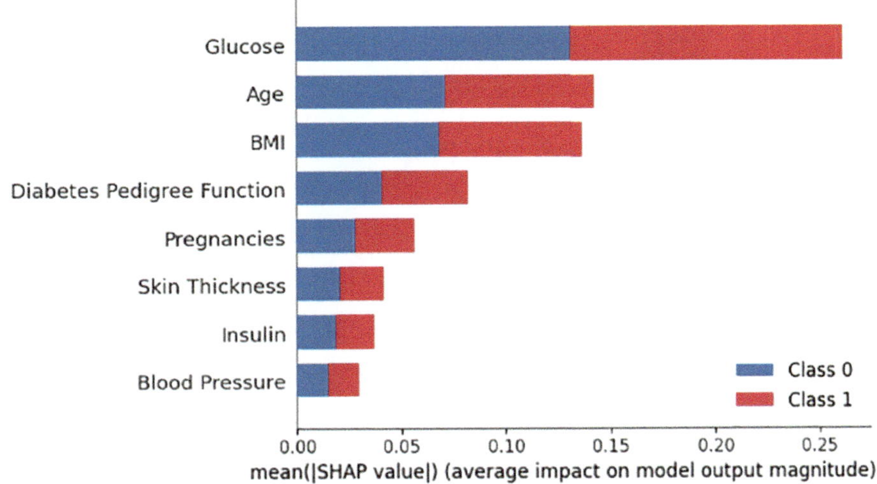

FIGURE 6.5 Global interpretation with feature importance plot using SHAP.

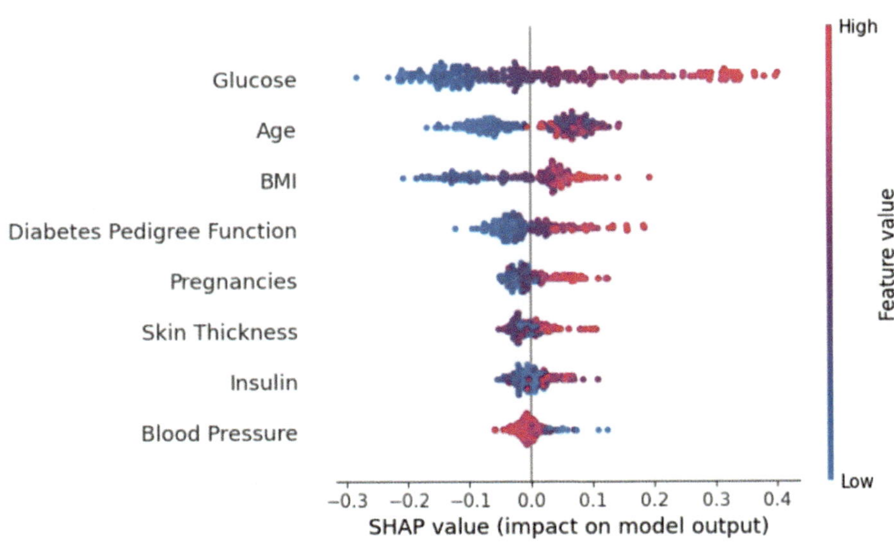

FIGURE 6.6 Shapley values with explanations for model output.

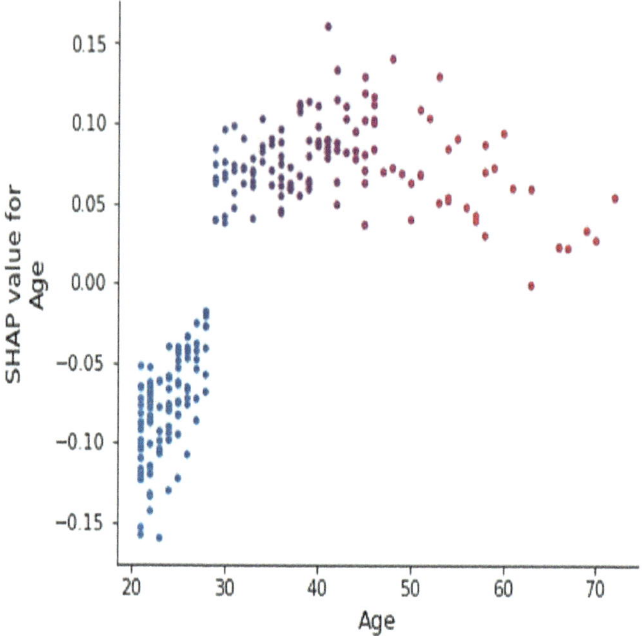

FIGURE 6.7 SHAP dependency plot for age in diabetes dataset.

of points on both sides of the axes, as shown in Figure 6.7. From Figure 6.7, one can interpret clearly that patients with age less than 30 do not have diabetes, whereas persons above 30 years of age are mostly found with diabetes.

Now, we use LIME as a local interpretable XAI method to understand the explainability of the prediction model outcome and feature relevance analysis. In this, two class levels with values 1 and 0 are represented as having diabetes or no diabetes, respectively. Only one sample from the testing dataset is used for explanations. Next, LIME is used to show the explanations for the model developed using the Random Forest algorithm on the diabetes dataset, and the results obtained are presented in Figure 6.8. Figure 6.8 reveals that the single patient sample with an age of 37 and a glucose level of 154 has a 69% model predictability of having diabetes symptoms with the following explanations: low insulin (126), low body mass index (BMI) (31.30), low Skin Thickness (29), pregnancies (4), and high blood pressure (72) in comparison to their respective thresholds as mentioned on the left side of the detailed feature values.

6.7 CONCLUSIONS

This chapter discusses the basics of explainable AI along with several types of explanations with examples. Further, it aims to discuss the XAI framework and methods available for post-hoc interpretation of the black box AI model to enhance the understanding of the predicted model for more transparency and acceptance. Next, the

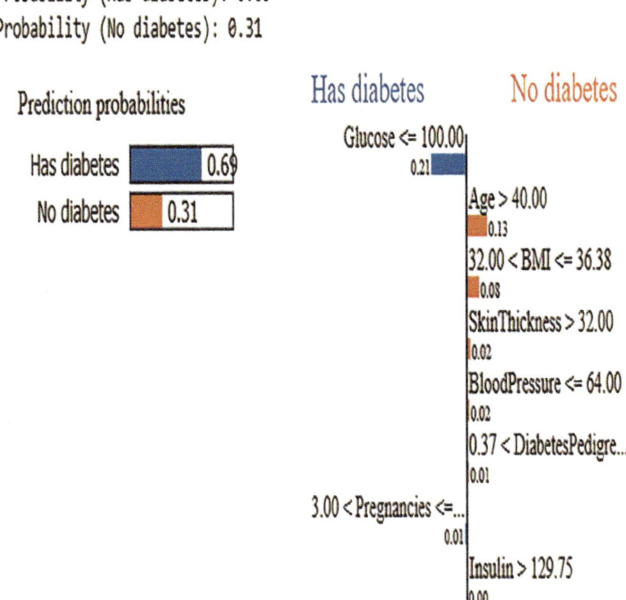

FIGURE 6.8 Explanation of random forest model using LIME for persons with diabetes.

trustworthiness of AI is discussed with various definitions of trust in model development. Out of several XAI methods available, the random forest machine learning model is used on a publicly available diabetes dataset for the prediction of diabetes, and then SHAP and LIME methods are used for its explainability. Feature significance and dependence plots for SHAP as a global interpretability method along with LIME, as a local interpretability method, are discussed, which provide more insights into the decision-making process. In the future, the aim is to explore further possibilities with other black box AI as well as explanation techniques to have better explanations in decision-making with trustworthiness.

REFERENCES

1. Clement, T., Kemmerzell, N., Abdelaal, M., and Amberg, M. 2023. XAIR: A systematic metareview of explainable AI (XAI) aligned to the software development process. *Machine Learning and Knowledge Extraction* 5(1): 78–108. https://doi.org/10.3390/make5010006.
2. Beaudouin, V., Bloch, I., Bounie, D., Clémençon, S., d'Alché-Buc, F., Eagan, J., Maxwell, W., Mozharovskyi, P., and Parekh, J. 2020. Flexible and Context-Specific AI Explainability: A Multidisciplinary Approach. *CoRR*. arXiv:2003.07703v1 [cs.CY] 13 Mar, 2020, pp. 1–65.
3. Mahbooba, B., Timilsina, M., Sahal, R., and Serrano, M. 2021. Explainable artificial intelligence (XAI) to enhance trust management in intrusion detection systems using decision tree model. *Complexity* 2021: 11 pages. https://doi.org/10.1155/2021/6634811.

4. Kim, J., Rohrbach, A., Akata, Z., Moon, S., Misu, T., Chen, Y.-T., Darrell, T., and Canny, J. 2021. Toward explainable and advisable model for selfdriving cars. *Applied AI Letters* 2021: 1–13. https://doi.org/10.1002/ail2.56.
5. Zolbanin, H.M., Delen, D., Crosby, D., and Wright, D. 2019. A predictive analytics-based decision support system for drug courts. *Information Systems Frontiers* 22: 1–20. https://doi.org/10.1007/s10796-019-09934-w.
6. Meske, C., Bunde, E., Schneider, J., and Gersch, M. 2022. Explainable artificial intelligence: Objectives, stakeholders, and future research opportunities. *Information Systems Management* 39(1): 53–63. https://doi.org/10.1080/10580530.2020.1849465.
7. Mane, S. and Rao, D. Explaining Network Intrusion Detection System Using Explainable AI Framework. https://arxiv.org/abs/2103.07110.
8. Omeiza, D., Webb, H., Jirotka, M., and Kunze, L. Explanations in Autonomous Driving: A Survey, IEEE, pp. 1–21. https://doi.org/10.48550/arXiv.2103.05154.
9. Kartikeya, A. Examining Correlation between Trust and Transparency with Explainable Artificial Intelligence, pp. 1–6, 2021. arXiv:2108.04770v1 [cs.HC] 10 Aug 2021.
10. Markus, A.F., Kors, J.A., and Rijnbeek, P.R. 2021. The role of explainability in creating trustworthy artificial intelligence for health care: A comprehensive survey of the terminology, design choices, and evaluation strategies. *Journal of Biomedical Informatics* 113: 103655. https://doi.org/10.1016/j.jbi.2020.103655.
11. Hussain, F., Hussain, R., and Hossain, E. Explainable Artificial Intelligence (XAI): An Engineering Perspective. arXiv:2101.03613v1 [cs.LG] 10 Jan 2021.
12. Jagatheesaperumal, S.K., Pham, Q.-V., Ruby, R., Yang, Z., Xu, C., and Zhang, Z. Explainable AI over the Internet of Things (IoT): Overview, State-of-the-Art and Future Directions. arXiv:2211.01036v2 [cs.AI] 7 Nov 2022.
13. Mankodiya, H., Jadav, D., Gupta, R., Tanwar, S., Hong, W.-C., and Sharma, R. 2022. OD-XAI: Explainable AI-based semantic object detection for autonomous vehicles. *Applied Sciences* 12: 5310. https://doi.org/10.3390/app12115310.
14. Javed, A.R., Ahmed, W., Pandya, S., Maddikunta, P.K.R., Alazab, M., and Gadekallu, T.R. 2023. A survey of explainable artificial intelligence for smart cities. *Electronics* 12: 1020. https://doi.org/10.3390/electronics12041020.
15. Renda, A., Ducange, P., Marcelloni, F., Sabella, D., Filippou, M.C., Nardini, G., Stea, G., Virdis, A., Micheli, D., Rapone, D., et al. 2022. Federated learning of explainable AI models in 6G systems: Towards secure and automated vehicle networking. *Information* 13: 395. https://doi.org/10.3390/info13080395.
16. Kim, H.-S., and Joe, I. 2022. An XAI method for convolutional neural networks: In self-driving cars. *PLoS ONE* 17(8): e0267282. https://doi.org/10.1371/journal.pone.0267282.
17. Madanu, R., Abbod, M.F., Hsiao, F.-J., Chen, W.-T., and Shieh, J.- S. 2022. Explainable AI (XAI) applied in machine learning for pain modeling: A review. *Technologies* 10: 74. https://doi.org/10.3390/technologies10030074.
18. Mehta, H., and Passi, K. 2022. Social media hate speech detection using explainable artificial intelligence (XAI). *Algorithms* 15(8): 291. https://doi.org/10.3390/a15080291.
19. DeYoung, J., Jain, S., Rajani, N.F., Lehman, E., Xiong, C., Socher, R., and Wallace, B.C. Eraser: A Benchmark to Evaluate Rationalized NLP Models. arXiv 2020, arXiv:1911.03429.
20. Vaishak, B. 2021. Papantonis ioannis, principles and practice of explainable machine learning, *Frontiers in Big Data* 4. https://doi.org/10.3389/fdata.2021.688969.
21. Lipton, Z.C. 2016. The Mythos of Model Interpretability. arXiv preprint arXiv: 1606.03490.
22. Goethals, S., Martens, D., and Evgeniou, T. 2022. The non-linear nature of the cost of comprehensibility. *Journal of Big Data* 9(30). https://doi.org/10.1186/s40537-022-00579-2.

23. Arrieta, A.B., Díaz-Rodríguez, N., Ser, J.D., Bennetot, A., Tabik, S., Barbado, A., García, S., Gil-López, S., Molina, D., Benjamins, R., et al. 2020. Explainable Artificial Intelligence (XAI): Concepts, taxonomies, opportunities and challenges toward responsible AI. *Information Fusion* 58: 82–115.
24. Das, A., and Rad, P. 2020. Oppertunities and challenges in explainable artificial intelligence (XAI): A survey. arxiv preprint arXiv:2006.11371.
25. Carvalho, D.V., Pereira, E.M., and Cardoso, J.S. 2019. Machine learning interpretability: A survey on methods and metrics. *Electronics*, 8: 832.
26. Kenny, E.M., et al. 2019. Predicting grass growth for sustainable dairy farming. In: *Case Based Reasoning Research and Development*. https://doi.org/10.1007/978-3-030-29249-2_12.
27. Laugel, T., et al. 2019. The dangers of post-hoc interpretability: Unjustified counterfactual explanations. In: *Proceedings of the 28th International Joint Conference on Artificial Intelligence (IJCAI-19).*
28. Kenny, E.M., and Keane, M.T. 2020. On generating plausible counterfactual and semi-factual explanations for deep learning. arxiv:2009.06399.
29. Hoffman, R.R., Mueller, S.T., Klein, G., and Litman, J. 2018. Measuring Trust in the XAI Context. Technical Report, DARPA Explainable AI Program.
30. Friedman, J.H. 2001. Greedy function approximation: A gradient boosting machine. *Annals of Statistics* 29: 1189–1232.
31. Gkolemis, V., Dalamagas, T., and Diou, C. 2023. DALE: Differential Accumulated Local Effects for efficient and accurate global explanations. In: *Asian Conference of Machine Learning (ACML).* https://arxiv.org/abs/2210.04542v1.
32. Slack, D., Sophie, H., Emily, J., Singh, S., and Lakkaraju, H. 2020. Foolong LIME and SHAP: Adversarial attacks on post hoc explanation methods. In: *Proceedings of the AAAI/ACM Conference on AI, Ethics, and Society*, pp. 180–186.
33. Lee, H.-T., Yang, H., and Cho, I.-S. 2021. Data-driven analysis for safe ship operation in ports using quantile regression based on generalized additive models and deep neural network. *Sensors* 21: 8254. https://doi.org/10.3390/ s21248254.
34. Främling, K. 2022. Contextual importance and utility: A theoretical foundation. In: Long, G., Yu, X., Wang, S. (eds.) *AI 2021: Advances in Artificial Intelligence*, AI 2022. Lecture Notes in Computer Science, vol. 13151. Springer, Cham. https://doi.org/10.1007/978-3-030-97546-3_10.

7 Explainable AI and Its Usefulness in the Business World

Shivam Sakshi, Nagalakshmi Vallabhaneni, Rajesh Mamilla, Prabhavathy Paneer, and Venkatesan M.

7.1 INTRODUCTION

Artificial intelligence (AI) simulates human intelligence in machines that think and act like humans [1]. These machines can learn, adapt, and problem-solve, and AI is used in healthcare, finance, education, and transportation. AI improves efficiency, accuracy, and cost-effectiveness. AI's ethical and social effects, job displacement, and other unintended consequences are also concerns. On the other hand, XAI means "solving stakeholder questions about AI system decision-making." AI systems need to be reliable, capable of handling failure, unbiased, accountable, and understandable by people. The problem is that most AIs are unable to justify their decisions. By providing complete, accurate, and human-understandable justifications for all system outputs, XAI helps build confidence in AI systems. The four common principles of XAI are shown in Figure 7.1 [2]. XAI allows humans to understand a machine learning (ML) system's decision-making process. A human-friendly explanation of a system's predictions, classifications, and other decisions is one of the capabilities of an XAI system. Users can better trust and comprehend the system's decisions due to this. AI systems that operate as black boxes cannot justify their choices.

The use of XAI and AI in business is increasing as technology progresses. XAI explains an AI model's operation, potential effects, and biases. It also helps assess AI-supported decision-making accuracy, fairness, transparency, and results. When using AI models in production, marketing managers frequently use purchase intentions to forecast sales, and XAI is essential for fostering confidence and trust. AI and XAI adoption are constantly changing. However, as AI becomes more prevalent, privacy concerns grow. XAI lets you choose which attributes are visible to end-users and customers while hiding sensitive information [3].

The purpose of this chapter is to educate the reader on the basics of AI and XAI, as well as their potential benefits. The chapter also discusses the objectives, applications, and challenges that consumers face in different applications. Additionally, it explores the risks and benefits of using AI and XAI in marketing, along with the legal and social implications of AI-enabled market structures. Sensitivity information may

DOI: 10.1201/9781003442509-7

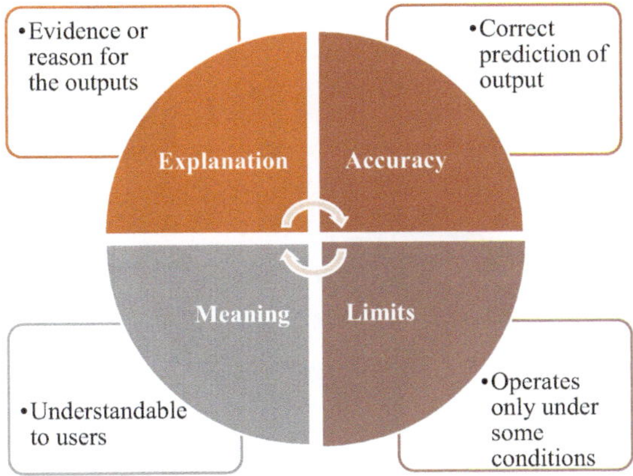

FIGURE 7.1 The four principles of XAI.

be exposed if feature selection for explainability rules is not followed when converting a black-box regression model to XAI.

7.1.1 Explainability and XAI

The authors in [4] state that "Explainability" in XAI refers to the capacity to explain an AI decision's justification in terms understandable to humans and relevant to a larger group of end users. The various end users focus on various explainability angles. Clinical inference and prediction, however, are of greater interest to doctors and other medical experts. Two related concepts are explainability and interpretability. Explaining the significance of an abstract idea is an example of a concept's interpretability. The former refers to how a model's predictions can be understood in the context of new data, while the latter refers to how that model can be understood after it has been trained.

XAI techniques can also be separated into post hoc and intrinsic groups. Without the need for additional information, an intrinsic method enables comprehension of a decision-making process or the fundamentals of a technique. Bayesian models, decision trees, rule-based learners, logistic regression, k-nearest neighbor, and linear regression are standard intrinsic techniques. AI is divided into ML and deep learning (DL). We also think that XAI is a subset of AI, and its fundamental techniques are ML. XAI, DL, and ML are just a few of the types of AI connected, as shown in Figure 7.2.

Some of the benefits that XAI can provide are as follows:

Minimizing the error costs:

Wrong predictions significantly impact decision-sensitive industries like medicine, finance, law, etc. Monitoring the results lessens the impact of incorrect results and helps identify the underlying causes, which improves the underlying model.

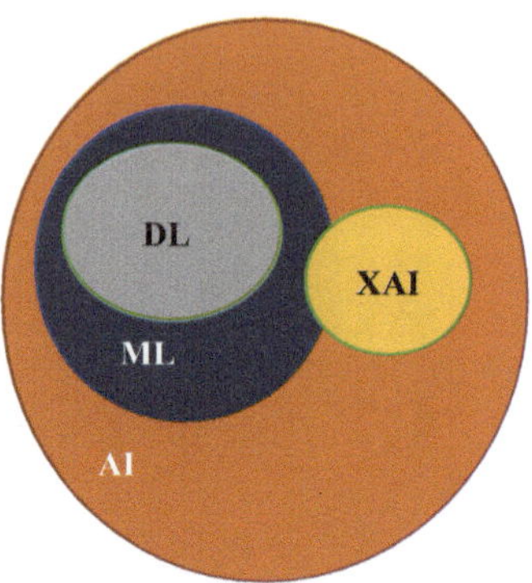

FIGURE 7.2 The relation between AI, ML, DL, and XAI.

Minimizing model bias:

Several instances of bias in AI models have been documented. Here are a few cases: racial bias in autonomous vehicles; gender bias in Apple's credit card system; racial bias in Amazon's Rekognition software; and so on. A system that can explain its decisions can mitigate the damage done by inaccurate forecasts.

Responsibility and accountability:

There is always room for error in the predictions made by AI models, and empowering a person who can take responsibility for and account for those errors can improve the system.

Code confidence:

The system's assurance grows with each inference and its justification. Autonomous vehicles, medical diagnosis, the financial sector, etc., are just a few examples of user-critical systems that require high confidence in the code from the user to be utilized to their full potential.

Code compliance:

Companies are under increasing pressure from regulators to adopt and deploy XAI to meet new regulations.

7.2 APPLICATIONS OF XAI

XAI can potentially significantly improve a wide variety of application areas for AI. Several potential applications of explainable models are explored and are shown in Figure 7.3.

Transportation:

Automated vehicles can improve mobility and reduce traffic fatalities, but explaining AI decisions is difficult. Autonomous vehicles must categorize objects quickly. Classification issues may cause self-driving cars to act strangely, and consequences may be severe. Anonymous sources said the car's software detected "wind-borne debris" in its path. Only an explicable system can resolve this confusion and hopefully prevent it. Transportation

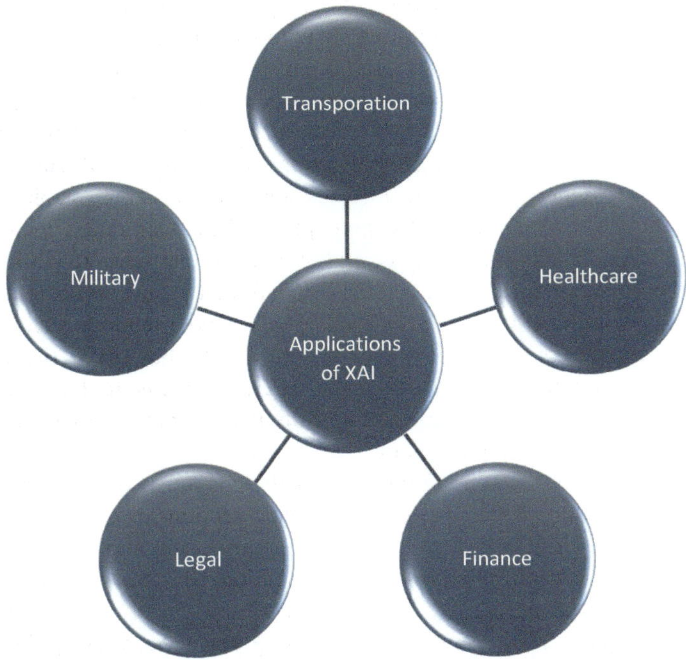

FIGURE 7.3 Applications of XAI.

could benefit from XAI. Many efforts have begun to illuminate autonomous vehicle behavior, but much work remains.

Healthcare:

The efficacy of medical diagnostic models is crucial to human survival. How can we put our faith in a black-box model and take its advice regarding a real patient? A mid-1990s ANN predicted pneumonia hospitalization and outpatients. Neural nets initially outperformed classical statistical methods. The results of the neural network's extensive testing indicate that patients with pneumonia and asthma have a lower mortality rate and do not require hospitalization. The training data showed that asthma and pneumonia patients were more likely to be admitted to the ICU for life-saving treatment, which defies medical logic. The project was scrapped because medical AI is risky. Avoiding this critical issue, identified only through model interpretation, is crucial. Researchers explain AI-based clinical systems. This literature shows the difficulty of XAI and the growing interest in healthcare.

Legal:

AI can potentially reduce crime and incarceration costs while also enhancing recidivism assessments. A criminal decision model must accurately, honestly, and without bias predict court recidivism. In the lawsuit, it was alleged that the program "Correctional Offender Management Profiling for Alternative Sanctions" infringes on procedural rights by considering a person's race and gender. The Judge was unsure of the causal audit procedure, and the algorithms were trade secrets. Although making the justifications for automated legal system decisions transparent is crucial, much research has not been done in this area.

Finance:

Financiers can benefit from AI in several ways, including improved customer service, investment advice, and wealth management. Concerns about data security and fair lending have been raised in light of these tools. Loan providers in the financial industry must follow the law and make fair lending decisions. Therefore, providing a "reason code" to borrowers becomes more challenging when AI-based credit scores and models are used to make lending decisions. This is especially true when the basis for denial is the result of a mysterious ML algorithm. Both Equifax and Experian are conducting promising studies to improve the auditability of AI credit-based score decisions and generate automated reason codes.

Military:

Military researchers created the current, well-known XAI initiative. The increased interest in the subject is mainly attributable to requests for research and Defence Advanced Research Projects Agency (DARPA) projects. XAI has applications in image recognition, education, entertainment, government, and other fields for automated decision-making.

7.3 IMPLICATIONS OF XAI

Organizations should consider many ways XAI can affect their operations.

- People's faith in these systems can be increased through XAI, which explains how AI models come to conclusions. This is of utmost importance to remember in areas like medicine and finance, where even a seemingly insignificant mistake can have dire consequences.
- With the help of XAI, organizations can gain insight into the reasoning behind AI models' decisions and hold them to a higher standard of accountability.
- Organizations can benefit from XAI because it allows them to understand the factors that go into an AI model's decision-making process.
- XAI can increase transparency by clearly outlining how an AI model makes decisions and the factors that affect those decisions.
- Organizational decision-making under the General Data Protection Regulation (GDPR) in the European Union must be open and accountable to its stakeholders.
- More businesses may be persuaded to adopt AI if they understand the reasoning behind the system's decisions, which could be made possible by XAI.
- Ethical issues arise when explaining AI because it may reveal private/personal details.

7.4 HISTORY OF XAI

XAI research began with expert systems in the early 1990s. By using a large set of rules, expert systems attempt to mimic human knowledge. The term "expert system" was coined for this reason. The implications of a rule are the new conclusions that can be drawn if the rule's premises are true. An explanation is a trail of rules with consistent outcomes and foundations. For example, "the system came to this diagnosis because it applied these rules in this order to these initial symptoms, and thus concluded that the patient has this sickness" would suffice as an explanation. These explanations, known as trace explanations, were the first to be developed [5].

Justification, strategy, and terminological explanations came later. These methods are domain-knowledge-enriched traces by providing definitions of domain-specific terms and supporting arguments for the expert system's rules, for instance. The results of expert systems are usually explained using a mixture of different types of explanations. Training systems, medical support systems, legal support systems, and educational systems all use similar methods to describe their operations [6].

Symbolic representations are the foundation of all the systems mentioned above. Symbolic systems are built to be processed by computers, but the languages they use are understandable by humans. To make decisions, the rules of "if.. then.." are used to model human logic. These specifications commonly take the form of human-created and -understood computer code. Moreover, understanding the reason for symbolic systems may be challenging for those who are not experts in the field [7]. As a result, XAI frequently focuses on determining the accuracy of the results. Therefore, interpretable ML is sometimes used instead of XAI to explain and present model behavior in human-comprehensible terms [8]. Non-symbolic AI systems store non-linear correlations rather than human-readable rules. Experts can validate the logic of symbolic systems, but non-symbolic systems require translation steps to be human-readable. Only post-hoc analysis can show results. Many of these algorithms constantly learn from their own decisions and new data. Self-learning algorithms will influence explanations. As a result, explanations must be updated as well [9].

Based on the statistical model learned by a machine, XAI intends to generate a symbolic model that humans can understand automatically. Since the model could be understood, it would be simple to account for the system's results. The process of explaining and interpreting machine-learned models has been extensively researched. Taylor decomposition and layer-by-layer relevance propagation are used in sensitivity analysis to show how machine-learned models work. The explanations provided by such models require the expertise of interpreters. Since this is the case, descriptions have meaning. When dealing with complex situations, explanations tend to be less clear [10].

Some researchers distinguish between model-centric explanations and subject-centric explanations. Variable changes in an algorithm cause it to reverse its decision, aiming to eliminate the negative consequences of AI-based decision-making. While promising, such techniques must be refined before they can be helpful and widely used [11]. According to available literature, XAI is a promising and growing research area, although no widely accepted techniques exist to explain the underlying AI models. Non-symbolic ML-based models are challenging to explain and interpret, particularly for beginners. Given these stumbling blocks to comprehension, the term "meaningful" is occasionally used in explanations. To be meaningful, an explanation must be understandable by all users. Given the variety of users, various justifications may be required. Many factors influence meaningful explainability, including the audience, the goals and objectives to be achieved, the complexity of the environment in which AI will be used, and the type of data used.

7.5 CLASSIFICATION OF XAI MODELS

In the literature, XAI models are divided into four groups, each defined by their generalizability and scope. These models include Model-Specific Global Explanation (MSGE), Model-Specific Local Explanation (MSLE), Model-Agnostic Global Explanation (MAGE), and Model-Agnostic Local Explanation (MALE).

- **MSGE** incorporates constraints on deep learning models' structure, learning rules, and parameters to make them more interpretable.
- **MSLE** uses tailored mechanisms to focus on high-dimensional input characteristics to explain DL models.
- **MAGE** makes alternative global models that can be understood, and that map model inputs to outputs based on how important and related they are.
- **MALE** seeks model-independent explanations for an instance or its vicinity.

7.5.1 Objectives of XAI

Figure 7.4 provides a concise summary of the overarching goals of the XAI.

i. Understand the system's behavior and avoid spurious correlations to find previously unknown vulnerabilities and flaws. We think explaining the system's behavior is crucial to assessing its performance.
ii. AI's inner workings and consequences are crucial for developers designing the algorithm. A system's accuracy and utility can benefit from being more explicable. As a result, XAI techniques offer the possibility and the likelihood of accomplishing improvement [12].

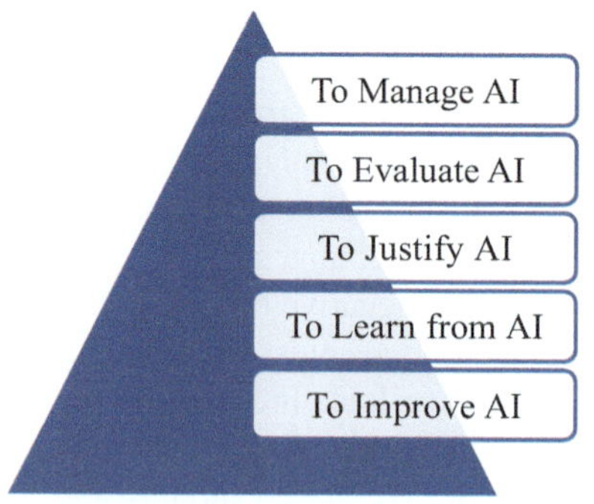

FIGURE 7.4 Generalizable objectives of XAI.

iii. "Deep knowledge" explains why specific rules were programmed into knowledge-based systems. AI explanations can find causal relationships in data without machine learning model programming. XAI learns from the algorithm's results [13]
iv. There has been a rise in the use of AI in complex settings. Legislation like Europe's GDPR may not have established formal "rights for an explanation." Still, it requires accountability and transparency in automated decision-making processes that may substantially affect particular individuals [14].
v. AI may alter work routines and results. Explainability aids AI evaluation, improvement, learning, and defense [15].

7.6 AI ETHICS

Scientists in a wide range of fields have begun paying more attention to questions of machine ethics and AI ethics as the number of applications for such systems grows. The increasing concerns about AI decisions' ethical, social, legal, and safety implications drive this interest. Both implicitly and explicitly, moral agents have been discussed in the literature on AI ethics. In both scenarios, human involvement is required for intelligent systems to determine what is and is not ethical. Conversely, rules established by their human designer prevent implicitly moral agents from engaging in immoral conduct. However, those who claim to be explicitly ethical take it for granted that they can evaluate their morality [16].

AI systems may hold both tacit and overt moral beliefs. The primary benefit of AIs with implicit ethics is that they are easy to implement and regulate because they are incapable of engaging in unethical behavior. However, this apparent minimalism implicitly reflects the designer's ethical stance and outlook. Explicit ethics systems claim to be able to appraise the normative status of actions independently and reason about what they consider unethical, thus resolving normative conflicts. In addition, they might break some rules in a way that helps them achieve their ultimate moral goals. However, their complexity and possibly unexpected behavior are their main drawbacks.

7.6.1 ETHICAL ISSUES WITH AI

Artificial intelligence researchers contend that moral considerations ought to be one of the main factors influencing the creation and use of AI. This subject is closely related to the accountability and openness of ML algorithms. Researchers have documented numerous instances of racial bias in AI algorithms. These cases showed racial discrimination against non-white mortgage applicants and the imposition of harsher prison sentences on black defendants. Because human lives are on the line, and driverless cars are a well-known example, researchers are urging the creation of principles that will guide AI decision-making. Other problems that highlight the need for ethical AI include a lack of accountability and transparency and the routine invasion of people's privacy [17].

According to ethical AI proponents, fairness, transparency, and privacy should come first in AI algorithms and models. Data, algorithms, and human characteristics must all adhere to discriminatory non-harm. AI systems capable of defending dubious AI algorithmic decisions are necessary for accountability. Sustainability guarantees the transformation of people and society through AI-enabled strategies. Finally, transparency allows the AI system to justify the ethical acceptability, non-discriminatory safety, and public trustworthiness of the outcomes and process by outlining in plain English the considerations that were made while acting in a certain way. These principles illustrate how to address data privacy, model confidentiality, fairness, and accountability when dealing with interpretability issues in AI models. The authors of [18] argue that developers and organizations must study these principles together to develop, adopt, and implement AI methods ethically.

In response to a request from the European Parliament's Committee on Legal Affairs, the policy division for "Citizens' Rights and Constitutional Affairs" commissioned, oversaw, and published the study that served as the foundation for the resolution. The report highlights the importance of a resolution urging the swift creation of robot and AI legislation that can foresee and accommodate any anticipated scientific advancements in the medium term. Figure 7.5 summarizes the main ethical and legal concerns related to the use of AI in healthcare settings.

7.6.2 Legal Issues

Applying XAI methods in a domain can have significant legal repercussions, even though the theoretical foundations of AI are typically regarded as legal. When working with sensitive data, it is essential. It is unknown if a model's explainability with slightly different data sets can be used to infer information about specific individuals. Information technology security analysts employ this method when working with anonymized data sets or only partially releasing sensitive information. Similar assaults on ML models [20,21] yielded a wealth of data, latent causalities, and correlations that could be used to infer private details.

Federated ML involves exchanging feature sets to be combined centrally while running machine learning algorithms locally on private data sets. These are distinct from running learning algorithms and centrally gathering all sensitive data. Federated ML must balance sensitive protective inferences and model robustness. Due to performance or quality issues, many anonymization techniques are inappropriate for

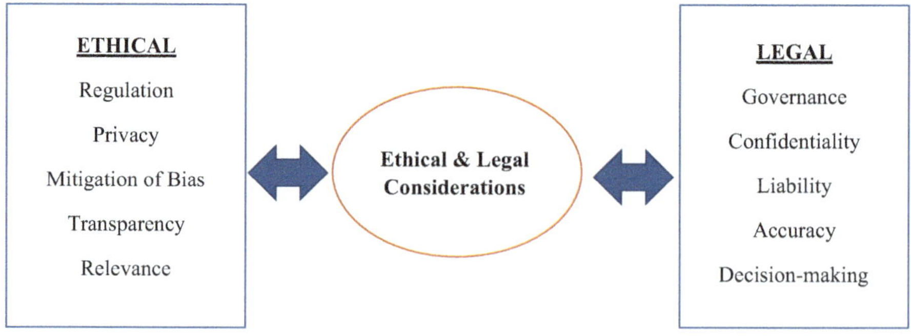

FIGURE 7.5 Ethical and legal issues of AI in healthcare [19].

application scenarios. The right to be forgotten is yet another legal issue. According to the statement, this "reflects a person's claim to have certain data deleted so that third parties can no longer trace them." The exercise of this right is made more difficult by technical issues like erasing entries in existing systems, determining specific individuals' information from aggregates, and excluding particular individuals from the aggregation process [22].

Despite the aforementioned issues, AI and machine learning explainability benefit the law. It can be challenging to address the issue of transparency, which is a crucial requirement of the GDPR. However, explainability might change that by offering in-depth information about the places, times, and levels at which a specific person's data was utilized in a data analysis workflow [23]. Even though it is not currently required by law, doing so could significantly impact how people perceive AI and how well their data is protected in today's data-driven society. Additionally, bias is being addressed as a significant issue in machine learning, especially since straightforward solutions have been ineffective. If data-driven systems' findings could be more clearly communicated to stakeholders and the general public, especially those used for societal or economic purposes, they might gain legitimacy and credibility.

7.6.3 Potential Issues with XAI

Data, the fuel for the algorithms, should be the primary focus of any discussion of AI transparency. Businesses should be transparent about the sources of the information used by their AI systems. Businesses often use and sell personal information about their customers without their knowledge or consent. Consumers should have the right to know how their data is being used. Since data is the backbone of AI, it's reasonable to wonder if and how it could explain the systems' inherent biases and seemingly irrational choices. On the side of algorithms, IBM and other tech giants' posturing around the concept of "explainable AI" is without virtue signalling with no basis in reality. For instance, I am not aware of anywhere where IBM has exposed the inner workings of Watson, such as how its algorithms function. We need to know why they give the advice or make the forecasts they do.

The concept of AI can be explained with two significant flaws. The first is definitional: What exactly do we mean when discussing explainability? Precisely what are our concerns here? What are the underlying algorithms/statistical models? What are some of the ways the parameters of learning have shifted over time? What form does a model take for a given forecast? There is a wide range of complexity involved with each of these options. Some are pretty simple; after all, somebody had to create the algorithms and data models, and that person undoubtedly remembers precisely what they did and why. We have a good understanding of what these models entail. One of the most welcome aspects of the current AI boom is that most research results are published in open, peer-reviewed papers that anyone can access.

Current methods, such as deep learning, outperform our definitions of success because they automatically identify relevant variables. Explaining this complexity is challenging because the system discovers variables and relationships that humans have not previously recognized or articulated. If we could, we would write it into a computer program.

Evaluating the trade-offs of "true explainable and transparent AI" is the second significant aspect to consider when thinking about explainable AI. Currently, performance and explainability, along with business implications, are traded off in some tasks. Intellectual property would cease to be a differentiator if the entire inner workings of an AI-powered platform were made available to the public. It would be like asking a company to reveal its source code if they were forced to detail every aspect of its proprietary AI system. If the intellectual property were worth anything, the business would be over as soon as the send button was pressed. Thus, pushing for such requirements would typically benefit market incumbents with deep pockets and substantial market share and stifle innovation among startups.

Don't take this as support for "black box" AI; that's not what the authors meant. Before we can fully accept the need for companies to be open with their data and explain their AI systems to those interested, we need to think about what that means for society, both in terms of what we can do and what kind of business environment we make. When it comes to artificial intelligence, I believe it will have a net positive effect, and I support open source and transparency [24]. We are placing an enormous burden on a new but potentially lucrative industry by placing such a premium on openness.

7.6.4 Miscellaneous Issues

XAI appears to be an appealing concept at first glance because explanations are valued. However, cultural values and norms are not always translated literally, and social acceptability varies by society. Even within a single community, cultural norms can vary. This is why policy decisions are thought to necessitate careful thought. Any policymaking process should strive to create standards that are universally accepted. Although the reasons for explainable AI's failure to deliver on its promise are complex and intertwined, Table 7.1 summarizes the significant XAI challenges. The issues are complex and intertwined [25].

TABLE 7.1
Typical Challenges of XAI

Challenge	Description
Lack of knowledge	The average person doesn't know enough about the subject at hand to evaluate whether or not the decision was made fairly.
Disputed justifications	Most people don't know how to evaluate the decision's fairness or understand its reasoning.
Causality does not influence decisions	A chain of events connecting inputs and outputs does not guarantee the algorithm's conclusion. There is a chance that our views on causality will evolve as well.
Dynamics of information and decision	As time passes, both data and decisions may have different explanations.
Problems' wickedness	A wicked problem lacks clear parameters, has multiple potential resolutions and is difficult to define. Algorithms produce a single answer that can be debated and evolved.
Context dependent	The algorithm doesn't use its explanation of inputs and outputs. Causality explanations may change.

7.7 CONCLUSION

The role of AI in shaping our everyday lives is expanding. The widespread adoption of AI-based solutions in industries as diverse as human resources, finance, law enforcement, healthcare, and education has far-reaching personal and professional implications. AI researchers and practitioners have shifted their attention to XAI to gain more confidence in and comprehend large-scale models. Recent research has shown that XAI is crucial to implementing ML techniques in real-life scenarios. This chapter clarifies model explainability concepts and demonstrates how various goals motivate the search for more understandable ML techniques. The chapter begins by defining some key terms before delving into XAI's fundamentals, applications, implications, and recent research in this field. Additional XAI model classifications and objectives are covered. Possible pitfalls and ethical concerns with using XAI in real-time are discussed. We believe it's critical to consider data privacy, model confidentiality, fairness, and accountability considerations when assessing a model's interpretability. Only by studying these XAI principles together can we ensure AI techniques' safe and ethical application in businesses and government agencies worldwide.

REFERENCES

1. Zhang, Bo, Jun Zhu, and Hang Su. "Toward the third generation artificial intelligence." *Science China Information Sciences* 66, no. 2 (2023): 1–19.
2. Phillips, P. J., Amanda C. Hahn, Peter C. Fontana, David A. Broniatowski, and Mark A. Przybocki. "Four principles of explainable artificial intelligence (draft)." (2020). https://doi.org/10.6028/NIST.IR.8312-draft.
3. Nwakanma, Cosmas Ifeanyi, Love Allen Chijioke Ahakonye, Judith Nkechinyere Njoku, Jacinta Chioma Odirichukwu, Stanley Adiele Okolie, Chinebuli Uzondu, Christiana Chidimma Ndubuisi Nweke, and Dong-Seong Kim. "Explainable artificial intelligence (XAI) for intrusion detection and mitigation in intelligent connected vehicles: A review." *Applied Sciences* 13, no. 3 (2023): 1252.
4. Doshi-Velez, Finale, and Been Kim. "Towards a rigorous science of interpretable machine learning." arXiv preprint arXiv:1702.08608 (2017).
5. Xu, Feiyu, Hans Uszkoreit, Yangzhou Du, Wei Fan, Dongyan Zhao, and Jun Zhu. "Explainable AI: A brief survey on history, research areas, approaches and challenges." In *Natural Language Processing and Chinese Computing: 8th CCF International Conference, NLPCC 2019*, Dunhuang, China, October 9–14, 2019, Proceedings, Part II 8, pp. 563–574. Springer International Publishing, 2019.
6. Conati, Cristina, Kaska Porayska-Pomsta, and Manolis Mavrikis. "AI in Education needs interpretable machine learning: Lessons from Open Learner Modelling." arXiv preprint arXiv:1807.00154 (2018).
7. Preece, Alun, Dan Harborne, Dave Braines, Richard Tomsett, and Supriyo Chakraborty. "Stakeholders in explainable AI." arXiv preprint arXiv:1810.00184 (2018).
8. Du, Mengnan, Ninghao Liu, and Xia Hu. "Techniques for interpretable machine learning." *Communications of the ACM* 63, no. 1 (2019): 68–77.
9. Jordan, Michael I., and Tom M. Mitchell. "Machine learning: Trends, perspectives, and prospects." *Science* 349, no. 6245 (2015): 255–260.
10. Du, Mengnan, Ninghao Liu, and Xia Hu. "Techniques for interpretable machine learning." *Communications of the ACM* 63, no. 1 (2019): 68–77.

11. Venkatasubramanian, Suresh, and Mark Alfano. "The philosophical basis of algorithmic recourse." In *Proceedings of the 2020 Conference on Fairness, Accountability, and Transparency*, Barcelona, Spain, pp. 284–293, 2020.
12. Gilpin, Leilani H., David Bau, Ben Z. Yuan, Ayesha Bajwa, Michael Specter, and Lalana Kagal. "Explaining explanations: An overview of interpretability of machine learning." In *2018 IEEE 5th International Conference on Data Science and Advanced Analytics (DSAA)*, pp. 80–89. IEEE, Turin, Italy, 2018.
13. Chandrasekaran, Bruce, Michael C. Tanner, and John R. Josephson. "Explaining control strategies in problem solving." *IEEE Intelligent Systems* 4, no. 01 (1989): 9–15.
14. Adadi, Amina, and Mohammed Berrada. "Peeking inside the black-box: A survey on explainable artificial intelligence (XAI)." *IEEE Access* 6 (2018): 52138–52160.
15. Rai, Arun, Panos Constantinides, and Saonee Sarker. "Next generation digital platforms: Toward human-AI hybrids." *MIS Quarterly* 43, no. 1 (2019): iii–ix.
16. Moor, James H. "The mature, importance, and difficulty of machine ethics." In Wendell Wallach, Peter Asaro (eds.) *Machine Ethics and Robot Ethics*, pp. 233–236. London: Routledge, 2020.
17. Leslie, David. "Understanding artificial intelligence ethics and safety." arXiv preprint arXiv:1906.05684 (2019).
18. Arrieta, Alejandro Barredo, Natalia Díaz-Rodríguez, Javier Del Ser, Adrien Bennetot, Siham Tabik, Alberto Barbado, Salvador García et al. "Explainable Artificial Intelligence (XAI): Concepts, taxonomies, opportunities and challenges toward responsible AI." *Information Fusion* 58 (2020): 82–115.
19. Naik, Nithesh, B. M. Hameed, Dasharathraj K. Shetty, Dishant Swain, Milap Shah, Rahul Paul, Kaivalya Aggarwal et al. "Legal and ethical consideration in artificial intelligence in healthcare: Who takes responsibility?." *Frontiers in Surgery* 9 (2022): 266.
20. Shokri, Reza, Marco Stronati, Congzheng Song, and Vitaly Shmatikov. "Membership inference attacks against machine learning models." In *2017 IEEE Symposium on Security and Privacy (SP)*, pp. 3–18. IEEE, SAN JOSE, CA, 2017.
21. Rudra Kumar, M., Rashmi Pathak, and Vinit Kumar Gunjan. "Diagnosis and medicine prediction for COVID-19 using machine learning approach." In *Computational Intelligence in Machine Learning: Select Proceedings of ICCIML 2021*, pp. 123–133. Singapore: Springer Nature Singapore, 2022.
22. Malle, Bernd, Peter Kieseberg, and Andreas Holzinger. "Do not disturb? Classifier behavior on perturbed datasets." In *Machine Learning and Knowledge Extraction: First IFIP TC 5, WG 8.4, 8.9, 12.9 International Cross-Domain Conference, CD-MAKE 2017*, Reggio, Italy, August 29–September 1, 2017, Proceedings 1, pp. 155–173. Springer International Publishing, 2017.
23. Zhang, Quan-Shi, and Song-Chun Zhu. "Visual interpretability for deep learning: A survey." *Frontiers of Information Technology & Electronic Engineering* 19, no. 1 (2018): 27–39.
24. Lakshmi, T. Naga, S. Jyothi, and M. Rudra Kumar. "Image encryption algorithms using machine learning and deep learning techniques-A survey." In *Modern Approaches in Machine Learning and Cognitive Science: A Walkthrough: Latest Trends in AI*, Volume 2, pp. 507–515. Cham: Springer International Publishing, 2021.
25. de Bruijn, Hans, Martijn Warnier, and Marijn Janssen. "The perils and pitfalls of explainable AI: Strategies for explaining algorithmic decision-making." *Government Information Quarterly* 39, no. 2 (2022): 101666.

8 Fair and Explainable Systems
Informed Decision Making in AI/ML

Parth Birthare, Maheswari Raja, and Sharath Kumar Jagannathan

8.1 INTRODUCTION

Trustable AI is a very recent area of research, and even the most reputed companies are either unfamiliar with this area or have just begun its implementation in their practices. Moreover, regulatory authorities like the Monetary Authority of Singapore and the European Commission High-Level Expert Groups have started to raise awareness on the FEAT - Fairness, Ethics, Accountability, and Transparency of AI, and have laid out several guidelines for financial institutions to follow. Leaders at the executive level from top companies even express uncertainty about whether the current methods for achieving Trustable AI are correct because there has not been much exploration in this area.

Accuracy and precision are just two of several parameters, but not the absolute deciding factors of the correctness of outcomes in an AI system. Deep learning models are black boxes: input goes in, and output comes out. What happens within the model is simply unknown. For example, an AI model predicts whether a loan will be sanctioned to a person based on several factors. What if their race is used as a factor in predicting the loan decision? Though it might increase accuracy to use such features, many organizations have faced financial losses and criticism for employing them.

Therefore, it is necessary to instill trust in AI. There has been only some progress in the area of Trustable AI. Moreover, most of the work focuses on categorical data, whereas Natural Language-based systems remain a challenge. Trustable AI involves not only a specific stage of the Machine Learning/AI workflow but is also crucial at every stage.

To provide more clarity on what Trustable AI really is, here is an example: there is an autonomous vehicle. It has a speeding truck coming towards it. But if it does not steer away, it will hit the truck. However, on the right is a little child and on the left an old man. What should the autonomous vehicle do to ensure it represents human values? Another example is of a loan providing agency that profiles people using AI to assess if they are a risk or not. However, they also use gender as a factor in

DOI: 10.1201/9781003442509-8

predicting if the person is a risk. This may increase the accuracy of their model, but ethically, it is a very wrong practice. Moreover, institutions have faced huge penalties and condemnation for using such AI models. Therefore, Trustable AI has its roots widely spread in the technological world, and the sky is the limit.

Explainability of AI models can occur both at global and local levels. Global explanations elucidate the entire model, whereas local explanations focus on explaining models at a local level. To elaborate further, global models provide insights into features that are highly correlated with a certain kind of attribute across the entire dataset, while local models consider only one data point and explain that by considering the features locally.

8.2 BACKGROUND STUDY

For AI systems to be trustworthy, they should be fair and transparent, reliable, cause no harm in any manner, robust with assured privacy, explainable, accountable, aligned with the key values they are designed for, and able to understand logical justifications behind a problem. Additionally, they should communicate uncertain outcomes when unsure of decisions. If these concerns are not handled aptly, severe consequences can arise: the risk of the programmer's bias affecting the outcome increases, resulting in a model based on previously known biases and stereotypes that reflects a narrow vision of the community. Furthermore, these consequences can stem from poorly chosen training data, hacking by malicious attackers to introduce bias, or developers' assumptions and stereotypes applied in the machine learning process.

Google's photo app once classified images of black people as gorillas, leading to significant scrutiny [1].

A similar case was observed with Nikon's S630 camera, which encountered racial bias in its AI model, miscategorizing Asian people as blinking [2]. All these incidents only contribute to the increasing hesitation and level of doubt placed in AI systems.

HP webcam facial tracking software was also unable to track people with darker skin tones, while it successfully tracked people with lighter skin tones [3].

According to a report published by ProPublica, Recidivism Risk Detection Software made predictions based on race and labeled incorrectly [4]. The software predicted a lower risk score of three for Dylan Fugett, who had one prior offense of attempted burglary and a lighter skin tone. Conversely, the same software predicted a higher risk score of ten for Bernard Parker, who had one prior offense of resisting arrest without violence. Subsequently, Dylan committed subsequent offenses of three drug possessions, while Bernard committed no subsequent offenses.

Furthermore, according to ProPublica's analysis of data from Broward County, Northpointe's assessment tool correctly predicts recidivism 61%of the time. However, blacks are almost twice as likely as whites to be labeled as higher risk but not actually re-offend. Conversely, it makes the opposite mistake among whites: they are much more likely than blacks to be labeled as lower risk but go on to commit other crimes [4]. Even predictive policing software identified wrong areas as crime hotspots and consequently made more arrests in those regions in Oakland [5].

In a similar vein, Amazon's same-day delivery service was inaccessible to black neighbourhoods [6]. Now, is race indirectly being used by the models to make predictions? Are the models prey to human biases?

In the work by Kruger and Wilson, they point out that the notion of trust is essentially human-originated, even within AI systems [7]. This means that the level of trust placed by humans is subjective, and that subjectivity is imparted to the machines. They refer to this as human bias. They conducted a study on the role and necessity of trust in AI, suggesting that it helps the government gain legitimacy and improves customer satisfaction. In conclusion, they argue that trust is essential, and since AI is already a social reality, better ways to trust it should be found. In the work by Reinhardt, the author researched different AI ethics papers [8]. They point out that trust in AI is linked to reliability, robustness, safety, and security. Additionally, they mention traceability and verifiability, as well as interpretability and explainability, as other characteristics of trustworthy AI. They further state, "AI systems should serve the needs of the user, or even enhance environmental and societal well-being and the good of humanity, individuals, societies and the environment and ecosystems." They further pointed out that notions of trust and trustworthiness have gained significant attention in AI research, especially after the publication of Europe's High-Level Expert Group's Ethics guidelines for trustworthy AI.

Alvarado, in his work, pointed out that the only trust that can be put into AI systems is epistemic trust [9]. They further point out that trusting AI otherwise may betray conceptual confusion. Also, they note that the level of trust placed on AI is dependent on the receiver. According to the work, it is better to have predictable outcomes in either case, whether dealing with a human or an AI machine. However, they argue that trusting AI is still not that easy and is epistemic. Trust is developed only according to the receiver. The concept of trust in AI was linked in three verticals in the work by Siau and Wang [10]. According to the authors, trust is not a result of a single trait of AI systems; rather, the notion should be involved and seen as a whole in the entire process. Trust is developed not only by having informed decisions and traceable outcomes but also by education and awareness among people. Whenever technology has replaced jobs, it has always created more than it replaced, and the demand is ever-increasing. In the work by Jan and others, the authors focused on implementing explainable solutions in business practices and processes [11]. The main focus is to explain the importance of interpretable models so that the results can be easily tracked down. In the work, they mentioned the importance of a tool called LIME, which is locally interpretable model-agnostic explanations, and which will be used in my work too, to explain the end results of the AI models. They also mentioned the need to maintain quality explanations that lead to value judgment.

In the work by Linardatos and the team, the authors started by pointing out the statistics of the term "Explainable AI" used in Google searches [12]. It showed that the popularity of the search term "Explainable AI" has grown exponentially over the past 10 years. The work concerned itself with the interpretability of models, firstly by starting with the relation between explainability and interpretability. Further, they discussed the tools available for explainable AI, for texts and images. They concluded by mentioning the taxonomy of the existing machine learning interpretability methods that allow for a multi-perspective comparison among them. Another work focused on the interpretability and explainability of AI, and how the term has been used interchangeably, with the help of a graph [13]. This work also discusses the same methods that are available for developing interpretable and explainable models. There is a lack of implementation in real-life scena rios like all the other works till now. The paper focused more on what has been done till now in statistics; the way to implement explainable AI is, however, not covered. Barbara's work, however, focuses on the use of AI in criminal justice reform [14]. This work also sheds light on formal fairness and the accuracy of the models; the latter results in a trade-off between interpretability again. The authors further discussed if the systems are having disparate results for punishment since the algorithms are being used widely in the crime and punishment department. These are just the discussions; however, the actual implementations of the algorithms and the approaches are not discussed in their work.

Belengueur's work is different from the previously discussed work in the sense that it deals with bias in AI [15]. With reduced bias, AI automatically becomes fair. It illustrates four phases of detecting bias in AI systems. It also includes the use of AIF360 – AI Fairness 360 toolkit by IBM for detecting possible biases. The author, however, proposes only the methods by which bias in AI can be dealt with. It does not discuss practical examples where it is implemented and therefore does not address the challenges that arise in the processes.

Delgado and team focused on works related to COVID-19 that were published between 2020 and 2021 [16]. Their aim was to identify any biases in the algorithms. They concluded that the main biases are related to data collection and management. The authors further identified that there were ethical problems related to privacy, consent, and lack of regulation, which led to bias-related health inequalities. The study is more focused on the COVID-19 side, and they faced a challenge of having high-quality data. They, however, found that there were biases related to apps used in healthcare. Whereas Johansen and others state that the origin of biases in algorithms arises from the dataset itself [17]. And since humans are involved in data generation, human biases have an important effect. They further state that despite this traditional way of bias arising in the systems, there are new sources of biases in AI systems: programmers' ethical standards and education, and the contextual programming environment. This is a very important point and a factor, which is also a reason to include multiple people in the development of an AI system, and from different backgrounds, to mitigate any possible chances of biases that can occur. They concluded with a flowchart of types of biases – long-term or short-term, strong and weak.

A new approach to explainable AI was discussed in Ribeiro and the team's work, where they introduced LIME, or locally interpretable model explanations [18]. This method builds a local model on top of the AI model to trace back the features that lead to a particular outcome. It is widely used to make models explainable. However, as the model complexity increases, the method is not as useful. LIME can only explain the models locally and not globally; it can explain individual results but not the entire model. They further focused only on one aspect of the issues of trustworthy AI.

In the work by Ntoutsi and others, they provide a broad multidisciplinary overview of the area of bias in AI systems, focusing on the technical challenges and solutions [19]. They divided the process of dealing with biases into three steps: bias detection, bias mitigation, and bias accounting. While discussing the individual parts, they mention that biases can be handled pre-process and in-process, during the AI model generation. However, changing the data in the pre-process to reduce bias may also cause legal issues. This work also discusses just the conceptual part of bias without providing any practical solutions to it. Panch and team, on the other hand, focused their area of research on the trust of AI systems in the healthcare sector [20]. Major challenges include the lack of a clear standard of fairness, lack of contextual specificity, and the black box nature of deep learning models. They further discuss the actions that can be taken to mitigate biases. These actions include establishing the context in which algorithms will be developed and deployed, processes to counter the risks of bias in algorithm development, and the balanced development of the discipline of health data science. The last point has been given considering AI in healthcare both as a science and an art. It also talks about transparency and conveys that the implications of trustworthy AI are vast.

Liu and others dealt with bias in textual data, which is the most difficult to deal with and mitigate [21]. They further discussed the two types of bias: implicit and explicit, which can occur in the process. Implicit bias is the major and most difficult to identify and mitigate. While focusing their work solely on textual data, they trained a Convolutional Neural Network (CNN) text classification model and proposed methods to mitigate bias from it. Again, they only took into account the biasness, and since CNN and Recurrent Neural Network (RNN) are complex deep learning models, the explainability of such systems is compromised, which is where the work lacks its superiority.

8.3 METHODOLOGY

Every type of data is dealt with in a different manner. What practices work for one kind of data may not work for another kind. When it comes to the creation of any AI model, the first and foremost aspect is the dataset. Data choice and its diversity are very important, and the model creation should not be biased in any sense. One way to remove possible personal biases from the models developed is by having developers from different backgrounds and cultures, each with different mindsets, to eliminate possible biases. Another aspect is for the training data to be as representative of the real scenario as possible. There is no universally accepted ethical system for AI. Furthermore, computers lack human morality – they have no feelings, therefore instilling ethical behaviours in them is a challenge.

There are many methods that can be used for different types of data. Choosing the right method is not an exhaustive combination of practices, but rather a mix and match. One solution is using Inverse Reinforcement Learning. In this method, the systems are allowed to observe how people behave in different scenarios, identify what people actually value, and then let the system make decisions based on underlying ethical principles.

Creating fair and explainable AI and ML models revolves around the notion of transparency. Transparency is instilled considering three factors:

1. Transparency in process

   For trust to be developed in AI, people should know what processes lead to a particular outcome. Deep learning models are black boxes, and their predictions cannot be traced back. However, some AI models provides excerpts from documents or knowledge bases that lead to a particular conclusion. This answers the "what" and "why" of a conclusion.
2. Transparency in use of information

   Developing trustworthy AI systems that are fair and explainable also involves understanding what information a system processes while interacting with people. For instance, consider an emergency app that has control over audio, video, and location features of a device; the user should be provided with adequate access to control the application's access to those features. This way, transparency can be ensured in those AI systems. However, this approach involves a trade-off between privacy and the efficient use of the system.
3. Transparency through education

   Misconceptions should be cleared about the emergence of AI systems. People should be made aware of what AI can and cannot do, and which jobs AI can and cannot affect. Furthermore, in scenarios where advisory work is involved, people do not want to consume precious faculty time. It has always been the case that where jobs are affected by technology, it is observed that new technology has only opened new jobs.
4. Multidisciplinary approach to AI.

   As the human brain is composed of interdependent elements, natural intelligence is similarly complex. It can only be better understood through multidisciplinary approaches.

Transparency, as discussed in the above text, is one way to make AI systems more trustworthy. Another metric used to measure the trustworthiness of AI is bias. One way of checking bias in data is through counterfactual texts. Counterfactual outcomes are those that do not actually occur but can occur under separate and different scenarios. For textual data, counterfactual texts are synthetically generated texts created differently by a conditional model. Therefore, counterfactuals provide a varied or alternate dataset and often serve as test cases for evaluating the fairness of any AI model. For example, consider sentence A: Frey eats lemons. Consider sentence B: Tim eats leaves. A counterfactual example here would be defined by flipping the original class label if Frey eats lemons instead of leaves. There are four important properties of counterfactual texts:

1. **Plausibility:** Ensuring that the generated test cases are real and can be used for retraining.
2. **Diversity:** Ensuring high input space coverage as defined by the goal.
3. **Goal-oriented:** Ensuring that the generated counterfactual text samples differ from the original text samples on particular aspects (Named Entity Recognition, Semantic Role Labeling, synonyms, fairness attributes) as required.
4. Effectiveness.

To ensure the desired properties of counterfactual texts are preserved during generation, loss functions are used. For generation, the Generative Pre-Trained Transformers – 2 (GPT-2) decoder can be utilized. It relies on specific proximity and diversity constraints to change input sequences in many ways. Bias is observed when there is a difference in the sentiments for different counterfactual texts. For example, let sentence A be "The trainer was a female" and sentence B be "The trainer was a male." Suppose the sentiment model assigns a positive label to sentence A and a negative label to sentence B; then the sentiment changes with a minimal change in text, such as gender. Therefore, the model is biased. Counterfactual samples can act as training data for data augmentation algorithms to de-bias the sentiment model. Furthermore, counterfactuals should maintain diversity and preserve the semantic and syntactic structure of the input sentence.

A standard supervised machine learning workflow involves three processes as displayed in Figure 8.1.

Process 1: The training dataset contains sequences of instances: each instance has two components – features and correct prediction.

Process 2: A machine learning algorithm produces a machine learning model after training on the training data.

Process 3: Validation data is used to test the accuracy of the model.

Biasness can trickle into the system in any of these three processes in the form of:

1. **Pre-processing:** The training data may be biased in a way that the outcomes are inclined towards a particular class.
2. **In-processing:** The training algorithms that create the model may be biased towards some particular classes.
3. **Post-processing:** Test data can be biased in a way where the concentration of the data is more for a particular class.

Therefore, bias is the inclination towards a particular group that shares common characteristics. Moreover, providing equal opportunity is not equivalent to being fair. Entities that are already at an advantage from the beginning, when provided equal weightage, will still be at an advantage against other entities. For instance, consider

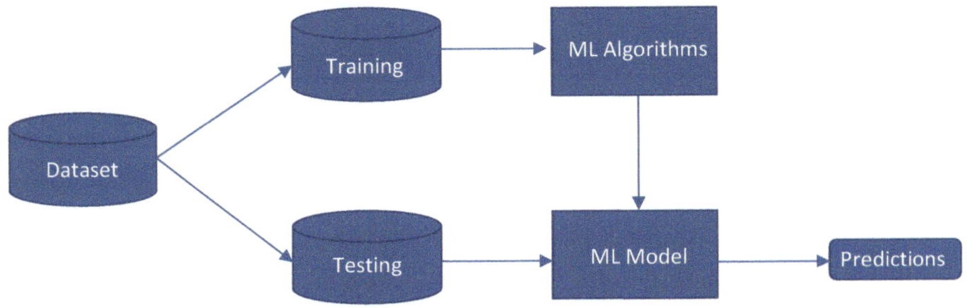

FIGURE 8.1 Standard machine learning workflow.

a fence that is 2 m high. There are three boys with heights of 1, 1.5, and 2 m. Now, to see what is behind the fence, they need to be at a level more than 3 m. So, providing a wooden stand to each of them of height 0.5 m will still make the shortest boy unable to see behind the fence. What can be done instead is to provide the shortest boy with two 0.5 m stands, the mid-heighted boy with 0.5 m stand, and none for the tallest boy. This will enable all of them to see what is behind the fence, and this is what fairness essentially is about.

Biasness is inevitable as long as there are humans. Therefore, since all the data is directly or indirectly generated by the involvement of humans, data tends to be biased. If biasness is not addressed in the initial stages, it will persist in the further stages of the pipeline, eventually leading to biased outputs.

8.3.1 Bias Identification

To determine if the data is biased, firstly, protected attributes must be identified. Protected attributes are those features that make our model biased. After that, the statistical parity difference (SPD) and disparate impact ratio are calculated. SPD is the difference in the ratio of favourable outcomes of privileged groups and the ratio of favourable outcomes of unprivileged groups. The disparate impact ratio is the ratio of favourable outcomes of unprivileged groups divided by the ratio of favourable outcomes of privileged groups. A negative value for SPD or a value of less than 0.8 for disparate impact ratio indicates biasness.

8.3.2 Bias Mitigation

Mitigation of bias works differently for different problems. In certain scenarios, it may be useful to remove certain bias-causing features from the dataset entirely. Whereas, it may be useful to adjust the weights assigned to the biased features to make them unbiased during the training process of the model. This is an important step if that feature is highly correlated with other features in the data.

Other methods to make AI trustworthy include using tools such as Local Interpretable Model-agnostic Explanations (LIME) and SHapley Additive exPlanations (SHAP). These are explainability tools that elucidate the models and individual examples, providing both global and local explanations for the model. These explainers assign a numerical value to each feature reflecting its importance.

8.4 CASE STUDY

This is a case study of customer complaints classification on a custom dataset. The dataset originally consisted of over 2 million records. Below is the procedural workflow diagram as depicted in Figure 8.2.

The pipeline starts with the pre-processing of the data – removing stop words, handling missing values, and vectorizing the data, which follows the data exploration stage, aimed at identifying patterns in the data. The next step is to resample the data using techniques such as SMOTE – synthetic minority oversampling technique.

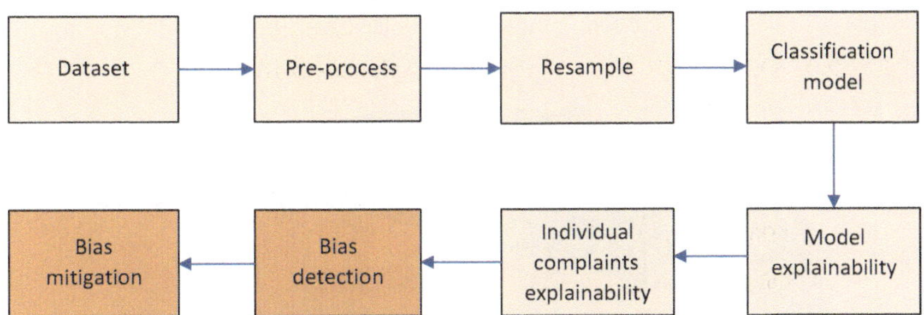

FIGURE 8.2 High-level procedural workflow.

The data distribution before pre-processing is displayed in Figure 8.3. The dataset size before pre-processing was approximately 800,000 records.

After pre-processing and resampling, the data distribution is depicted in Figure 8.4. The dataset size after resampling became 1,500,000.

The SMOTE algorithm up-sampled the records for each class so that they are in equal proportion. It did so on the tokenized data. That means it increased the quantity of certain tokens and used them to generate new complaints. However, one thing to keep in mind while using SMOTE with textual data in this case is that the new data or the complaints generated by the algorithm will not have any contextual meaning, as the tokens are randomly distributed to generate a sentence. At the same time, each token in the text is already assigned a score, as is the role of the Term Frequency-Inverse Document Frequency (TF-IDF) vectorizer, so it is not possible to retrieve the original text. However, in this case, it will not affect us, as we do not need

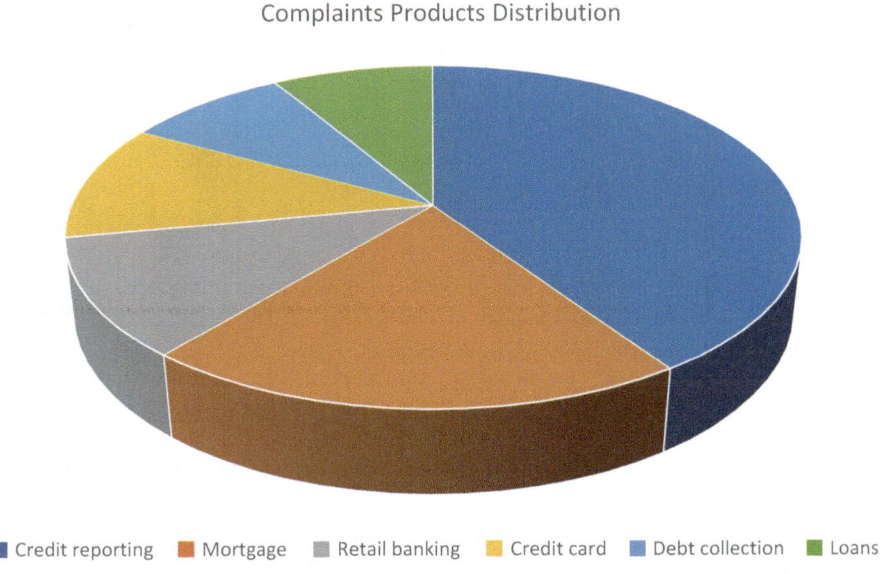

FIGURE 8.3 Data proportion before resampling.

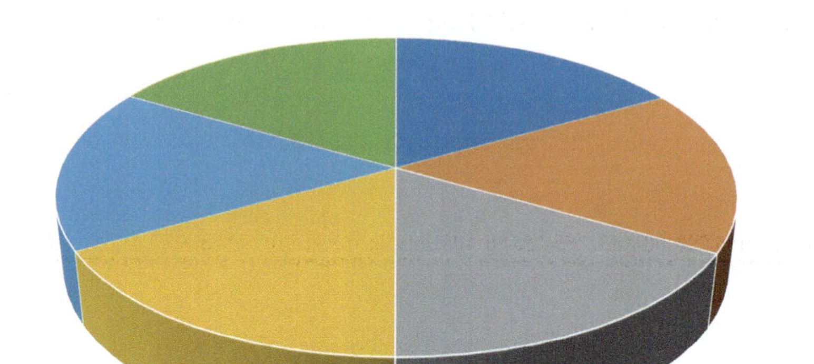

FIGURE 8.4 Data proportion after resampling.

to retrieve the original complaints, and we would need to convert the text to numerical format for further processing anyway.

Now that the data is ready to be processed with algorithms, various multi-class classification models are tested. Their parameters are varied to determine the best algorithm, and optimal parameter tuning is performed to achieve the most accurate solution. Below are the classification algorithms used:

1. K-nearest neighbours,
2. Decision trees,
3. Naïve Bayes,
4. Random Forest,
5. Support vector machine, and
6. Logistic regression.

For the K-Nearest Neighbour (KNN) algorithm, the obtained accuracy was approximately 86.3%. Decision trees yielded an accuracy of 85.1%. Naïve Bayes algorithm achieved an accuracy of 78%, whereas Random Forest reached 85.8%. Moreover, Logistic Regression resulted in an accuracy of 88%, while Support Vector Machine (SVM) achieved 89.4%. These accuracies are all from testing and can be observed in Table 8.1.

Observing the accuracy using the above models, it was noted that Linear Support Vector Classifier (Linear-SVC) has better accuracy on the given dataset than any other algorithms. Apart from that, using other kernel functions has an impact on the explainability of the model, which is a desired outcome of the problem and is much needed. Therefore, only Linear SVC is used and not any other kernels.

TABLE 8.1
Model Accuracies

Model	Accuracy (%)
KNN	86.3
Decision trees	85.1
Naïve Bayes	78
Random Forest	85.8
Logistic regression	88
SVM	89.4

8.5 SELECTION OF BEST MODEL

Selecting the best model depends on accuracy. However, at this stage, if we aim to explain the model and understand which features are indicative of a particular class, we should opt for a less complex model.

As the model complexity increases, the explainability or interpretability of the model decreases. A linear regression model is the easiest to explain. However, as more layers are added to the model, its complexity increases, making it difficult to trace back to the original features. Therefore, Linear SVC is chosen in this case, which not only achieves higher accuracy with the data but is also comparatively less complex than other kernels in SVM.

The entire procedure for choosing the best model and working with it was as follows:

1. Use TF-IDF vectorizer to convert complaints to vector form.
2. Employ Linear SVC considering unigrams and bi-grams.
3. Explore different combinations to determine the maximum number of features: here, it was found to be 10,000. 5000 resulted in lower accuracy, and more features generated useless words and overfitting issues.
4. Set the minimum document presence for features to 10.
5. Set the maximum document presence for features to 50%.
6. Achieve a resting accuracy of 89.4%.
7. Achieve a training accuracy of 90.1%.

From the testing and training data, it is also concluded that since there is not much difference between the two, the model is well-fitted, meaning it is neither underfit nor overfit. If a large difference is observed, then it suggests that there is some problem with the model or the parameters are not optimal.

8.6 MODEL EXPLAINABILITY

Deep learning models are essentially black boxes. An input goes in, and an output comes out, as shown in Figure 8.5. It becomes very important to trace the stages of the data from the input until it generates the output to achieve transparency.

FIGURE 8.5 Deep learning: a black box.

Now that we have our model trained, we need to identify which features have a higher weightage towards a particular class and how they affect the model's behaviour. If a feature is considered bad, meaning it does not contribute to predicting the classes – as noticed during manual evaluation – the score of that feature has to be reduced or the feature itself has to be removed.

Based on the weights assigned by the classifier to features during training, we identify the top features for each of the classes. Here, we not only consider features that positively influence or point towards a particular class, but also those that negatively influence a complaint belonging to a particular class. Positively influencing features indicate that the presence of a particular word or feature in the sentence suggests that the complaint belongs to a particular class, and the associated score reflects the confidence of the feature for that class. Similarly, negatively influencing features suggest that the presence of a particular word or feature in the complaint indicates it definitely does not belong to that class, and the score indicates the intensity of the complaint not belonging to that particular class.

The most positively and negatively influential words from each of the classes, specifically the top eight, were displayed in the form of an interactive graph, as shown in Figures 8.6 and 8.7.

The way to interpret the graph is that words/features on the right are positively influential, while those on the left are negatively influential. Additionally, the scores assigned can be traced along the scale provided on the side. For instance, if the word

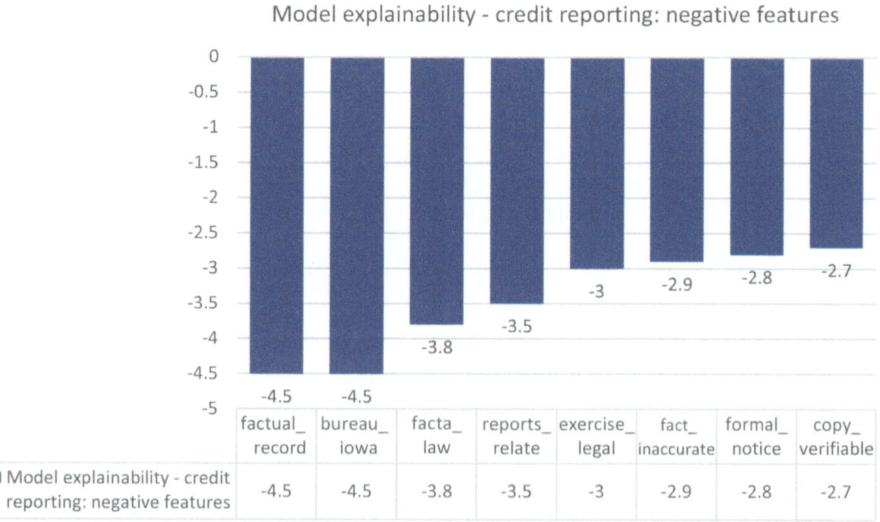

	factual_ record	bureau_ iowa	facta_ law	reports_ relate	exercise_ legal	fact_ inaccurate	formal_ notice	copy_ verifiable
■ Model explainability - credit reporting: negative features	-4.5	-4.5	-3.8	-3.5	-3	-2.9	-2.8	-2.7

FIGURE 8.6 Top negatively influencing features for credit reporting class.

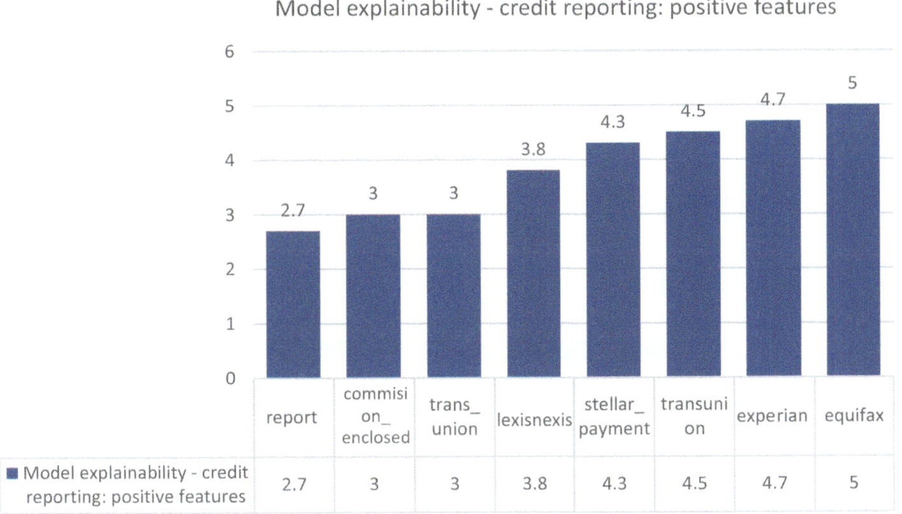

FIGURE 8.7 Top positively influencing features for credit reporting class.

"Equifax" is present, there is a high likelihood for the complaint to belong to the credit reporting category, with a score of around 5. Conversely, if the word "factual record" is present, there is a very low likelihood that the complaint belongs to the credit reporting class, with a score of approximately −4.

Similarly, by observing the features in this manner, the transparency of the model is realized. If any feature is deemed undesired, it can be either removed or assigned a lower score during modelling to have less weightage on the class.

8.7 INDIVIDUAL COMPLAINTS EXPLAINABILITY

At this stage, the model has already been developed. The test data is used to test the model and to examine the individual explainability of the predicted category of complaints. This allows us to specifically observe which features are leading to a particular product class, along with their probabilities or confidence scores.

One tool utilizes LIME, which stands for Local Interpretable Model-agnostic Explanations. Initially, the model is used to make predictions. Using these predictions, a local interpretable model is trained on the output of the model, taking it as an input for the local model. Based on this, the local model determines the probabilities of individual features belonging to a particular class.

Features are plotted alongside their influence value, indicating their contribution to a specific class. Positive and negative affecting words are displayed on either side. An example of this is provided in Figure 8.8 for the predicted class being credit reporting, along with its probability. The model is configured to use features such as addresses, report, lived, accounts, and opened.

On the left side, the graph shows the confidence percentage, ratio, or probability of each class for the complaints belonging to that class. It is predicted from the

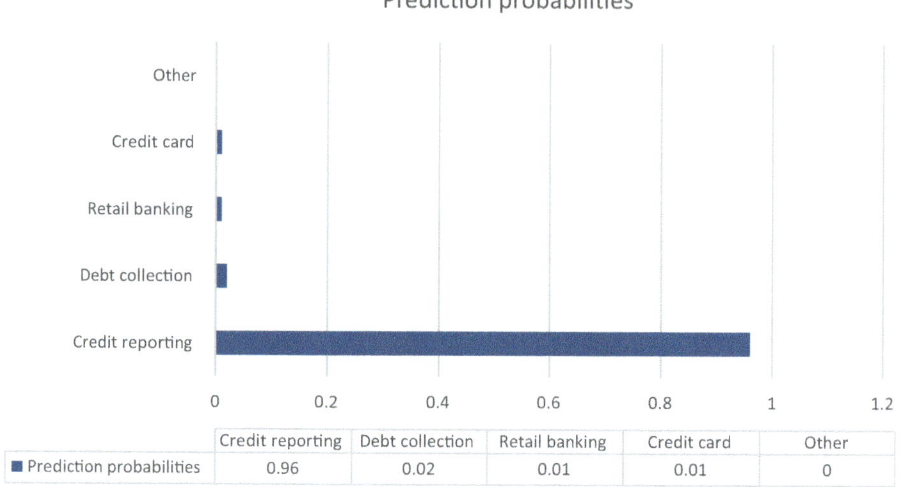

	Credit reporting	Debt collection	Retail banking	Credit card	Other
■ Prediction probabilities	0.96	0.02	0.01	0.01	0

FIGURE 8.8 Class prediction probability for individual complaints.

originally trained model and not the local model. Therefore, our Linear SVC model predicted that there is a 96% chance that the complaint belongs to the credit reporting category.

The individual complaint explainability, along with the feature influence, can be observed in Figure 8.9 for the credit reporting class.

The influence of features on the prediction of debt collection by the model is depicted in Figure 8.10.

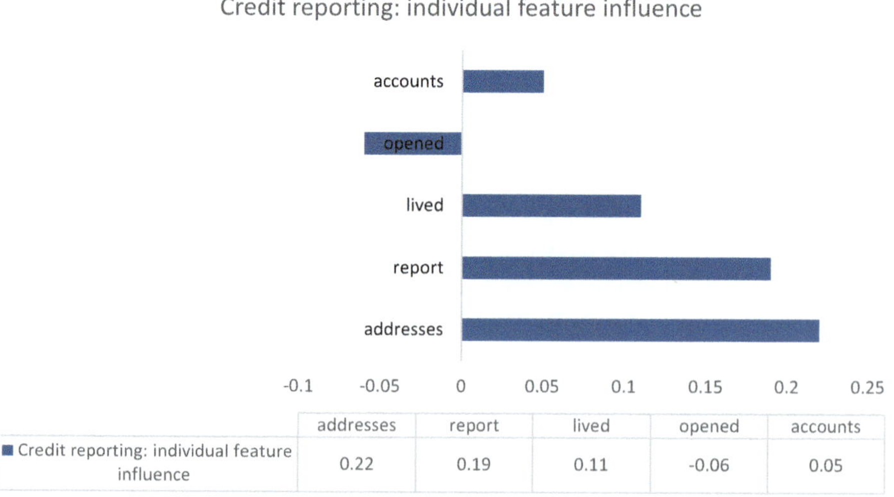

	addresses	report	lived	opened	accounts
■ Credit reporting: individual feature influence	0.22	0.19	0.11	-0.06	0.05

FIGURE 8.9 Individual complaints explainability for credit reporting class.

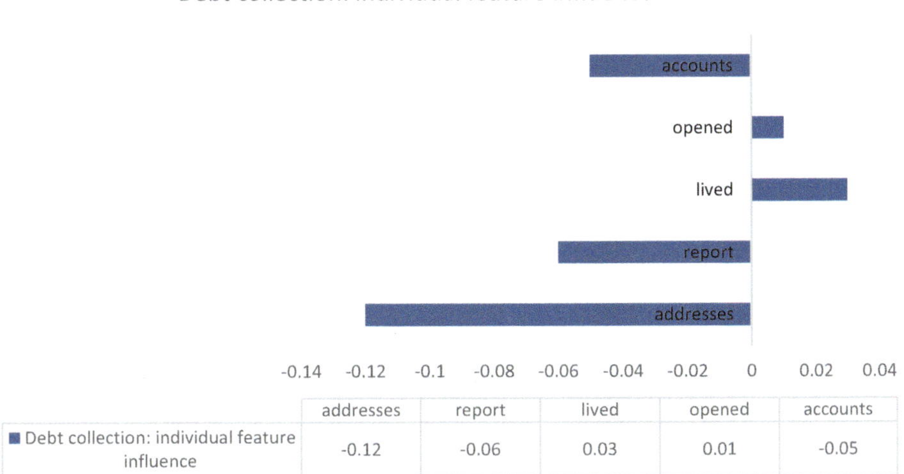

FIGURE 8.10 Individual complaints explainability for debt collection class.

Another thing to note is that the complaint above has already been pre-processed to remove stop words and undergone other processing. Therefore, the text contains only specific words, and no connectors or stop words are present. On the right side, for each class, the influence value of words in that complaint is shown. In this case, for the class credit reporting, the presence of the word "addresses" indicates a 22% chance that the complaint would belong to the credit reporting class. Conversely, due to the presence of the word "opened," there is a 0.06% chance that the complaint does not belong to the credit reporting class.

Similarly, the number can be varied for the total number of classes displayed in the graph. This helps to understand the percentage of confidence our model has for a particular class, and within that complaint, which words have confidence that the complaint belongs to the particular class along with a numerical score associated with that confidence. The individual words in the complaints also indicate the words and their score, indicating why they do not think that the complaint belongs to a particular class.

It is very important for non-technical analysts to see which features or words are actually influencing a decision by the model and to make relevant and appropriate decisions in case of any doubts or concerns.

8.8 FAIRNESS OF MODEL

There might be some features that support a particular product class but are actually biased towards them. These are undesired. An example of bias is using gender as a deciding factor to grant a loan by a bank. The model may have higher accuracy that way, but it is unethical to judge someone's eligibility for a loan based on their gender. Therefore, bias must be reduced, and many algorithms can measure and mitigate bias from the workflow. Even if some protected attributes or variables that might cause bias are identified, neglecting them won't solve the issue. Moreover, providing every feature with the same weightage does not guarantee that our algorithm will be unbiased.

To identify bias in our model, we consider the top positive affecting features as protected attributes. These features are likely to cause bias in the system. Additionally, the classes with complaints containing the protected attributes are designated as favourable classes. This means that the model favours these classes whenever those protected features are present in the complaints. Moreover, records where at least one of the features is present constitute the privileged group. Since the model favours certain classes (favoured) due to the presence of certain features (protected), it deems those complaints or groups as privileged.

Records where none of the features are present are unprivileged groups. This means that these are the complaints being discriminated against simply because they do not contain any of the protected attributes. For bias detection, two metrics are used.

1. Statistical parity difference
2. Disparate impact ratio

To calculate it, first compute the following:

I. Calculate the probability of having a favourable attribute for unprivileged groups.
   a. It is calculated by dividing the number of complaints belonging to the unprivileged groups and containing favourable attributes by the total number of complaints belonging to the unprivileged groups.

II. Calculate the probability of having a favourable attribute for privileged groups.
   a. It is calculated by dividing the number of complaints belonging to the privileged groups and containing favourable attributes by the total number of complaints belonging to the privileged groups.

I–II gives the statistical parity difference. If a negative value is observed, the model is biased towards the feature. SPD difference should never be negative for the model to be fair. I/II is the disparate impact ratio. As per the standards set by industry experts, a ratio of two-fifths is considered unbiased. That means the ratio should be more than 0.8 or more than 80% for the model to be considered unbiased. After calculating the score using these metrics, the result can be plotted for easy visual understanding and is depicted through Figure 8.11.

The graph shows that the Disparate Impact ratio is around 0.45, which is less than 0.8. This means that possible bias has been obtained in the model. On the right side of the graph, the SPD is around −0.12. It is a negative value, and SPD should never be negative for fair models. Therefore, it is evident from these metrics that the model is biased. To solve the biasness problem, these words can be either removed or their scores changed during the training stage to lessen their effect.

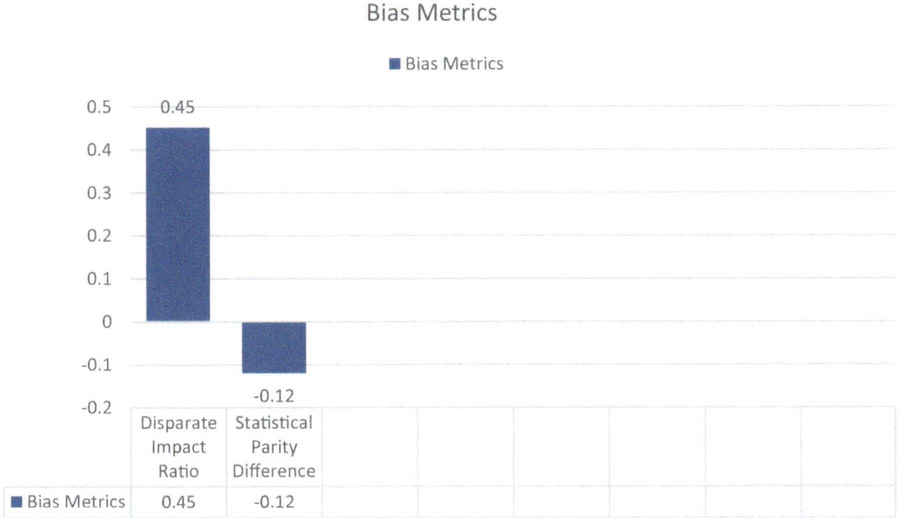

FIGURE 8.11 Metrics for fair model – DIR and SPD.

8.9 CONCLUSION

It is observed that there is no fixed set of practices and methodologies to be followed for a particular kind of dataset to make AI systems fairer and more explainable. There is always a mix and match of practices and models to arrive at the best metrics for creating bias-free and explainable AI systems. Furthermore, feature selection plays a crucial role in model creation. Choosing the right features with the right weightage ensures unbiased model generation and thereby fairer systems. There are tools such as the AIF360 toolkit by IBM, SHAP, and LIME to help with bias, fairness, and explainability of ML models, enabling informed decision-making.

8.10 FUTURE WORKS

Due to the limited computational capacity of the systems, selecting a larger training set was not possible. Future work will involve training on larger datasets, as larger training sets decrease the likelihood of bias. Furthermore, having high computational availability enables the utilization of a larger feature matrix and more combinations of features during the training stage.

REFERENCES

1. A. Barr, "Google mistakenly tags black people as 'gorillas,' showing limits of algorithms," *The Wall Street Journal*, https://www.wsj.com/articles/BL-DGB-42522 (accessed May 28, 2023).
2. G. Sharp, "Nikon camera says Asians: People are always blinking - sociological images," *Sociological Images*, https://thesocietypages.org/socimages/2009/05/29/nikon-camera-says-asians-are-always-blinking/ (accessed May 28, 2023).

3. B. X. Chen, "HP investigates claims of 'racist' computers," *Wired*, https://www.wired.com/2009/12/hp-notebooks-racist/ (accessed May 28, 2023).
4. J. Angwin, J. Larson, L. Kirchner, and S. Mattu, "Machine Bias," *Machine Bias-ProPublica*, https://www.propublica.org/article/machine-bias-risk-assessments-in-criminal-sentencing (accessed May 28, 2023).
5. K. Lum and W. Isaac, "To predict and serve?" *Significance*, 13(5), 14–19, 2016.
6. D. Ingold and S. Soper, "Amazon doesn't consider the race of its customers. should it?" *Bloomberg*, https://www.bloomberg.com/graphics/2016-amazon-same-day/ (accessed May 28, 2023).
7. S. Krüger and C. Wilson, "The problem with trust: On the discursive commodification of trust in AI," *AI Society*, 38, 1753–1761, 2022.
8. K. Reinhardt, "Trust and trustworthiness in AI Ethics," *AI and Ethics*, 3, 1–10, 2022.
9. R. Alvarado, "What kind of trust does AI deserve, if any?" *AI and Ethics*, 3, 1–15, 2022.
10. K. Siau and W. Wang, "Building trust in artificial intelligence, machine learning, and robotics," *Cutter Business Technology*, 31, 47–53, 2018.
11. S. Jan, V. Ishakian, and V. Muthusamy, "AI trust in business processes: The need for process-aware explanations," *Innovative Applications of Artificial Intelligence*, 2020. doi: 10.1609/aaai.v34i08.7056.
12. P. Linardatos, V. Papastefanopoulos, and S. Kotsiantis, "Explainable AI: A review of machine learning interpretability methods," *Entropy*, 23, 18, 2021.
13. D. Carvalho, E. Pereira, and J. Cardoso, "Machine learning interpretability," *Electronics*, 8, 832, 2019.
14. C. Barabas, "Beyond bias: "Ethical AI" in criminal law," *The Oxford Handbook of Ethics of AI*, pp. 736–753, 2020. doi: 10.1093/oxfordhb/9780190067397.013.47.
15. L. Belenguer, "AI bias: Exploring discriminatory algorithmic decision-making models and the application of possible machine-centric solutions adapted from the pharmaceutical industry," *AI and Ethics*, 2, 771–787, 2022.
16. J. Delgado and A. Manuel and I. Parra et al., "Bias in algorithms of AI systems developed for COVID-19: A scoping review," *Journal of Bioethical Inquiry*, 19, 407–419, 2022.
17. J. Johansen, T. Pedersen, and C. Johansen, "Studying human-to-computer bias transference," *AI Society*, 38, 1659–1683, 2021.
18. M. T. Ribeiro, S. Singh, and C. Guestrin, "Why Should I Trust You?": Explaining the Predictions of Any Classifier, arXiv, 2016, arXiv:1602.04938.
19. E. Ntoutsi and P. Fafalios, and U. Gadiraju, "Bias in data-driven artificial intelligence systems-An introductory survey," *Wires*, 2020. doi: 10.1002/widm.1356.
20. T. Panch, H. Mattie, and R. Atun, "Artificial intelligence and algorithmic bias: Implications for health systems," *Journal of Global Health*, 9, 020318, 2019.
21. H. Liu, W. Jin, H. Karimi, Z. Liu et al., "The authors matter: Understanding and mitigating implicit bias in deep text classification", ACL-IJCNLP, 2021. doi: 10.18653/v1/2021.findings-acl.7.

9 Interpretation of Deep Network Predictions on Various Data Sets Using LIME

Sengul Bayrak

9.1 INTRODUCTION

Artificial intelligence (AI) refers to the simulation of human intelligence processes by computer systems. These processes include learning, reasoning, and self-correction. The goal of AI research is to create systems that can perform tasks that normally require human intelligence, such as visual perception, speech recognition, decision-making, and language translation. ML is a subfield of AI that uses statistical and computational techniques to enable computers to learn from data and make predictions or take actions without being explicitly programmed. Machine learning (ML) is used in a wide range of applications, including signal processing, natural language processing (NLP), computer vision, and robotics. In signal processing, ML has been used to analyse, process, and understand different types of signals, such as audio and video (Bayrak et al. 2019). In NLP, ML is used for tasks such as language translation, speech recognition, and text summarization (Bayrak et al. 2022). In computer vision, it is used for image and video analysis, such as object recognition and tracking, face recognition, and scene understanding (Selwyn et al. 2023). In robotics, it is used for tasks such as navigation, grasping, and manipulation (Uslu et al. 2022). In ML, a black box model is a model whose inner workings cannot be seen or interpreted by the user. These models are often complex and can contain multiple layers of non-linear transformations, making it difficult to understand how the model arrived at a particular decision or prediction.

Interpretable ML (IML) is a field of study that focuses on developing ML models and techniques that can be easily understood and explained by humans. It is particularly important in safety-critical applications such as healthcare and finance, where the consequences of errors or biases can be severe. The goal of IML is to create models that are transparent and accountable so that users can understand how the model arrived at its predictions and identify any errors or biases that may be present.

Several methods and techniques are commonly used in IML such as feature importance, local interpretable models, rule-based models, and model visualization.

DOI: 10.1201/9781003442509-9

Feature importance measures the importance of each input feature in the model's predictions, helping identify which features are most important and how the model is using them. Local interpretable models are simple, interpretable models that approximate the behaviour of a complex, non-linear model in the vicinity of a specific prediction. Rule-based models represent the decision-making process of the model as a set of explicit, human-readable rules, aiding in understanding how the model is making its predictions. Model visualization provides a visual representation of the internal structure and the parameters of the model, helping understand how the model is making its predictions.

One of the main challenges with these models is that it can be challenging to understand and trust the decisions or predictions they make. This can be particularly problematic in areas where risks are high, and the decisions made by the model can have significant consequences. LIME, one of several methods developed to make black box models more interpretable, attempts to explain the predictions of any classifier in an interpretable and faithful way by learning a locally interpretable model around the prediction.

LIME is a method for interpreting and understanding the predictions of ML models, particularly those that are complex and non-linear. It creates a simplified, interpretable model that approximates the behaviour of the complex model near a specific prediction. This allows the user to understand why the complex model made a particular prediction and can be used to identify and diagnose problems with the model. LIME is "model-agnostic" because it can be applied to any ML model, regardless of its specific algorithm or architecture. In this way, LIME can provide a human-intelligible description of the predictions made by a black box model, highlighting the features that are most important in the prediction. In the literature, LIME is a technique that is used to explain the predictions of ML models. The main idea behind LIME is to approximate the complex, non-linear decision boundary of the ML model in the vicinity of a specific prediction using a simple, interpretable model. This allows the user to understand why the complex model made a particular prediction and can be used to identify and diagnose problems with the model. LIME can be applied to any ML model, regardless of the specific algorithm or architecture used, which is why it is called "model-agnostic" (Ribeiro et al. 2016).

Interpreting deep network predictions can be challenging because deep networks are complex, non-linear models with many layers and parameters. Recent studies have proposed various extensions and modifications to the original LIME algorithm to enhance its performance on different types of data with DNN (deep neural networks) models. This study outlines recent developments in the field of LIME with deep networks and its applications in different domains, such as image, signal, and unstructured text data.

The rest of the study is structured as follows: Section 9.2 presents LIME explanation models for image, signal, and unstructured text data. Section 9.3 provides literature reviews about image, signal, and text data. Section 9.4 discusses future studies on the LIME model. Finally, Section 9.5 concludes the study.

9.2 LIME EXPLANATION MODEL FOR THE IMAGE/SIGNAL/TEXT DATA

The original image/signal/text data (x) is segmented into samples by the algorithms. The disclosure model is defined as $g \in G$, where G represents the set of interpretable models that can be graphically represented to a user such that $\pi_x(z)$ is then used to represent the proximity between instance z and x and to define locality around x. An aim function $\xi(x)$ is then defined, and the L-function on $\xi(x)$ is provided as a metric describing how the infidelity g in the local definition approximates f via $\pi_x(z)$. The L-function is minimized for human comprehension to obtain the optimal solution of the objective function when $\Omega(g)$, which expresses the explanatory model complexity, is sufficiently low. The description function $\xi(x)$ obtained with the LIME algorithm is represented as (Equation 9.1).

$$\xi(x) = \arg\min_{g \in G} L(f, g, \pi_x(z)) + \Omega(g) \tag{9.1}$$

Equation (9.2) is used to calculate the degree of similarity $\pi_x(z)$.

$$\pi_x(z) = \exp\left(-\frac{D(x, z)^2}{\sigma^2}\right) \tag{9.2}$$

The definition of the degree of similarity $\pi_x(z)$ in (Equation 9.2) and the original aim function are given in (Equation 9.3). $f(z)$ is the estimated value of the original feature of the perturbed sample in d-dimensional space, and the response of this estimated value. $g(z')$ is the estimated value in d-dimension. The similarity weight values are optimized by linear regression analysis of the objective function. Perturbed z' samples can be obtained by turning off /on the features.

$$\xi(x) = \sum_{z, z' \in Z} \pi_x(z)\left(f(z) - g(z')\right)^2 \tag{9.3}$$

The perturbed z' samples back into the original input and get z as the new input of f. The estimates of z are the respective labels of the perturbed samples. Then, z' and $f(z)$ make up the dataset. LIME then uses the perturbed z' samples extracted around the neighbour sample x to train the interpretable model $g(z')$ to always satisfy the objective that $g(z') \approx f(z)$ in the local interval. The described deep network model can be augmented with a locally trained interpretable model.

9.3 LITERATURE REVIEW

Perturbation-based explanation methods measure the significance of a feature by probing the f model and obtaining patterns from it. In this study, a systematic review of the latest developments in DNN and LIME and their applications in different fields such as image, signal, and unstructured text data is conducted.

9.3.1 LIME Applications with Image Data

LIME explains the decisions made by DNN models, which are widely used in image-based decision systems. By masking, deleting, or blurring different regions of the input data, it is important to determine the impact of the region of interest on the output. The method works by creating a locally interpretable model that fits the prediction of the DNN model while considering the influence of each feature (pixel) in the image on the prediction. This is achieved by creating input-output sampling pairs and perturbing features to estimate their importance. The resulting explanations provide insight into the factors driving the model's decisions, making it easier to understand and trust the model's output. Explaining model decisions from image inputs is essential for the deployment of deep network-based models as decision systems. It is common in practice to support decision-making from images used in different domains. Images capture different aspects of the same underlying areas of interest. Therefore, explaining DNN decisions for images is an important challenge. So, LIME, a perturbation-based method, is used to explain DNN decisions and uses input-output sampling pairs to estimate feature importance. Figure 9.1 represents the flow chart of deep networks for image data using the LIME algorithm.

Trend studies for image data applied to LIME models are summarized in Table 9.1.

Bansal and Jain used the heatmap to present a localized area of X-ray images that influences the LRP model's prediction of the likely localized area of COVID-19 infection. The original images were displayed first, followed by the LRP and LIME heatmap sequentially. LRP partitioned the entire image into red and blue pixels. Each pixel represented as a feature with specific weights. Red pixels contributed positively to the prediction while the blue pixels made a negative contribution. The perspective of the presentation of the LRP was masked with the original image for the results. On the other hand, LIME gave the superpixels results in green and red colours. The green region indicated the part of the image that mostly contributed to this prediction, while the red represented the area that mostly contributed against the predicted class (Bansal and Jain 2022).

LIME can be used to improve the quality of synthetic aperture radar (SAR) images. The method estimates the contribution of each pixel to the model's prediction by perturbing the image. By comparing the significance scores of the original SAR image and the synthetic image of the target, it can be concluded whether the synthetic image

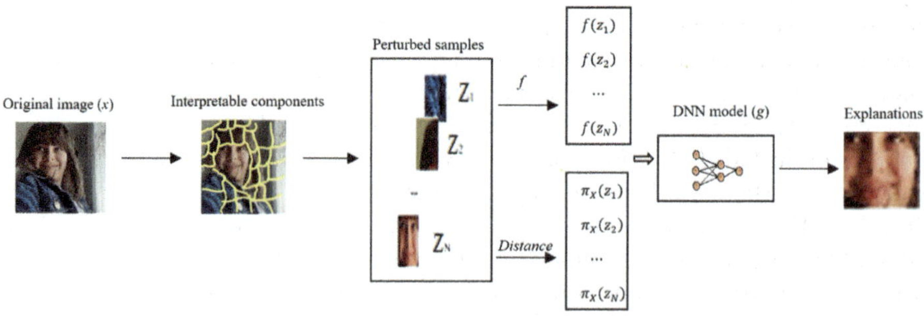

FIGURE 9.1 Flow chart on deep networks for image data using LIME.

TABLE 9.1
Trend LIME Models for Image Data

Methods	Image Explanations	Deep Learning Models
Bansal and Jain (2022)	X-ray images for the severe acute respiratory syndrome corona virus-2	Layer-wise Relevance Propagation (LRP)
Zhu et al. (2022)	Photo-realistic radar synthetic aperture radar images	Convolutional Neural Networks (CNN) and Generative Adversarial Networks (GANs)
Onler (2022)	Cassava leaf images	Transfer learning, CNN
Chayan et al. (2022)	Glaucoma dataset	ResNet50, VGG-19
Huo et al. (2022)	Remote sensing images	YOLOv4
Saarela et al. (2022)	Dermatoscopic images	CNN
Zhang et al. (2022)	Spam images	CNN
Lee et al. (2022)	Facial ultrasonography images	15 different CNN models
Jin et al. (2023)	3D images to classify glioma	Gradient-based
Abir et al. (2023)	Deep fake images	CNN

correctly represents the target's key features. This approach allows the assessment of SAR image quality and the optimization of synthetic image generation techniques for specific applications. Zhu et al. used an innovative way to assess if a SAR image created the most representational characteristics of a given target type. First, LIME was used to illustrate the useful visualization of the positive contributions of the input SAR image to the accurate estimation of the classifier. Then, the representative SAR images could be easily selected by evaluating the degree to which the positive contribution region corresponds to the target. The empirical results showed that the proposed method could largely remove the "Clever Hans" phenomenon induced by the pseudo-relation between the generated SAR images. GAN test scores were obtained before selection and after selection, respectively 71.67%, and 80.62% (Zhu et al. 2022).

Onler modelled the classification of five different types of cassava leaf disease using data consisting of leaf images with transfer learning. The LIME library was used to investigate which parts of the image were affected by the CNN (convolutional neural network) model. With LIME, the image was divided into superpixels and then visualized to decide. The clear conclusion of the study was that visual descriptions of the model predictions were important for decision-making. LIME produced reliable results for identifying and summarizing the region of interest of the superpixels in the image (Onler 2022).

In the context of glaucoma diagnosis, LIME can help identify which features of eye images are most important in predicting whether a patient has glaucoma or is at risk of developing glaucoma. For example, an ML model could use various features of an eye image, such as the size of the optic nerve or the thickness of the retinal nerve fiber layer, to predict whether a patient has glaucoma. LIME can be used to build a local interpretable model around a particular prediction made by the larger

ML model. So, doctors and clinicians can better understand how ML models make predictions about glaucoma diagnosis and progression. This can help them make more informed decisions about patient care and treatment. This local model can then be used to highlight the features that are most important in making the prediction. Chayan et al. conducted a comprehensive comparison of the effectiveness of pre-trained models along with glaucoma diseases based on transfer learning. LIME was applied to the models for comparison. The verification accuracies obtained from ResNet50 and VGG-19 models were 94.7% and 93.3%, respectively (Chayan et al. 2022).

Huo et al. modelled remote sensing images with the YOLOv4 method for effective automatic detection of vehicles in unstable environments. LIME optimized interpretation method was used to interpret the results of the model established for the recognition of remote sensing images of vehicles and the recognition results were found to be reliable (Huo et al. 2022).

In the context of skin lesion recognition, LIME could be used to detect specific features of a skin lesion that are most indicative of a particular skin condition. For instance, LIME can emphasize the presence of irregular borders, asymmetric shapes, or a distinct colour pattern, which are common characteristics of malignant skin lesions. Using LIME to describe the predictions of a skin lesion detection model can help improve the transparency and interpretability of the model's decision-making process, which can be especially crucial in medical environments where accurate diagnoses are essential. Saarela et al. studied a well-performing deep CNN classification model that classified seven different types of skin lesions. Both XAI/IML techniques adhered 100% to the original CNN model. However, the integrated gradients were clearly better achieved in terms of stability and robustness measures. In contrast, LIME produced a different annotation in each run and was less accurate than the integrated gradients annotation. Furthermore, unlike integrated gradients, which depend on the model's internal gradient, the LIME descriptor was model-independent and therefore more portable and applicable when the classification model was changed (Saarela et al. 2022).

Zhang et al. modelled spam and normal images with CNN in the context of detecting spam image data in email filter systems. They were able to highlight certain colours and repetitive patterns in the image with heatmap annotation to identify certain characteristics or features of the image that were most indicative of spam content. By using LIME to annotate the predictions of the spam image detection model, they helped increase the transparency and interpretability of the model's decision-making process. This has been particularly important in avoiding false positives or false negatives in spam detection. With the CNN model developed based on LIME, spam images could be classified as 97.16% (Zhang et al. 2022).

Lee et al. developed 15 different CNN models for 9 different slices from ultrasound images obtained from the faces of 86 people. The most successful model was VGG-16 with 93% accuracy. In the deep learning models, it was observed that muscles, blood vessels and nerves, which lack contrast in face segmentation, could be easily ignored in US images. In the characteristics of low-quality US images, modelling with VGG-16 showed little performance difference from the analysis using LIME. In their study, LIME helped increase the transparency and interpretability of the technique by providing insights into the complex relationships between the features of facial ultrasonography images and the predictions of the deep learning model (Lee et al. 2022).

Jin et al. developed a multimodal heatmap description scaled to [−1, 1] by using two cases of high-grade glioma and low-grade glioma. In their implementation, they fed perturbation-based annotation methods with a superpixel mask of the inputs. This drastically reduced the number of features and maintained the computation time within an acceptable tolerance. This visualization presented an interpretable explanation of the model's decision, enabling clinicians and researchers to better understand the factors directing the model's output and streamlining the diagnosis or treatment of medical conditions (Jin et al. 2023).

Abir et al. proposed an approach using deep learning models to detect deep fake images by training them on a dataset of real and manipulated images. These models learned to identify patterns and characteristics of manipulated images, such as changes in facial expressions, body posture, or background. While his approach could be effective in detecting deep fake images, the models could be considered a black box, making it difficult for humans to understand the logic behind the model's decisions. To address this issue, they enhanced LIME to use deep learning models to generate deep fake images and then detect manipulation. They verified 70k real images manipulated with InceptionResNetV2 software, one of the CNN models, using the LIME method with 99.87% accuracy (Abir et al. 2023).

9.3.2 LIME Explanation Model for the Signal Data

LIME is developed to explain important signal data by considering the temporal dependence between time series features, as represented in Figure 9.2.

In recent years, studies for signal data applied to LIME models are summarized in Table 9.2.

Pandey et al. used a one-dimensional convolutional neural network (1D CNN) for automatic damage detection using the Lamb wave response of a thin aluminium plate to identify damage. Using LIME, the 1D CNN model was interpreted in terms of damage feature contributions and the interpretation was found to correspond to an insightful damage signature. Finite element simulations of the Lamb wave response were obtained using a commercial package (ABAQUS) for different frequencies (100, 125, 150 kHz). Experimental data were generated for a 1.6 mm thick aluminium plate using piezoelectric wafer transducers. The 1D CNN architecture provided approximately 100% accuracy. Using the key features affecting the classification obtained

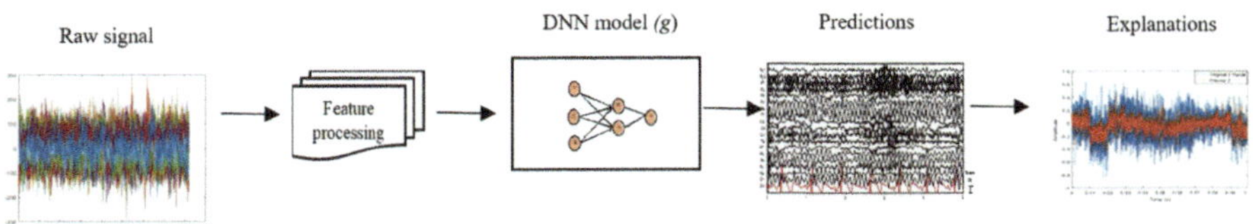

FIGURE 9.2 Flow chart on deep networks for signal data using LIME.

TABLE 9.2
Trend LIME Models for Signal Data

Methods	Signal Explanations	Deep Learning Models
Pandey et al. (2021)	Lamb wave	CNN
Khan et al. (2022)	Time series-based attack from Industrial Internet of Things (IIOT)	CNN and Recurrent Neural Network (RNN)
Cui et al. (2022)	32-channel EEG signals	CNN
Zhang et al. (2023)	Ultrasonic guided waves	CNN
Abdullah et al. (2023)	Electrocardiogram (ECG) signals	CNN – Gated Recurrent Unit (GRU)

from LIME, damage localization was obtained with about a 7% deviation from the true damage localization (Pandey et al. 2021).

Industrial Internet of Things (IIoT) service quality, in terms of time collected from heterogeneous sensing devices, typically depends on data integrity and accuracy, which can be exploited by injecting malicious events such as false data injection and data poisoning attacks. Therefore, anomaly recognition and annotation are critical for ensuring quality services and enabling security managers to interpret the causal justification of prediction decisions and evidence of key data. Khan et al. proposed an autoencoder-based detection framework that uses convolutional and recurrent networks to discover and model seven different cyber threats in IIoT networks. A two-stage sliding window (SW) was adopted to better learn the underlying representations of the data features. The malicious points in the raw time series were converted into constant-length series. Each series was transformed into a continuous time-dependent sub-series through another small SW to learn the hidden representations of malicious events. The fully connected networks utilize the extracted temporal and spatial features for classification and annotation of attack events. Experimental results revealed that this framework effectively extracts features that contain the contexts of malicious patterns with the modelling LIME with 99.35% accuracy. They proved this model to be robust in detecting malicious events and performed better than contemporary state-of-the-art methods (Khan et al. 2022).

To use LIME with EEG (electroencephalogram) data, a deep learning model first needs to be trained on the EEG data. This can be a model for classification tasks, such as classifying different types of brain activity or detecting neurological disorders, such as predicting brainwave amplitudes or frequencies. Once the model is trained, LIME can be used to explain its predictions. LIME functions by creating local approximations of the model at each point in the data and then using these approximations to identify the most important features for making a given prediction. In the cases of EEG data, these features include the location of the electrodes used to record the frequency of specific brain waves. LIME's output provides valuable information about the neural activity underlying the deep learning model's predictions. It could reveal that a type of brainwave is particularly important for predicting a particular neurological condition, or that activity in a particular brain region is particularly relevant to a particular cognitive task. Overall, using LIME with EEG data can help provide a deeper understanding of brain function and the neural activity underlying neurological conditions and can be a powerful tool to advance our understanding of the brain. Cui et al. used LIME to explain the predictions of a simple CNN in an EEG-based emotional brain-computer interface by modelling which features in the EEG better discriminate target emotions. They applied two types of segmentation, one of which segmented the data into top, middle, and bottom regions, while the other segmented it into left, middle, and right regions. The results showed that the model provides insights into the neural processes involved in learned behaviours about valence and arousal (Cui et al. 2022).

Zhang et al. proposed small-sample damage detection for ultrasonic-guided waves by modelling an attention-based interpretable prototypical network. This prototypical network was used as a framework to compute the prototype of each category and calculate the similarity between samples based on the metric space. Feature extraction was built with the channel attention module, which enables the network to highlight valid features. By modelling with LIME to explain the damage in the intrinsic mechanism, the identification performed by the network in terms of the critical feature was successfully achieved. In practice, effective damage detection based on small data modelled the problem with four classes: intact, pinhole, crack, and corrosion damage. This classification performance demonstrated that the proposed network could serve as an effective model. Overcoming the lack of limited data for damage detection and LIME analysis significantly improved the interpretability of the network, achieving 97.92% accuracy (Zhang et al. 2023).

Abdullah et al. proposed Bootstrap-LIME as an improvement of the LIME method to produce a robust and significant description of electrocardiogram (ECG) signal data. In order to take into account the temporal dependence of the signals and compare the performance with LIME, they first generated a random set of data points that were locally faithful to the neighbourhood of the prediction to produce reliable descriptions. They obtained a visual representation of LIME on the ECG dataset by applying a heat map to highlight important areas in the heartbeat signal that the CNN-GRU (gated recurrent unit) model used for prediction. They developed a hybrid 1D CNN-GRU model that combined two powerful deep learning algorithms to classify four (F, N, S, V) arrhythmia types. Comparing the BLIME and LIME modelling results, the null hypothesis was rejected with a p-value of 8.544500×10^{-35} between the two methods based on the calculated classification accuracies, providing evidence of a significant difference (Abdullah et al. 2023).

9.3.3 LIME Explanation Model for the Unstructured Text Data

LIME has been implemented to explain unstructured text data by considering the dependence between text features, as illustrated in Figure 9.3.

In recent years, important studies for text data applied to LIME models are summarized in Table 9.3.

Wang et al. studied the textual data of malware used in Advanced Persistent Threat (APT) threats. In the article, LSTM and CNN were used as deep learning methods. LIME, one of the black box models, was used to improve the interpretability of the model by explaining the reasons for the model's decisions. LIME was

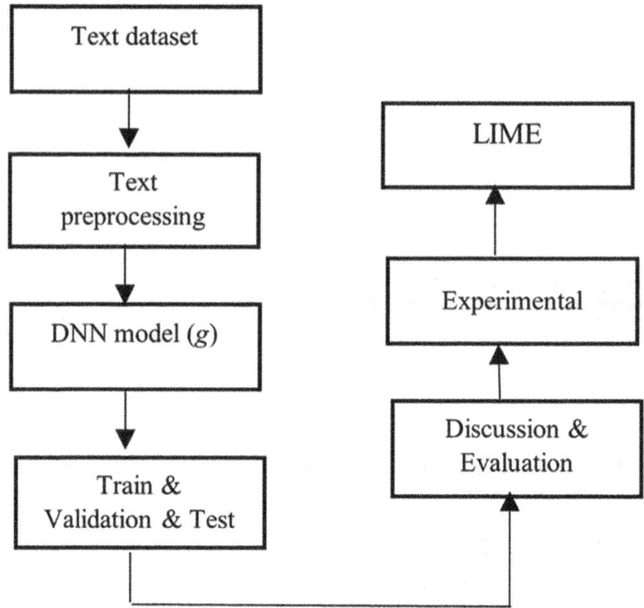

FIGURE 9.3 Flow chart on deep networks for unstructured text data using LIME.

TABLE 9.3
Trend LIME Models for Text Data

Methods	Text Data Explanations	Deep Learning Models
Wang et al. (2021)	Code features and string features for APT attribution	LSTM, CNN
Datta et al. (2021)	Machine translation	LSTM
Shakil and Alam (2022)	Hate speech	CNN
Bell et al. (2022)	Donor Text Narratives	LSTM, GRU
Abdelwahab et al. (2022)	LASIK Surgeries	LSTM
Kanneganti (2023)	Fake news	Transfer learning

used to explain the decisions of LSTM and CNN models. In this study, it was stated that a balance needs to be struck between the performance of deep learning models and the ability of LIME to provide interpretability. While deep learning models had high accuracy rates, they were poor in terms of interpretability. LIME was used to improve the interpretability of the model's decisions while giving some accuracy. As a result, the use of the methods in the study resulted in higher accuracy rates for APT attacks, and at the same time, the interpretation of the reasons for the models' decisions was made more informative. This provided a better understanding in terms of detection and analysis of attacks and helped in the development of defence mechanisms (Wang et al. 2021).

Datta et al. studied texts translated from English into German. In the study, the sequence-to-sequence (seq2seq) model was primarily used. This model worked in an encoder-decoder structure and was widely used in language models. Then the transformer model was used to solve the long-term dependency problem in seq2seq models. The LIME method was used to improve the translation performance and explainability of the model. LIME is a method used to understand how a prediction is made and is used to explain predictions in deep learning models, so-called black boxes. The results between deep learning modelling and LIME modelling were compared. The results showed that LIME improved the explainability of the outputs of deep learning models. Moreover, LIME modelling identified modifications that were made to improve the performance of the model. The results obtained using deep learning models and LIME modelling were evaluated in terms of human comprehensibility (2021).

Shakil and Alam studied hate speech tweets collected from social media platforms. Since the CNN algorithm was used as a black box model, LIME was used to improve the interpretability of the model and to understand which features were more important for the model's predictions. The comparison between deep learning methods modelling and LIME modelling showed that LIME was more effective in explaining the specific features that cause the model's predictions, increasing the interpretability of the model. LIME helped to understand which words were most important for the model's predictions and which words had less influence on the predictions. They developed a method for classifying hate speech tweets using ML algorithms in combination with NLP and CNN. Furthermore, by using LIME, the interpretability of the model was improved, and it was determined which features were more important for the predictions of the model. The results showed that the developed method provided high accuracy for hate speech classification (Shakil and Alam 2022).

Bell et al. studied text data from the medical records of potential liver transplant donors. This text data included clinical notes, laboratory results, radiologic reports, images, and other data. Deep learning algorithms such as LSTM and GRU were used in the study. LIME was a method that explained how model predictions were made and which features affected the output of the model most informatively. The study showed that deep learning algorithms such as LSTM and GRU achieve the highest accuracies. LIME modelling was used as an important tool for interpreting the decisions of the models. A classification model was built using deep learning and ML algorithms on text data of potential donors for liver transplantation. By analysing the data, this model predicted whether potential donors were suitable for liver transplantation. Moreover, the model also helped to predict the post-transplant outcomes of the donor (Bell et al. 2022).

Abdelwahab et al. demonstrated the effectiveness of LIME in sentiment analysis of Arabic text based on the comments of patients who have undergone Laser-Assisted in Situ Keratomileusis (LASIK) surgery. As a result of the study, it was observed that the LSTM model had high accuracy, but it was insufficient to understand why and how the model made decisions. Therefore, the use of explainability tools such as LIME helped in understanding the model's decisions (Abdelwahab et al. 2022).

Kanneganti used Bidirectional Encoder Representations from Transformers (BERT) for cross-domain text classification with various Covid-19, mixed data, and cross-domain datasets. An interpretable linear model was trained on 500 examples to predict and represent sample behaviour, and the LIME model was used to design a new explainable approach for cross-domain levels. Given the known high performance of the pre-trained BERT model in detecting fake news, the researchers experimented using a variety of machine models and achieved good accuracy by fine-tuning the model with a SoftMax layer for the classification task. Other researchers have performed comparative analyses and achieved better results by combining other ML models and developing a fine-tuned model. LIME modelling predicted the behaviour of a sample of 500. An interpretable linear model has been trained to represent and interpret the LIME modelling results, although they were not explicitly stated. However, theoretically LIME has proven to provide a good local approach. Four-level (low to high) cross-domain stars were analysed using the local interpretation model, and the ideal pair that provided a better explanation was identified. As a result of the experiments, it was found that LIME performed well in low-level cross-domain classification. In the last experiment, the high-level cross-domain LIME model showed consistency in correctly classifying news (Kanneganti 2023).

9.4 FUTURE STUDIES

Black box research is tasked with analysing and understanding complex systems or processes whose inner workings cannot be easily observed or understood. While it is difficult to predict specific future work related to black boxes, some areas of potential interest can be explored based on current trends and challenges. Future work in the field of Explainable AI could focus on developing techniques and methodologies to make AI algorithms more explainable and enable people to understand the decision-making processes of black-box AI systems. Ethical implications arise from black box systems, raising concerns regarding accountability, fairness, and bias. Forthcoming work could explore ways to address these concerns by developing ethical frameworks and guidelines for the design, deployment, and regulation of black box systems to ensure transparency and prevent unintended consequences. Black box systems can be vulnerable to cyber-attacks and manipulation. Future work will be able to explore methods to improve the security and trustworthiness of black box technologies. Increasing the transparency of models in the financial and economic domain enables regulators and stakeholders to better understand and assess their impact on markets and the economy. Future work focuses on developing methods to interpret and explain the logic behind these models to increase trust in machine learning used in medical diagnosis and treatment, improve patient outcomes, and facilitate collaboration between healthcare professionals and AI systems. Social media platforms can investigate the impact of these algorithms on information dissemination, user behaviour, and potential biases they reveal in order to promote transparency and address issues related to filter bubbles and misinformation for black box models used to curate content and personalize user experiences. Indeed, these are only a few potentially promising directions for future studies on black boxes. As technology advances and social needs change, researchers and professionals will explore new ways to make such complex systems more transparent, accountable, and beneficial for humanity.

9.5 CONCLUSION

Deep learning has recently gained immense popularity. They are often considered as black box models as they cannot explain the causal rationale of the prediction decisions for the behaviour of the model developed with deep learning. In recent years, studies with image, signal, and textual data have investigated, the validity of the prediction decisions and the behaviour of the deep learning model, which can be explained thanks to the LIME method. Using LIME has been found to be important for achieving confidence in the paradigm for different data type classifications.

REFERENCES

Abdelwahab, Y., Kholief, M., and Sedky, A. A. H. (2022), "Justifying Arabic text sentiment analysis using explainable AI (XAI): LASIK surgeries case study," *Information*, 13(11), 536. https://doi.org/10.3390/info13110536.

Abdullah, T. A., Zahid, M. S. M., Ali, W., and Hassan, S. U. (2023), "B-LIME: An improvement of LIME for interpretable deep learning classification of cardiac arrhythmia from ECG signals," *Processes*, 11(2), 595. https://doi.org/10.3390/pr11020595.

Abir, W. H., Khanam, F. R., Alam, K. N., Hadjouni, M., Elmannai, H., Bourouis, S., Rajesh, D., and Khan, M. M. (2023), "Detecting deepfake images using deep learning techniques and explainable AI methods," *Intelligent Automation & Soft Computing*, 35(2). https://doi.org/10.32604/iasc.2023.029653.

Bansal, V., and Jain, A. (2022), "Interpreting the predictions of deep network build to identify early detection of COVID-19 in X-ray images," *Journal of Applied Material Science & Engineering Research*, 6(1), 2689–1204.

Bayrak, S., Yucel, E., and Takci, H. (2019), "Classification of extracranial and intracranial EEG signals by using finite impulse response filter through ensemble learning," In *2019 27th Signal Processing and Communications Applications Conference*, IEEE, Sivas, Türkiye, pp. 1–4. https://doi.org/10.1109/SIU.2019.8806334

Bayrak, S., Yucel, E., and Takci, H. (2022), "Epilepsy radiology reports classification using deep learning networks," *CMC-Computers Materials & Continua*, 70(2), 1–15. https://doi.org/10.32604/cmc.2022.018742.

Bell, K., Hennessy, M., Henry, M., and Malik, A. (2022), "Predicting liver utilization rate and post-transplant outcomes from donor text narratives with natural language processing," In *2022 Systems and Information Engineering Design Symposium*, IEEE, pp. 288–293. https://10.1109/SIEDS55548.2022.9799424.

Chayan, T. I., Islam, A., Rahman, E., Reza, M., Apon, T. S., and Alam, M. D. (2022), "Explainable AI based Glaucoma Detection using Transfer Learning and LIME," *arXiv preprint arXiv:2210.03332*. https://doi.org/10.48550/arXiv.2210.03332.

Cui, C., Zhang, Y., and Zhong, S. (2022), "Explanations of deep networks on EEG data via interpretable approaches," *In 2022 IEEE 35th International Symposium on Computer-Based Medical Systems (CBMS)*, IEEE, pp. 171–176. https://10.1109/CBMS55023.2022.00037.

Datta, G., Joshi, N., and Gupta, K. (2021), "Study of NMT: An Explainable AI based approach," In *2021 5th International Conference on Information Systems and Computer Networks (ISCON)*, pp. 1–4. https://10.1109/ISCON52037.2021.9702396.

Huo, Y., Ai, Y., Zhao, C., and Li, Y. (2022), "Interpretable analysis of remote sensing image recognition of vehicles in the complex environment," In *Second International Symposium on Computer Technology and Information Science*, vol. 12474, pp. 666–672. https://doi.org/10.1117/12.2653489.

Jin, W., Li, X., Fatehi, M., and Hamarneh, G. (2023), "Generating post-hoc explanation from deep neural networks for multi-modal medical image analysis tasks," *MethodsX*. https://doi.org/10.1016/j.mex.2023.102009.

Kanneganti, D. (2023), "A New Cross-domain Strategy based XAI Models for Fake News Detection," *arXiv preprint arXiv:2302.02122*. https://doi.org/10.48550/arXiv.2302.02122.

Khan, I. A., Moustafa, N., Pi, D., Sallam, K. M., Zomaya, A. Y., and Li, B. (2021), "A new explainable deep learning framework for cyber threat discovery in industrial IoT networks," *IEEE Internet of Things Journal*, 9(13), 11604–11613. https://10.1109/JIOT.2021.3130156.

Lee, K. W., Lee, H. J., Hu, H., and Kim, H. J. (2022), "Analysis of facial ultrasonography images based on deep learning," *Scientific Reports*, 12(1), 16480. https://doi.org/10.1038/s41598-022-20969-z.

Onler, E. (2022), "Explaining image classification model (CNN-based) predictions with LIME," *World Journal of Advanced Engineering Technology and Sciences*, 7(2), 275–280. https://doi.org/10.30574/wjaets.2022.7.2.0176.

Pandey, P., Rai, A., and Mitra, M. (2022), "Explainable 1-D convolutional neural network for damage detection using lamb wave," *Mechanical Systems and Signal Processing*, 164, 108220. https://doi.org/10.1016/j.ymssp.2021.108220.

Ribeiro, M. T., Singh, S., and Guestrin, C. (2016), "Why should I trust you?" Explaining the predictions of any classifier," In *Proceedings of the 22nd ACM SIGKDD International Conference on Knowledge Discovery and Data Mining*, San Francisco, CA, USA, pp. 1135–1144. https://doi.org/10.1145/2939672.2939778.

Saarela, M., and Geogieva, L. (2022), "Robustness, stability, and fidelity of explanations for a deep skin cancer classification model," *Applied Sciences*, 12(19), 9545. https://doi.org/10.3390/app12199545.

Selwyn, N., Campbell, L., and Andrejevic, M. (2023), "Autoroll: Scripting the emergence of classroom facial recognition technology," *Learning, Media and Technology*, 48(1), 166–179. https://doi.org/10.1080/17439884.2022.2039938.

Shakil, M. H., and Alam, M. G. R. (2022), "Hate speech classification implementing NLP and CNN with machine learning algorithm through interpretable explainable AI," In *2022 IEEE Region 10 Symposium* (*TENSYMP*), IEEE, pp. 1–6. https://10.1109/TENSYMP54529.2022.9864421.

Uslu, A. N., Yavuz, G. O., and Koçak Usluel, Y. (2022), "A systematic review study on educational robotics and robots," *Interactive Learning Environments*, 1–25. https://doi.org/10.1080/10494820.2021.2023890.

Wang, Q., Yan, H., and Han, Z. (2021), "Explainable apt attribution for malware using NLP techniques," In *2021 IEEE 21st International Conference on Software Quality, Reliability and Security* (*QRS*), Hainan Island, China, pp. 70–80. https://10.1109/QRS54544.2021.00018.

Zhang, Z., Damiani, E., Al Hamadi, H., Yeun, C. Y., and Taher, F. (2022), "Explainable artificial intelligence to detect image spam using convolutional neural network," *In 2022 International Conference on Cyber Resilience* (*ICCR*), Dubai, United Arab Emirates, pp. 1–5. https://10.1109/ICCR56254.2022.9995839.

Zhang, H., Lin, J., Hua, J., Zhang, T., and Tong, T. (2023), "Attention-based interpretable prototypical network towards small-sample damage identification using ultrasonic guided waves," *Mechanical Systems and Signal Processing*, 188, 109990. https://doi.org/10.1016/j.ymssp.2022.109990.

Zhu, M., Zang, B., Ding, L., Lei, T., Feng, Z., and Fan, J. (2022), "LIME-based data selection method for SAR images generation using GAN," *Remote Sensing*, 14(1), 204. https://doi.org/10.3390/rs14010204.

10 Comprehensive Study on Social Trust with XAI

Techniques, Evaluation, and Future Direction

G. K. Panda, Diptimaya Mishra, and Subhadeep Nayak

10.1 INTRODUCTION

Social trust refers to the belief that people can rely on each other to behave honestly and cooperatively. It has been associated as an important concept throughout human history. Aristotle (2400 years ago) discussed the concept of trust in his work "Nicomachean Ethics". He argued that trust is an essential component of a well-functioning society and that it is built through repeated interactions between individuals. Social trust has been quantified by several key measures in different ages. During the Middle Ages, trust was primarily based on social and religious hierarchies and who held positions of authority. During the Enlightenment period, social trust began to be seen as a product of reason and science rather than religion. With the rise of industrialization, trust became increasingly important for economic growth and stability. The development of complex markets and financial systems required trust in the reliability of contracts and institutions. In the 20th century, social trust became a subject of empirical research. Social scientists began to study the factors that contribute to trust, such as social networks, cultural norms, and government institutions. The current age is one of significant global change, characterized by the post-COVID-19 pandemic, geopolitical uncertainty (due to war and conflicts of interest among superpowers), and socio-economic challenges. In this era, social trust has become a critical issue for humankind. This has led to a renewed focus on building trust through transparency, accountability, and democratic participation. Additionally, the rise of social media and digital communication has also raised new questions about how trust is formed and maintained in an increasingly interconnected and complex world.

Intelligence can be very useful for fostering social trust, particularly in situations where trust is challenged or where critical thinking, problem-solving, communication, and emotional intelligence are required. The transformation of cognitive intelligence from human-centric to machine-centric has opened up new possibilities for solving complex problems and making more informed decisions.

The use of artificial intelligence (AI) in the study of social trust is a relatively interesting dimension of research. AI has the potential to provide insights into the

DOI: 10.1201/9781003442509-10

dynamics of social trust, as well as to develop tools to measure and predict trust. The most common AI techniques used in social trust research are machine learning (ML) and natural language processing (NLP). ML techniques are used to analyze large datasets to identify patterns and correlations between trust-related variables. NLP techniques are used to analyze text-based data to identify the sentiment and meaning of trust-related statements. Both techniques can be used to develop predictive models of social trust. Evaluation of AI-based social trust research is typically done through a combination of quantitative and qualitative approaches. The first type of approach includes statistical analysis and predictive modelling while the latter approach is based on surveys and personal interviews. Evaluation of AI-based social trust research is important for assessing the accuracy and reliability of the results.

The recent development of XAI may be viewed as an extension of AI that focuses on creating algorithms and models that can be more easily interpreted and understood, leading to deduced or predicted conclusions. The interpreted goal of XAI is to make AI more transparent and accountable, enabling humans to trust and use AI more effectively [1,2]. In other words, XAI aims to provide users and designers of AI systems with a better understanding of their predictions, including how and why the systems arrived at them. Traditional ML models utilize complex calculations inherited from the system to make predictions or classifications, but interpreting these calculations in most cases is observed to be difficult and not straightforward. This lack of transparency can lead to mistrust and skepticism of the technology, especially when it is used in critical applications, including social trust. To address these issues, researchers develop techniques and methods that enable humans to understand how AI models make decisions [3,4]. These techniques include visualizations, natural language explanations, and interactive interfaces that allow users to explore and manipulate the data and model. By making the core workings of AI more transparent, XAI can help users understand why a model made a certain decision, identify potential biases, and improve the accuracy and reliability of the technology. However, the development of XAI systems is necessary to ensure that AI is transparent, interpretable, and accountable, and to ensure that it can be trusted by humans to make decisions that are fair and reliable.

The organization of the remaining parts of the chapter is divided into four sections. The first section, which is the preliminary section, offers some conceptual ideas related to AI models and services, XAI organization and models, as well as social networks and similarity measures. In Section 10.3, an overview of trust is provided, along with the role of social trust in XAI and the effects of social bias on trust. The subsequent section focuses on building social network applications and evaluating trust with XAI. This is done by examining measures of centrality and closeness in social networks, and by using empirical analysis and explainable techniques, such as "LIME" and "SHAP", with ETs. Finally, the last section proposes a trust prediction model that uses explainable techniques, followed by the conclusion and references to the text.

10.2 PRELIMINARIES

10.2.1 AI Models and Services

Most of the research works pertaining to machine intelligence are based on domain-specific solutions. The use of domain-specific solutions and core AI models is essential for building effective machine intelligence systems. In most solutions, the set of core AI models exhibit decisions based on the patterns and relationships they learn from data. These models typically consist of mathematical functions that map input data to an output prediction or decision. The specific algorithm used depends on the type of problem being solved, but some common techniques include "neural networks", "decision trees", and "support vector machines". In Figure 10.1, we outline a broader spectrum of such models, commonly seen in machine intelligence-based solutions.

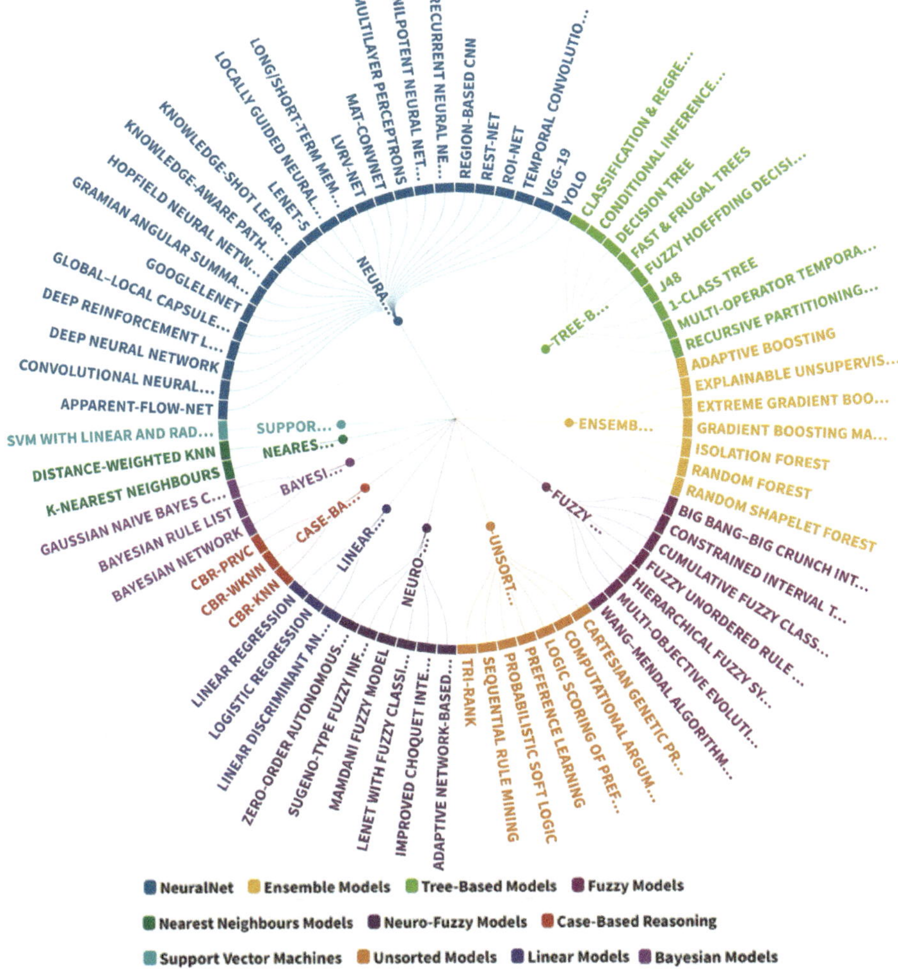

FIGURE 10.1 AI-based some major models.

The AI community which includes researchers, technology developers, and service providers, has noted that ensuring trust in AI solutions is a significant concern. Transparency and accountability at all levels are the dominant credentials of trust in AI products. Table 10.1 provides a broader overview of AI service fact sheets encomprising such issues. Therefore, there is a requirement for methods and techniques that enable us to comprehend the reasoning behind a particular decision made by an AI model, detect potential biases, and improve the accuracy and reliability of the technology.

10.2.2 XAI: Organisation and Models

XAI models make decisions in a similar way to traditional AI models, but with an added layer of transparency and interpretability. They are designed to produce not only a prediction or decision for the problem but also an explanation for how they arrived at that decision. Although there are many associated models and techniques, choosing the right model or explainable technique (ET) depends on the specific problem being solved and the level of interpretability required. Figure 10.2 outlines the basic organizational aspects of XAI.

There have been many discussions in the literature about ETs, but the question arises as to how to select a specific ET and determine the appropriate circumstances

TABLE 10.1
AI Services Related to Trust

AI Services: Fact Sheet			
Conventional AI: Objectives			
• System Operations • Testing Methodologies	• Algorithmic Training Data • Fairness & robustness checks	• Test setups & results • Intended user maintenance	• Performance benchmarks • Training & updates
AI Trust: Objectives			
• Data Governance	• Data Tracking	• Dataset De-biasing	• Resource optimization
AI Trust: Critical Components			
• Safety • Reliability	• Fairness • Robustness	• Accountability • Explainibility	• Ability to measure performance levels
AI Trust: Top Challenges			
Exposure to Bias	**Lack of Explainability**	**Susceptibility**	**Lineage**
Biased algorithms or datasets contribute to the unfair treatment of certain groups	Opacity of intelligent explainable systems leading to trust deficit	Due to Adversarial attack Due to Manipulation	Algorithm audit during development, deployment, and maintenance life cycle

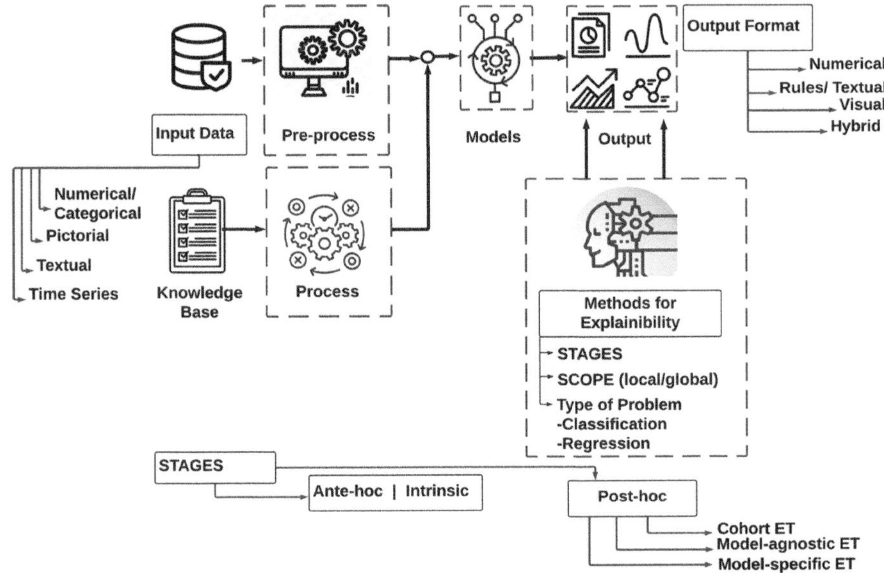

FIGURE 10.2 Organisational aspects of explainable-AI.

for its use. The following sections will provide information on some selected ETs in accordance with the categories of ante-hoc, intrinsic, model-specific, model-agnostic, and post-hoc techniques [5].

Ante-hoc techniques involve building models with interpretability in mind from the beginning while intrinsic techniques focus on building models that are inherently interpretable. This can include using interpretable model architectures, such as decision trees or rule-based models, or incorporating constraints into the optimization process to promote interpretability. Model-specific and model-agnostic refer to types of ETs used to build models, and the choice of adaptation depends on the problem at hand and the type of data being used. Model-specific ETs may provide better performance for certain types of models but may not be as flexible. Model-agnostic ETs may be more versatile but may not be optimized for any specific model architecture [1,6].

Model-specific ETs contribute to working with a specific type of model architecture, such as decision trees or neural networks. For example, random forests and gradient boosting are model-specific techniques that work well with decision trees. These techniques are optimized for the strengths and weaknesses of the model architecture and may not perform as effectively on other types of models. A disadvantage to this effect is that if we focus on providing explanations that are specific to a particular model, we might restrict our choice of neural networks, potentially excluding a neural network that could provide a better fit between the input data and output.

The model-agnostic ETs can be used with any type of model architecture. They focus on the data and the way it is processed rather than the specifics of the model architecture. Examples of model-agnostic techniques include cross-validation, feature selection, and permutation importance. These post hoc explanation techniques have the potential to apply to a broad spectrum of models, allowing for more flexibility in the modeling process.

Cohort explanation focuses on medical data analytics and aims to evaluate the model's generality before deployment (model validation phase), enabling testing of its accuracy using new or unknown data. It is also helpful in identifying the specific variables that may be responsible for the reduction in accuracy within a specific subset. It also allows for the measurement and comparison of model accuracy among different cohorts or subsets of data. Additionally, this method can aid in detecting bias in the model and identifying areas where datasets need improvement. Local explanations are useful for understanding specific predictions and identifying potential biases or errors in the model, while global explanations are useful for understanding the model's overall behavior and performance and comparing it to other models or benchmarks.

The methods like Class Activation Maps (CAMs), Gradient-weighted Class Activation Mapping (Grad-CAM), Integrated Gradients, and Saliency Maps belong to local-interpretable Model-specific Explainable Techniques (ETs) because they explain the specific decision made by the model at a local level (i.e., for a specific input). These methods [7] highlight the important regions of the input image that contributed to the model's prediction.

On the other hand, methods like Layer Visualization and Feature Importance belong to global-interpretable Model-specific ETs because they provide an explanation for the model's behavior at a global level (i.e., for all inputs). Layer Visualization allows us to visualize the activations of different layers of the model, which can help us understand how the model processes information. Feature Importance methods like Permutation Feature Importance (PFI) and Feature Importance Scores (FIS) highlight the features that are most important for the model's predictions across all inputs.

Methods like Local Interpretable Model-Agnostic Explanations (LIME) [8], SHapley Additive exPlanations (SHAP) [9] and Contrastive Explanations Methods (CEM) [10] belong to local-Model-Agnostic ETs because they can be used to explain the predictions of any black-box model for a specific input. LIME generates an interpretable model locally around the prediction for a specific input, while SHAP assigns a feature importance score to each feature of the input. CEM tries to find the smallest perturbation to the input that would change the prediction while still being semantically meaningful.

Methods like Partial Dependence Plots (PDP), Surrogate Models, Tree Interpreter, and Exploratory Data Analysis (EDA) belong to global-Model-Agnostic ETs [11] because they provide an explanation for the model's behavior across all inputs without making any assumptions about the model's internal workings. PDP plots the marginal effect of one or two features on the model's predictions, while Surrogate Models are simpler models that can be used to approximate the behavior of a black-box model, such as decision trees, random forests, and gradient boosting machines. The Tree Interpreter algorithm (like decision trees, random forest, and gradient boosting machines) computes the contribution of each feature to the final prediction by traversing the tree and accounting for the feature splits at each node. EDA is a data exploration technique that can help us understand the relationships between the input features and the model's predictions.

The **XAI:Local Interpretable Model-agnostic Explanations** (LIME) model [8] works by creating a simpler, more interpretable model that approximates the behavior of the AI model that can be used to explain the decisions made by the AI model. In other words, it fits a local interpretable model to each instance of data, approximating the behavior of the global model in the vicinity of the instance.

The **XAI:SHAP** (SHapley Additive exPlanations) [9] model is a popular model-agnostic ET that facilitates explaining the output of any ML model using tabular data. It employs a game-theoretic approach to assign an importance score to each input feature, indicating how much it contributes to the output prediction. The overall functionality of SHAP is described below which is not exhaustive regarding its original content.

The **XAI:Decision Tree** is a simple, interpretable model that can explain the decisions made by a more complex AI model. It operates by breaking down a decision into a series of smaller, simpler decisions, which are easily visualized and understood.

The **XAI:Rule-Based** models are another type of interpretable model that can be used for XAI. These models use a set of if-then rules to make decisions, making them easy to understand and interpret.

The **XAI:Anchor** is an interpretable machine learning system that generates locally faithful and human-understandable explanations for black box models. It helps identify relevant input features and determine where the model's predictions are influenced.

XAI:Counterfactual Explanations are used to explain the decisions made by a model by generating alternative input data points that would have resulted in a different decision. These explanations help users understand why the model made the decision it did and how the decision could have been different.

10.2.3 Social Networks and Similarity Measures

Social Network (SN) is described as a collection of nodes (actors/vertices) and the links (ties/edges) that connect them. Here, the nodes include elements related to individuals, organizations, places, websites, diseases, and many other related elements. SN analysis, with its social connections through graph structure, is the first-hand choice of researchers in the domain.

SNs expand by adding nodes with associated edges, making them highly dynamic objects. As new edges are added, the underlying social structure of the network changes swiftly, indicating the emergence of new relationships. With the basic goal of accurately identifying (predicting) edges that can fit into the network even at the outset of a brief change of time, link prediction has been extensively investigated in the SN sector. Inferring new and missing linkages that are hidden in an observable network is another way it manifests itself. To infer the link predictions, proximity measurements, shortest-path lengths, and volumetric neighborhoods are still used as some of the intuitive approaches. In this context, the developments in link prediction algorithms are Common Neighbors (CN, $O(Nk^2)$, 2007) [12], Hierarchical Structure Model (HSM, O(eN), 2008) [13], Local Path (LP, $O(Nk^3)$, 2009) [14], Stochastic Block Model (SBM, O(eN), 2009) [15], DeepWalk (DW, $O(\Upsilon\ NLogN)$, 2014) [16], Structural Perturbation Method (SPM, $O(dN^2)$, 2015) [17], Low Rank (LR, $O(dN^2)$,

2017) [18], Length Three (L3, $O(Nk^3)$, 2019) [19], Ensemble-Model-based Link Prediction (EMLP, 2020) [20], Derivation of Mapping Entropy (DME, $MO(N^3)$, 2021) [21], and Dempster–Shafer evidence theory (DS, $O(n^2\, m + n^2)$, 2022) [22].

In order to organize groupings of abstract items into classes of related objects, researchers have developed different clustering approaches for use in data analysis based on Data Mining and Machine Learning. To categorize the heterogeneous source into separate sub groups, numerous clustering techniques have been developed, with scaling the "similarity of measures" at their core. Model-based [23], Hierarchical [24], Constraint-based [25], Grid-based [26], Partition-based [27], and Density-based [28] techniques are a few of the common ones. The social graph shown in Figure 10.3 is made up of nodes and corresponding edges (whether directional or not), and the number of edges is measured by the degree of each node.

Centrality is a concept used in graph theory to identify the most important vertices or nodes within a graph. It has numerous applications, including identifying the most influential person or group of people in a social network. Degree centrality is a type of centrality that measures the total number of direct links that a node has with other nodes in the graph. *Eigenvector Centrality*, on the other hand, is a measure of the influence of a node in the network, which assigns scores to all nodes based on the concept that connections to high-scoring nodes contribute more to the score of the node in question than equal connections to low-scoring nodes. *Closeness Centrality* is another measure of centrality that is defined based on the distance metric between nodes in a connected graph. The farness of a node is calculated as the sum of its distances to all other nodes, and its closeness is defined as the reciprocal of the farness. Therefore, the more central a node is, the lower its total distance to all other nodes.

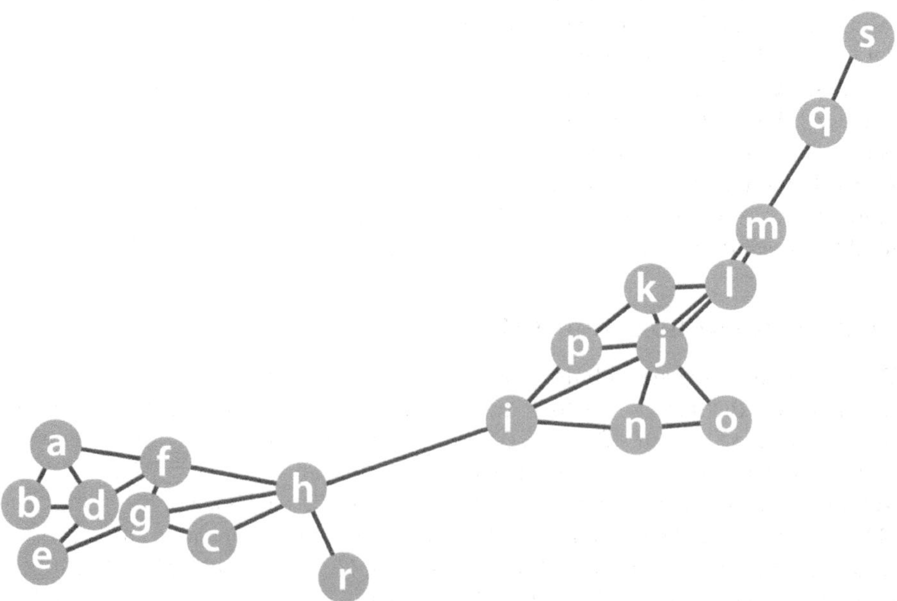

FIGURE 10.3 Social Graph with 19 nodes and their linkages.

Finally, *Betweenness Centrality* is a measure of centrality that quantifies the number of times a node acts as a bridge along the shortest path between two other nodes in a graph.

$$\text{Centrality}\left(\text{Degree}\right)\left(v_D\right) = \frac{dv}{\left(|N|-1\right)} \tag{10.1}$$

where, dv = degree of node v, N = number of nodes (sets) in the graph.

$$\text{Centrality}\left(\text{Eigen}_v\right)x_i = \frac{1}{\lambda}\sum_{j=1}^{n} a_{ij}x_j \text{ and } x = \frac{1}{\lambda}Ax \tag{10.2}$$

where, A is a similarity matrix representing the values of 1 and 0 for the existence of an edge between two nodes or not respectively.

$$\text{Centrality}\left(\text{Closeness}\right)\left(v_C\right) = \frac{|R(v)|}{|N|-1} * \frac{|R(v)|}{\sum_{u \in R(v)} d(u,v)} \tag{10.3}$$

where $R(V)$ = set of all nodes v which can be reached.

$$\text{Centrality}\left(\text{Betweenness}\right)\left(v_B\right) = \sum_{s \neq v \neq t} \frac{\sigma_{s,t}(v)}{\sigma_{s,t}} \tag{10.4}$$

where $\sigma_{s,t}$ = total no of shortest paths from node 's' to node 't' and $\sigma_{s,t}(V)$ = no. of those paths that pass through v. Because large nodes are more readily available in larger graphs, it is necessary to divide all node pairs aside from the present node to achieve normalization. Such normalization(s) in the case of undirected graphs and directed graphs are defined in Equations (10.5 and 10.6) respectively.

$$\text{No. of node pairs} = \left(\frac{1}{2}\right) * \left(|N|-1\right) * \left(|N|-2\right) \tag{10.5}$$

$$\text{No. of node pairs} = \left(|N|-1\right) * \left(|N|-2\right) \tag{10.6}$$

10.3 TECHNIQUES FOR BUILDING SOCIAL TRUST WITH XAI

10.3.1 Overview of Trust

The theory of social learning (cognitive reasoning) considers trust as a complex mental attitude that involves a set of beliefs, goals, and decisions. It is not just a simple emotion or feeling but rather a cognitive process that involves evaluating the reliability, competence, and intentions of another person or entity. The process of evaluating trust involves assessing the perceived risk and potential benefits of relying on the other person and making a decision about whether or not to trust them. The decision to trust is based on a combination of factors such as past experiences, social norms, and the specific situation at hand. Trust is a dynamic and context-dependent process that can vary depending on the specific situation and the individuals involved.

The above cognitive outcomes highlight that trust is a goal-oriented phenomenon. In a simple example, let us clarify the objectives with two agents (say, A_1 and A_2). Agent A_1 trusts agent A_2 only insofar as A_2 is perceived to have the ability (competence) and willingness (disposition) to help A_1 achieve a particular goal. In addition, A_1 must perceive some level of dependence on A_2 or believe that relying on A_2 will lead to a better outcome than relying on oneself or other agents. Finally, A_1 must believe that A_2 has the opportunity to perform the goal (fulfillment).

This formulation of trust implies that trust is not an all-or-nothing phenomenon but rather a matter of degree. Depending on the level of confidence that A_1 has in A_2's competence, disposition, dependence, and opportunity, A_1 may trust A_2 more or less. Additionally, trust is specific to a particular goal, so A_1 may trust A_2 to help achieve one goal but not another, depending on the relevant beliefs.

10.3.2 Social Trust

Researchers have studied trust as a foundation of the "human-automation" partnership in line with the advancement of technology. According to the theory presented in [29], trust plays a crucial role in guiding individuals to rely on automation. However, inappropriate reliance on automation can result in disastrous consequences. There are three main categories of research on trust in human-technology relationships, as identified in existing literature. These include "credentials-based trust", "experience-based trust", and "cognitive trust" [30–33]. The first category of research on trust is security technology that serves as a substitute for traditional methods of determining trustworthiness. The purpose of this approach is to evaluate whether a user or agent can be trusted based on a set of credentials and security policies. On the other hand, the second category of research on trust is centered around determining an agent's trust value based on their own past experiences, particularly in predicting the likelihood of a certain action being executed by another agent. Several methods are also available to calculate trust values, using different approximations of distributions. This approach is often used in reputation-based trust systems for e-commerce and peer-to-peer applications. The third category of research on trust captures the social (human) notion of trust, which is the most applicable type of trust in human-machine relationships [33] and referred to as a "mental state", that is, a belief of a cognitive agent about the achievement of a desired goal through another agent or through itself. The aspect is thus a key to expressing trust-based decisions in ML applications.

Luhmann [34] and Möllering [35] categorized social trust into interpersonal trust and institutional trust. Kei [36] and Zhou [37] used two variables to measure individuals' interpersonal trust: trust in neighbors and trust in villagers. Institutional trust was measured using trust in village officials and trust in government. To measure trust levels, rural households were asked to rate the trustworthiness of four categories of people: neighbors, villagers, village cadres, and government. Respondents used a Likert scale, with options ranging from very distrustful to very trustful, and values of 1–5 were assigned to each option, following previous studies [38,39].

Trust in information T_{info} is mathematically defined with a metric X, which is a range of values between −1 and 1, inclusive. The value '−1' corresponds to the category of distrust, indicating a negative perception of the trustworthiness of the information object. The value '0' indicates a state of uncertainty, ignorance, or insufficient information to make a decision about trusting the information object. Finally, the value '+1' represents a point on the range where there is sufficient evidence to trust the information object. Additionally, the total of X is calculated by taking into account all relevant attributed information details [40]. The variables W_{info} and R_{info} indicate the worth of information and reputation (changes in link credibility δL), respectively. Meanwhile, I, R, H, and Time refer to the significance, usefulness in a particular setting, information modelling value, and time (whether information is fixed or changeable), respectively.

$$T_{\text{info}} = \frac{w_{\text{info}} + R_{\text{info}}}{\text{Time}} \tag{10.7}$$

$$W_{\text{info}} = 1/n\left[\left(\sum_{x=1}^{n} X\right) + I + U\right] H \tag{10.8}$$

$$R = \frac{s + \delta L}{K} E \tag{10.9}$$

The use of XAI is to identify and analyze trust-related behaviors, as well as to develop tools for measuring and predicting trust. Additionally, there is a need for more research into the ethical implications of XAI-based social trust research.

10.3.3 Effect of Bias on Social Trust

Bias refers to the limitation of a model in accurately capturing the actual connection between variables. When a model has high bias, it may fail to identify important correlations between input features and output targets, resulting in underfitting.

In the context of social model solutions, bias can refer to the potential for the classification algorithm to exhibit discriminatory behavior against certain groups of people. This can happen if the training data used to develop the model contains imbalanced representations of different groups, leading the algorithm to make inaccurate predictions or decisions based on faulty assumptions or stereotypes. To address bias in social classification, it is important to ensure that the training data is diverse and representative of all groups in the population, and that the algorithm is tested for fairness and accuracy across different demographic categories.

It is important to note that while AI experts and novices share a goal of mitigating bias in machine learning systems, their needs and requirements may differ. AI experts may be more focused on identifying and mitigating bias in the training data used to develop machine learning models, while novices may be more interested in understanding how these models make decisions in real-world scenarios.

XAI can play a crucial role in identifying and mitigating bias in machine learning systems. XAI techniques allow for the interpretation and explanation of how machine learning models make decisions, enabling humans to understand the reasoning behind them. One common use case for XAI is in social network-based predictions and solutions. For example, XAI can help identify and mitigate bias in predictive policing algorithms that are used to identify individuals who are more likely to be involved in public disturbances (mobilize to the highly influential nodes having a high volume of social ties). By providing explanations for how these algorithms make decisions, XAI can ensure that they do not discriminate against certain individuals or groups. Another application of XAI is in assessing the fairness of automated decision-making systems, such as those used in hiring or referral/recommendation processes. XAI can help identify patterns of bias in the data used to train these systems, as well as provide explanations for how they arrive at their decisions. This can help ensure that these systems are fair and do not discriminate against protected groups. Here are some examples of how bias can be categorized:

- Algorithmic bias occurs when algorithms used in social networking analysis are not designed to be neutral and can lead to inaccurate results. This can be caused by a lack of diversity in the data used to train the algorithms, or by the algorithms themselves being designed to favor certain outcomes.
- Sampling bias occurs when the data used to analyze social networks is not representative of the population being studied, leading to skewed results and inaccurate conclusions.
- Data bias occurs when the data used to train a machine learning model is biased. For example, if a facial recognition algorithm is trained on a dataset that includes mostly light-skinned individuals, it may perform poorly when trying to identify people with darker skin tones.
- Selection bias occurs when certain groups or individuals are excluded from the analysis due to their characteristics or beliefs, resulting in an incomplete or distorted picture of the population being studied.
- Confirmation bias occurs when researchers focus on data that confirms their existing beliefs or hypotheses, rather than considering all available evidence, leading to inaccurate conclusions and a lack of objectivity in the analysis.
- Measurement bias occurs when the data used to train a machine learning model is not measured accurately or consistently. For example, if a sentiment analysis algorithm is trained on social media posts that use sarcasm or irony, it may not accurately reflect the true sentiment of those posts.

10.4 BUILDING SOCIAL NETWORK APPLICATIONS AND EVALUATING TRUST WITH XAI

10.4.1 Measures of Centrality and Closeness in SN

As discussed in Section 1.2.4, SN-community prediction is another term for the problem of clustering in graphs. Formally, communities can be defined as collections of network nodes that are intimately tied to one another and communicate with other nodes in the network only occasionally. A SN can be made easier to understand and study with the aid of community discovery and graph clustering. The community detection process is fraught with issues. The effect of participation of each node in inter- or intra-communities is studied in various threat detection models. It also offers potential solutions in the case of a sudden surge of the density in a community due to high usage. Relationships within the community and relationships outside it are depicted in social graphs, respectively, by internal edges and exterior edges. Within a society where there are few outward edges, the number of internal edges is high. There isn't an internal edge between two distinct groups; however, there are a few external edges. It is necessary to distinguish between the social media communities that are active and more significant on the graph. Analyses and statistics required for suggestions, searches, and trend analytics can be aided by determining the most active communities. The first established technique involves grouping vertices according to how similar they are. The second technique is based on sparse cut graph partitioning. A formal process to this effect is presented in Algorithm 10.1.

Algorithm 10.1: SN-based Community Prediction

1 Input: Nodes and corresponding Edges of Graph)

2 Output: Set of communities $\{C_1, C_2,..C_i\}$

3 for each e_i ($e_i \in$ Edges) do

4 Evaluate S_i using betweenness

5 end for

6 for each e_i ($e_i \in$ Edges) do

7 Evaluate $Similarity_{t1}$, $Similarity_{t2}$, $Similarity_{t3}$ w.r.t. e_i

8 $Sim = \sum_{i=1}^{3} Similarity_{ti}$

9 $C_{i+1} = C_i + Sim$

10 if($C_{i+1} > 1$) then $\{C_{i+1}=1\}$

11 else{ $C_{i+1}= 0$; POP the edges from G}

12 end if

13 end for

As the volume of data is one of these issues, identifying such communities with greater accuracy is a challenging effect.. We identify four communities in the social network displayed in Figure 10.3 using the Python libraries community and netgraph on the Colab platform, as illustrated in Figure 10.4. Furthermore, the volume of data produced by social networks is enormous. Community detection is an NP-hard problem [41].

In this context, it could also indicate a concern about privacy infringement if individuals (nodes) are recognized in publicly available social networks.

According to research findings in references [42,43], a social network satisfies the "k-anonymity" standard when the likelihood of identifying any given node is no more than *1/k*. We take into consideration the implementation and analysis of the straightforward social network synthesis shown in Figure 10.3 Using Equations (10.1–10.6) along with the Python NetworkX module, parametric measures concerning to SN, centrality were evaluated and are shown in Table 10.2. A graph is thus

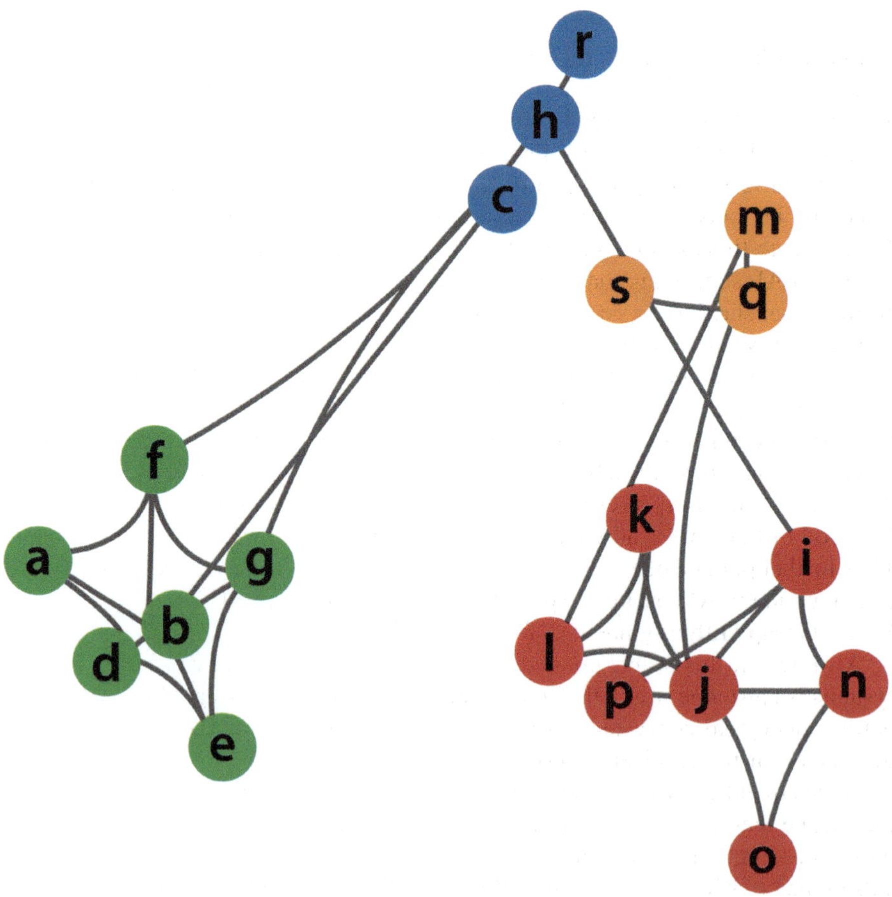

FIGURE 10.4 Social graph based community detection: Four derived communities coloured separately from the 19 nodes and their linkages.

TABLE 10.2
Centrality Analysis (V_D, X_i, V_C and V_B) of Corresponding Nodes in the Social Network with 19 Nodes and Their Linkages

		Centrality			
SN	Entity/Node	Degree (V_D)	Eigen (X_i)	Closeness (V_C)	Betweenness (V_B)
1	A	0.167	0.286	0.290	0.002
2	B	0.222	0.360	0.295	0.005
3	C	0.111	0.181	0.346	0.026
4	D	0.333	**0.465**	0.305	0.037
5	E	0.167	0.304	0.290	0.000
6	F	0.222	0.355	0.360	0.107
7	G	0.278	0.429	0.367	0.188
8	H	0.278	0.283	0.450	**0.575**
9	I	0.222	0.133	**0.462**	0.533
10	J	**0.389**	0.117	0.429	0.435
11	K	0.167	0.060	0.316	0.003
12	L	0.167	0.054	0.327	0.010
13	M	0.167	0.044	0.333	0.209
14	N	0.167	0.072	0.375	0.033
15	O	0.111	0.046	0.310	0.000
16	P	0.167	0.075	0.375	0.033
17	Q	0.111	0.011	0.261	0.111
18	R	0.056	0.069	0.316	0.000
19	S	0.056	0.003	0.209	0.000

Bold denotes the highest factor.

modeled to characterize the two important SN parametric measures (centrality and betweenness) and represented in Figure 10.5.

We further extend the analysis to compare it with high-density data using four real-world SN datasets. The first experimental dataset in our analysis (DS-1) is based on Facebook users and their communications in the social network. Identification of users is formulated in a text file with four-digit numbers. Similarly, descriptions of DS-2, DS-3, and DS-4 are shown in Table 10.3 and related to the Cora dataset [44], Flickr, and WebKB-wisc [45] respectively.

Some nodes are common between degree centrality (which measures the degree) and betweenness centrality in the social graph as seen through the employed dataset (which controls the information flow). It is normal for nodes that are more related to one another to be on the shortest paths connecting them. Using Equations (10.1–10.6), parametric measures were evaluated. The significance of nodes in SN is based on the essentiality of all centrality measurements, and Table 10.3 represents the outcomes of all datasets. In the context of space constraints, the output of the DS-1 graph with parametric measures is presented in Figure 10.6.

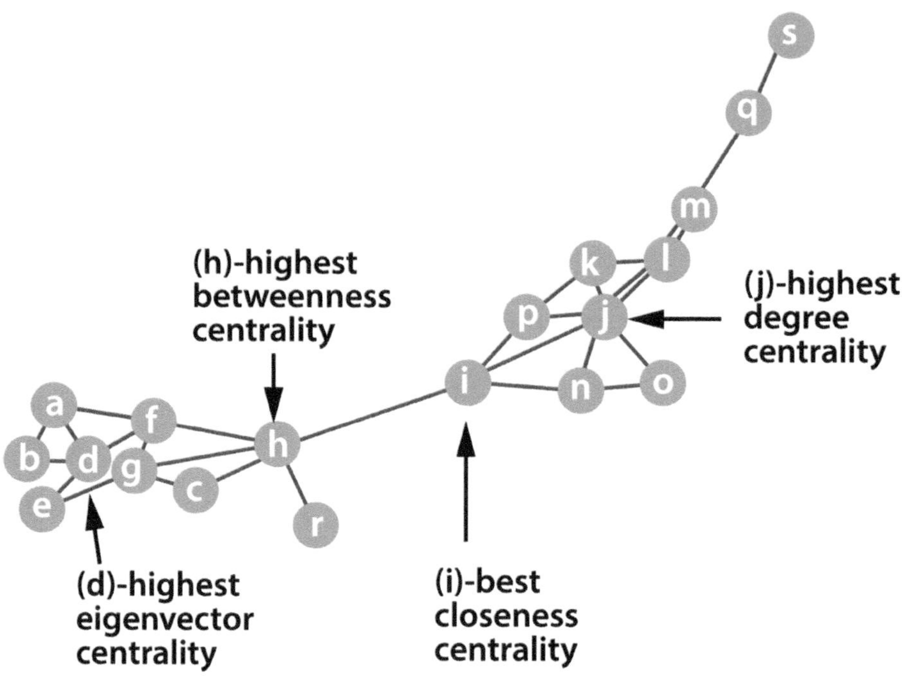

FIGURE 10.5 Parametric measures: representation of centrality and betweenness.

TABLE 10.3
Node Significance of Four Real-World Social Network Satasets

Dataset	Network Size		Node Analysis	
	Nodes	Edges	Top Five	Most Significant
DS-1	4039	88234	107, 1684, 3437, 1912, 1085	1,912
DS-2	2708	5278	163, 565, 427, 747, 523	163
DS-3	23323	2537992	3453, 7967, 11232, 19544, 21690	11,232
DS-4	7575	239738	754, 2456, 3523, 5443, 6412	754

10.4.2 Empirical Analysis and Explainable Measures with ETs

Suppose a social media company wants to predict (using an ML model) whether a user will click on a particular advertisement. The company has collected data on users' age, gender, location, interests, and past click behavior. To build a model, the company uses a random forest classifier, which achieves high accuracy on the test set. However, the company wants to ensure that the model is fair and unbiased, especially with regard to protected attributes such as gender and race. To investigate the model's behavior or measure its basic characteristics, one can use any of the following ETs. The explainer instance follows a basic flow where the desired XAI ET, ML-model, and instances are passed as arguments, and a set of explainable outputs is produced.

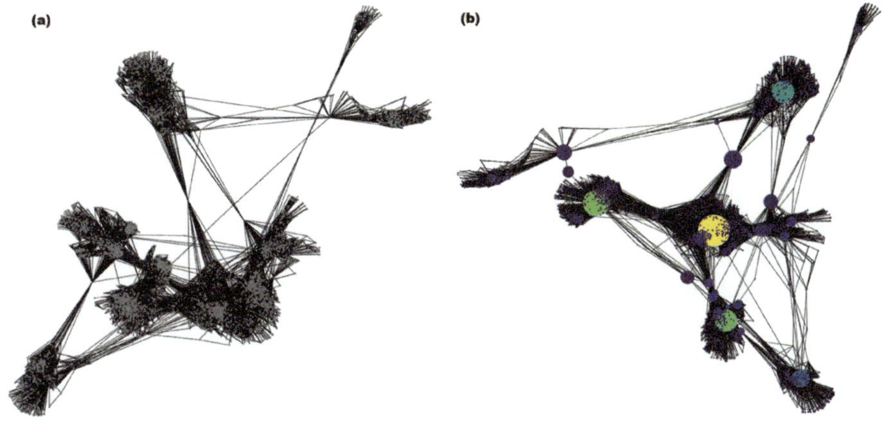

FIGURE 10.6 Parametric measures of facebook Dataset: (a) SN with 4039 nodes and 88234 edges (b) Centrality and betweenness with color weightings.

These outputs can include various explanations, such as the importance of specific features in the model's predictions, the process by which the model reached its conclusion/decision for a specific example/instance, and any potential biases in the model's partial forecasts or predictions. Additionally, the explainer instance can provide visualizations or summaries of the model's decision-making process, allowing for a more intuitive understanding of how the model is making its predictions. The working principles for two such ETs (XAI:LIME and XAI:SHAP) are provided below along with Python code for better comprehension.

Algorithm 10.2: XAI_Explainer_Instance (XAI-ET)

1 Input:{Set of Machine LearningModels, Instances}

2 Output: Set of explainable outputs {summary plot, textual description,interactive interface}

3 for each $inst_i$ ($inst_i \in$ Instances) do

4 S[i] = SHAP(feature)

5 end for

6 while each ($S_{value} \in$ S or user-input-features NOT NULL) do

7 model-prediction = SummaryPlot (most important features)

8 textual description = NLG (model-prediction)

9 Display textual description in an interactive interface

10 user-input-features = Interactive-interface(input features)

11 end while

12 return(summary-plot, textual-description)

10.4.2.1 Explainable Measures with XAI:LIME

This discussion is based on the LIME ET model scenario (outlined in Section 10.2.2). Here, we use the dataset concerning the financial positions of 190 member countries of the International Monetary Fund (IMF). The dataset used in this scenario is based on the participant countries data available in [46] as of March 28th, 2023. The data is categorized based on each country's GDP, with eight characterized attributes. Descriptions of these attributes include "GRA Credit Outstanding", "PRGT Credit Outstanding", "RST Credit Outstanding", "IMF's Holdings of Currencies", "Quota", "Reserve Tranche Position", "SDR Holdings" and "SDR Allocation". Our experiments on this dataset are conducted using the Python Google Colab platform, and we offer the Python code to elucidate the methodology in a more accessible manner, particularly for those who are new to XAI and interested in learning more.

```
#Train the Model with preliminary analysis
  import numpy as np
  import pandas as pd
  from sklearn.model_selection import train_test_split
  from sklearn.ensemble import RandomForestClassifier

#Data pre-processing.
  fin_pos =pd.read_csv('/content/sample_data/IMF-data.csv')

#Anonymising participant countries column
  fin_pos.drop('Member', axis =1, inplace =True)
  fin_pos.head()
#Split: Train/Test
  X =fin_pos.drop('Position', axis=1)
  y =fin_pos['Position']
  X_train, X_test, y_train, y_test =train_test_split(
    X, y, test_size=0.2, random_state=42)

#Model Training: Random forest-from ScikitLearn
  model =RandomForestClassifier(random_state=42)
  model.fit(X_train, y_train)
  score =model.score(X_test, y_test)

# The RF-classifier achieves 94.73% accuracy score
```

```
# Install LIME Python messaging library (for the first time)
pip install lime

import lime
from lime import lime_tabular

explainer =lime_tabular.LimeTabularExplainer(
    training_data=np.array(X_train),
    feature_names=X_train.columns,
    class_names=['cat2', 'cat1'],
    mode=              'classification'
)

exp_technique =explainer.explain_instance(
    data_row=X_test.iloc[5],
    predict_fn=model.predict_proba
)

exp_technique.show_in_notebook(show_table=True)
exp_technique =explainer.explain_instance(
    data_row=X_test.iloc[3],
    predict_fn=model.predict_proba
)
exp_technique.show_in_notebook(show_table=True)
```

For a sample test case of 5, the model has the top seven predictors aligned with cat2 (as shown in Figure 10.7) and is completely confident in this alignment. Similarly, for a sample test case of 3, the model has the top one predictor aligned with cat1 (as shown in Figure 10.8) and is completely confident in this alignment as well. Thus, the above example illustrates how the model demonstrates its ability to predict scores/accuracy using the classifier while also exhibiting its capability to explain the associated features of the dataset.

10.4.2.2 Explainable Measures with XAI:SHAP

SHAP can also be employed to produce explanations for each prediction made by the ML model. The prediction is swayed towards the baseline by red marks and

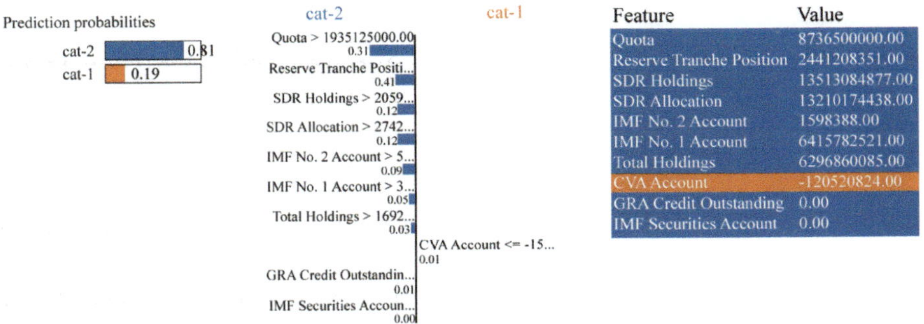

FIGURE 10.7 Output (Cat-2): Predicted explainer-xLIME.

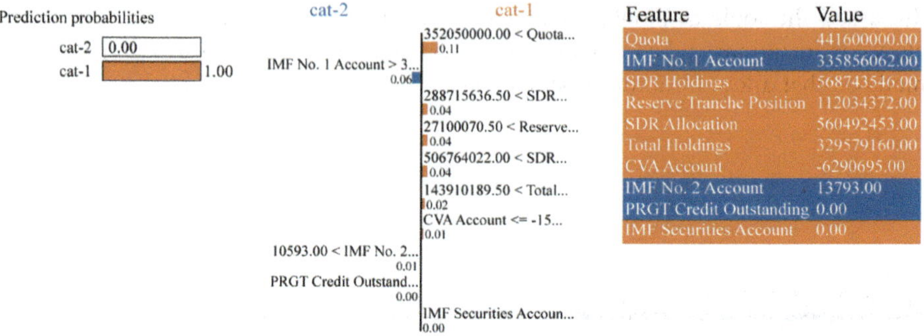

FIGURE 10.8 Output (Cat-1): Predicted explainer-xLIME.

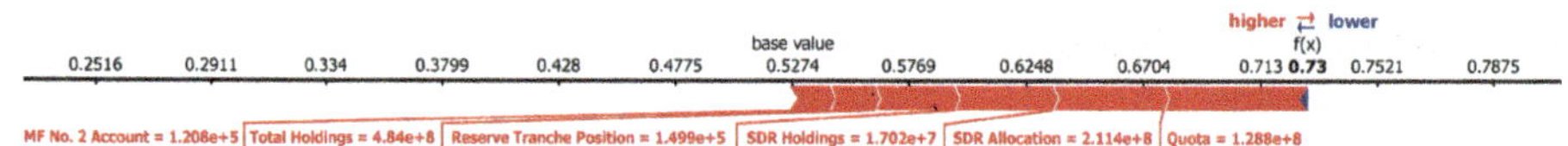

FIGURE 10.9 Output (IMF Dataset): Predicted explainer-xSHAP.

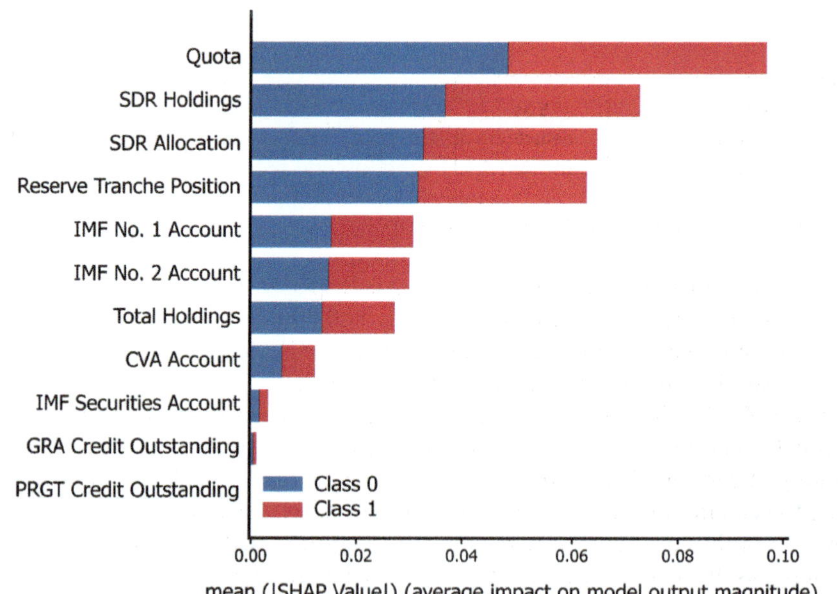

FIGURE 10.10 Output (IMF Dataset): Predicted explainer-xSHAP.

away from it by blue marks. In the given dataset, if red marks are more than blue marks, the dataset is categorized as cat1, and vice versa. The SHAP summary graph (Figures 10.9 and 10.10) offers a straightforward method to visualize the significance of each feature and its effect on the prediction.

```
# Install SHAP Python library (for the first time)
pip install shap

import shap
shap.initjs()
shap_explainer = shap.TreeExplainer(model)
shap_values = shap_explainer.shap_values(X)
test_1 =X_test.iloc[5]
shap.force_plot(shap_explainer.expected_value[0], shap_
values[0][0, :], test_1, link="logit")


#Summary Plot of the dataset
shap.summary_plot(shap_values, X)
```

10.4.3 Trust Prediction Model with Explainable Technique

It is necessary to classify objects in SN models based on both their links and their attributes to uncover distinct communities that are contained in the SN. The first step in applying conventional clustering methods to a social network graph would be to establish a distance metric. From hierarchical (agglomerative) to k-means (point-assignment) and spectral, SN-clustering methods have evolved.

Combining a couple of nodes that are connected by an edge is the first step in the "hierarchical clustering" of an SN network. To combine the clusters to which their two nodes belong, successive edges that are not between two nodes of the same cluster would be selected at random. Because every distance represented by an edge is the same, the decisions would be made at random. When using a "point-assignment method" to cluster SN, edges that are close together present a number of random variables that may cause some nodes to be assigned to the incorrect cluster. A particularly well-liked technique called "spectral clustering" includes iteratively partitioning the graph according to several criteria, such as the primary eigenvector of the adjacency matrix. Overlapping clusters are not allowed in any of these methods.

In Algorithm 10.3 the proposed trust prediction model is outlined. Here, the first objective is to check predicted labels in the model for consistency with the instance features and true label. If the predicted label is significantly distant from the closest label, it indicates a classifier issue under bias and allows the determine the Trust Score. Then the Counterfactual Explanation ET is used to explain the measures in detail.

Algorithm 10.3: XAI_Trust_Evaluation(XAI-ET)

1 Input: Input Description = (ML-Models, Dataset (T), Instances)

2 Output: Set of explainable outputs {summary plot, textual description, interactive interface}

```
3  /*Estimate alpha-high-density-set  */
   Sample_p (T)={t_1, t_2, .., t_n} derived from continuous-density-function f_cont
4  /*Estimate alpha-high-density-set [45] */
   alpha_high density (f_cont)=Estimate-HD-Set(Sample_p (T), f_cont)
5  /*Evaluate Trust-score [Equation 10.7] */
   S_Trust=classifier(training_data, test_example)
6  for each inst_i (inst_i∈ Instances) do
7     S[i]=DiCE(feature, S_Trust)
8   end for
9   while each (S_value∈ S or user-input-features != NULL) do
10     model-prediction=SummaryPlot (most important features)
11     textual description=NLG (model-prediction)
12     Display textual description in an interactive interface
13   user-input-features=Interactive-interface(input features)
14  end while
15 return(summary-plot, textual-description)
```

The methodology for evaluating social trust in ML models involves using XAI and measuring a Trust Score (S_{trust}) along with the counterfactual ET technique. The S_{trust} is derived from assessing the reliability of the predictions generated by an ML classifier based on the concepts presented in a seminal paper [47]. By leveraging the S_{trust}, models can obtain a deeper understanding of the relationships between data points, which may not be readily apparent when relying solely on the model's confidence level, especially when the model is trained using optimization algorithms in machine learning.

The first step of S_{trust} identification approach involves independent preprocessing of the training dataset to identify $alpha_{\text{high density}}$ set of each class. This requires removing the $\text{alpha}_{\text{fraction}}$ of samples with the lowest density (distance measures) say, in using nearest neighbors distance, centroid distance, or social network, which could be considered as outliers, to obtain the remaining training samples within the class.

Data-driven distance measures can include methods like nearest neighbors. Although distance measurements like Euclidean, Cosine, Jaccard, and edit distance are used to describe cluster similarity, SN-based distance measures depend on factors such as the shortest path between two nodes, breadth-first search, and eccentricity.

Next, to identify the S_{trust} with testing samples, the approach follows with a calculation of the proportion of the space between the test sample and the $alpha_{high\_density}$ cluster of the closest non-predicted category, to the distance between the testing sample and $alpha_{high\_density}$ set of the predicted class. The rationale behind this is that if the predicted label is significantly farther away than the closest label, it may indicate a potential mistake made by the classifier. The source [47] provides mathematical and comprehensive information that can assist in achieving a better comprehension of the subject matter.

The counterfactual explanation exhibits its property in explaining users understand why the model made the decision it did and how the decision could have been different. In this context, the Diverse Counterfactual Explanations (DiCE) ET is used to identify alternative events to a given model output. Here, the basic aim is to identify subsets and for which the approach of determinant point processes (DPP) is used. It is termed as the diversity constraint. DPP utilizes a matrix M with $M_{i,j}= 1/(1+\text{dist}(ce_i, ce_j))$, where distance (ce_i, ce_j) is a distance-metric between counterfactual explanations (represented by ce_i). The determinant of this matrix, denoted as *DPP* (diversity), captures diversity. The proximity between the c_i features and the input values is measured as the negative distance. Authors in [48] provide a thorough explanation on hyper parameters, loss functions, and model closeness/similarity related to the counterfactual explanation, and the Python *dice-ml* library provides DiCE functionalities.

10.5 CONCLUSION

This chapter provides an overview of the current state of AI-based research, including associated models and services. Although Explainable AI (XAI) is still in its early stages, the chapter highlights the organizational aspects of XAI and discusses the currently available models. The importance of trust in any application domain is emphasized, and the chapter explores social trust and bias in the context of intelligent and computable models. Two real-life datasets are used to outline parametric evaluation measures related to social networks, and the findings are presented graphically. Real-life data is also analyzed to examine the core concept of explainable parameters using a Python-based narration to explain the concepts in detail. The chapter concludes by outlining a trust prediction model that uses a trust score and the DiCE explainable technique. The findings from both theoretical and empirical analyses indicate that this approach may have significant practical expansion not only for predictions derived from data but also for graph-based investigations. Overall, the chapter emphasizes the significance of XAI in fostering social trust in social networks and offers a comprehensive framework for achieving this objective.

REFERENCES

1. Adadi, A., and Berrada, M., Peeking inside the black-box: A survey on explainable artificial intelligence (XAI), *Access*, 6, pp. 52138–52160 (2018).
2. Guidotti, R., Monreale, A., Ruggieri, S., Turini, F., Giannotti, F., and Pedreschi, D., A survey of methods for explaining black box models, *Comput. Surv. (CSUR)*, 51, pp. 1–42 (2018).
3. Vilone, G., and Longo, L., Notions of explainability and evaluation approaches for explainable artificial intelligence, *Inf. Fusion*, 76, pp. 89–106 (2021).
4. Zeng, Z., Miao, C., Leung, C., and Chin, J.J., Building more explainable artificial intelligence with argumentation. In *Proceedings of the 32nd Conference on Artificial Intelligence*, USA, 2–7, pp. 8044–8046 (2018).
5. Chen, Z., Xiao, F., Guo, F., and Yan, J., Interpretable machine learning for building energy management: A state-of-the-art review, *Adv. Appl. Energy*, 9, p. 100123 (2023).
6. Murdoch, W.J., Singh, C., Kumbier, K., Abbasi-Asl, R., and Yu, B., Definitions, methods, and applications in interpretable machine learning, *Proc. Natl. Acad. Sci. USA*, 116, pp. 22071–22080 (2019).
7. Zafar, M.R., and Khan, N., Deterministic local interpretable model-agnostic explanations for stable explainability, *Mach. Learn. Knowl. Extr.*, 3(3), pp. 525–541 (2021).
8. Ribeiro, M. T., Singh, S., and Guestrin, C., Why should I trust you? Explaining the predic- tions of any classiier. In *Proceedings of the 22nd ACM SIGKDD International Conference on Knowledge Discovery and Data Mining*, San Francisco, California, USA, pp.1135–1144 (2016).
9. Lundberg, S.M., and Lee, S.I., A unified approach to interpreting model predictions, *Adv. Neural Inf. Process. Syst.*, 30, pp. 4765–4774 (2017).
10. Dhurandhar, A., Chen, P.Y., Luss, R., Tu, C.C., Ting, P., Shanmugam, K., and Das, P., Explanations based on the missing: Towards contrastive explanations with pertinent negatives, *Adv. Neural Inf. Process. Syst.*, 31, pp. 1–22 (2018).
11. Pekala, K., Woznica, K., and Biecek, P., Triplot, Model agnostic measures and visualisations for variable importance in predictive models that take into account the hierarchical correlation structure, pp. 1–13 (2021). doi: 10.48550/arXiv.2104.03403.
12. Liben-Nowell, D., and Kleinberg, J., The link-prediction problem for social networks, *J. Assoc. Inf. Sci. Technol.* (2007). doi: 10.1002/asi.20591.
13. Clauset, A., Moore, C., and Newman, M.E., Hierarchical structure and the prediction of missing links in networks, *Nature*, 453(7191), pp.98–101 (2008).
14. Lü, L., Jin, C.H., and Zhou, T., Similarity index based on local paths for link prediction of complex networks, *Phys. Rev. E*, 80(4), p. 046122 (2009).
15. Guimerà, R., and Sales-Pardo, M., Missing and spurious interactions and the reconstruction of complex networks, *Proc. Natl. Acad. Sci.*, 106(52), pp. 22073–22078 (2009).
16. Perozzi, B., Al-Rfou, R., and Skiena, S., Deepwalk: Online learning of social representations. In *Proceedings of the 20th ACM SIGKDD International Conference on Knowledge Discovery and Data Mining*, New York, USA, pp. 701–710 (2014).
17. Lü, L., Pan, L., Zhou, T., Zhang, Y.C., and Stanley, H.E., Toward link predictability of complex networks, *Proc. Natl. Acad. Sci.*, 112(8), pp.2325–2330 (2015).
18. Pech, R., Hao, D., Pan, L., Cheng, H., and Zhou, T., Link prediction via matrix completion, *EPL (Europhys. Lett.)*, 117(3), p. 38002 (2017).
19. Kovács, I.A., Luck, K., Spirohn, K., Wang, Y., Pollis, C., Schlabach, S., Bian, W., Kim, D.K., Kishore, N., Hao, T., and Calderwood, M.A., Network-based prediction of protein interactions, *Nat. Commun.*, 10(1), pp. 1–8 (2019).
20. Li, K., Tu, L., and Chai, L., Ensemble-model-based link prediction of complex networks, *Comput. Netw.*, 166, p. 106978, (2020).

21. Hu, H., Wang, Y., Li, Z., Tian, Y., and Ren, Y., Link prediction based on the derivation of mapping entropy, *Complexity* (2021). doi: 10.1155/2021/4156832.
22. Liu, M., Wang, Y., Chen, J., and Zhang, Y., Link prediction model for weighted networks based on evidence theory and the influence of common neighbours, *Complexity*, 2022, pp. 1–16 (2022).
23. Yeung, K.Y., Fraley, C., Murua, A., Raftery, A.E., and Ruzzo, W.L., Model-based clustering and data transformations for gene expression data, *Bioinformatics*, 17(10), pp. 977–987 (2001).
24. Zhang, T., Ramakrishnan, R., and Livny, M., BIRCH: An efficient data clustering method for very large databases, *ACM Sigmod Record*, 25(2), pp. 103–114 (1996).
25. Tung, A.K., Han, J., Lakshmanan, L.V., and Ng, R.T., Constraint-based clustering in large databases. In *International Conference on Database Theory*, Springer, Berlin, Heidelberg, pp. 405–419 (2001). doi: 10.1007/3-540-44503-X_26.
26. Liao, W.K., Liu, Y., and Choudhary, A., A grid-based clustering algorithm using adaptive mesh refinement. In *7th Workshop on Mining Scientific and Engineering Datasets of SIAM International Conference on Data Mining*, Lake Buena Vista, Florida, USA, 22, pp. 61–69 (2004).
27. Boley, D., Gini, M., Gross, R., Han, E.H.S., Hastings, K., Karypis, G., Kumar, V., Mobasher, B., and Moore, J., Partitioning-based clustering for web document categorization, *Decis. Support Syst.*, 27(3), pp. 329–341 (1999).
28. Xu, X., Ester, M., Kriegel, H.P., and Sander, J., Clustering and knowledge discovery in spatial databases, *Vistas Astron.*, 41(3), pp. 397–403 (1997).
29. Lee, J. D., and See, K. A., Trust in automation: Designing for appropriate reliance, *Hum. Factors: J. Human Factors Ergonomics Soc.*, 46(1), pp. 50–80 (2004).
30. Muller, G., Secure communication trust in technology or trust with technology? *Interdiscip. Sci. Rev.*, 21, pp. 336–347 (2013).
31. Nielsen, M., Krukow, K., and Sassone, V., A bayesian model for event-based trust, *Electron. Notes Theor. Comput. Sci.*, 172, pp. 499–521 (2007).
32. Lahijanian, M., and Kwiatkowska, M., Social trust: A major challenge for the future of autonomous systems. In *AAAI Fall Symposia*, Arlington, Virginia (2016).
33. Falcone, R., and Castelfranchi, C., Social trust: A cognitive approach. In *Trust and Deception in Virtual Societies*. Springer, pp. 55–90 (2001). doi: 10.1007/978-94-017-3614-5_3.
34. Luhmann, N., *Trust and Power*, John Wiley and Sons: New York (1979).
35. Möllering, G., The nature of trust: From georg simmel to a theory of expectation, interpretation and suspension, *Sociology*, 35, pp. 403–420 (2001).
36. Kei, H.E., Junbiao, Z., Zhang, L., and Wu, X., Institutional trust and rural households' willingness to participate in environmental governance: The case of agricultural waste resourceization, *Manag World*, 5, pp. 75–88 (2015).
37. Zhou, Y.C., and Ao, D., A study of the differences in social capital between the self-employed and the employed, *Sociol. Study*, 5, pp. 198–224 (2011).
38. Chen, Z., Chen, F., and Zhou, M., Does social trust affect corporate environmental performance in China? *Energy Econ.*, 102, p. 105537 (2021).
39. Smith, E.K., and Mayer, A. A social trap for the climate? Collective action, trust and climate change risk perception in 35 countries, *Glob. Environ. Change*, 49, pp. 140–153 (2018).
40. Williams, T. A., and Marsh, S., Towards a computational model of information trust, In *IFIP Advances in Informationand Communication Technology Book Series (IFIPAICT)*, 528, pp. 124–136 (2018). doi: 10.1007/978-3-319-95276-5_9.
41. Mehdi, A., Rhouma, D., and Romdhane, L.B., Community detection in large-scale social networks: State-of-the-art and future directions, *Soc. Netw. Anal. Mining*, 9(1), pp. 1–32 (2019).

42. Sweeney, L., K-anonymity: A model for protecting privacy, *Int. J. Uncertaintly Fuzziness Knowl. Based Syst.*, 10(05), pp. 557–570 (2002).
43. Tripathy, B.K., and Panda, G.K., A new approach to manage security against neighborhood attacks in social networks. In *2010 International Conference on Advances in Social Networks Analysis and Mining*, IEEE, Odense, Denmark, pp. 264–269 (2010).
44. Renchi, Y., Jieming, S., Xiaokui, X., Yin, Y., Juncheng, L., and Bhowmick, S.S., Scaling attributed network embedding to massive graphs, *Proc. VLDB Endowment*, 14(1), pp. 37–49 (2021).
45. A graph and network repository facilitates for real-world networks and benchmark datasets. https://networkrepository.com/.
46. International Monetary Fund (IMF) Financial Data Query Tool. https://www.imf.org/external/np/fin/tad/queryoutput.aspx.
47. Jiang, H., Kim, B., and Guan, M.Y., To trust or not to trust a classifier. In *Proceedings of 32nd Conference on Neural Information Processing Systems (NIPS 2018)*, Montreal, Canada, pp. 1–25 (2018).
48. Mothilal, R.K., Sharma, A., and Tan, C. Explaining machine learning classifiers through diverse counterfactual explanations. In *Proceedings of the 2020 Conference on Fairness, Accountability, and Transparency*, Barcelona, Spain (2020).

11 Fuzzy Clustering for Streaming Environment with Explainable Parameter Determination

Subhadip Boral, Koustav Pal, and Ashish Ghosh

11.1 INTRODUCTION

A sequence of potentially infinite, non-stationary data (where the probability distribution of the data may change over time) arriving continuously where random access to the data is not feasible, and storing all the arriving data is impractical, is generally known as a data stream. Data streams are common in online trading, financial analysis, e-commerce and business, smart home, healthcare, transportation systems, global supply logistics chains, smart grids, industrial control, cybersecurity, and many other areas. Data stream has become readily available due to advances in data acquisition technology, and these massive data gathered as a continuous flow are often in need of real-time processing, which brings about unique challenges not easily handled by many of the current computational intelligence and machine learning methods operating in batch off-line mode. The main characteristics of streams include:

- Continuous flow,
- Rapid arrival rate,
- Huge data volumes,
- Change in data distribution over time.

Clustering is a technique of grouping a collection of objects into clusters such that objects within the same cluster are similar in a certain sense, and objects from different clusters are dissimilar. Objects are usually represented by a set of numeric features from which similarity degrees between pairs of objects are calculated, using a kind of similarity or distance measure. Unsupervised analysis of online data clustering has emerged as an essential data stream mining task because of its ability to capture natural structures from unlabelled, non-stationary data and its ability to reduce the complexity of the data by replacing a group of observations (cluster) with a representative observation (prototype). Clustering of data streams is performed to study time-changing grouping of data. The problem of data stream clustering is that of identifying an optimal clustering model in the presence of variations. Ideally,

DOI: 10.1201/9781003442509-11

learning methods should adapt to changing environments (such as concept drifts and concept evolutions). Concept drift denotes the way data distribution changes gradually over time, whereas concept evolution refers to the emergence of new concepts or clusters in the data distribution. The impact of concept drift and evolution on learning algorithms is enormous. While the effect of concept drift can be attenuated using parameter adaptation techniques, concept evolution may require a search in the underlying hypothesis space, which may be distinct. Computational models should be equipped with incremental learning algorithms to be able to evolve and deal with such changes. The nature of evolving data streams implies the following requirements for stream clustering:

- **No assumption on the number of clusters:** The number of clusters is often unknown in advance. Furthermore, in an evolving data stream, the number of clusters often changes.
- **Discovery of clusters with arbitrary shape:** The formation of clusters, or in other words, the distribution of the data is not known in prior.
- **Robustness to outliers:** Isolated observations or outliers are defined as the points that do not fit well into the clusters identified so far (Barbara 2002). As data streams evolve, new clusters emerge and old clusters fade out with time in unpredictable ways. Hence, a stream clustering algorithm must incorporate a mechanism to discriminate emerging clusters from outliers efficiently without impacting the quality of the clustering scheme.
- **Tracking model changes:** To maintain relevance, the clustering model should be able to adapt to concept drift and evolution.
- **The balance between robustness and tracking ability:** Insensitivity to outliers and reactivity to model changes are two competing objectives. This balancing act depends on the nature of the data stream and the choice of model parameters.

Now, clustering methods may be broadly classified into two categories partitional and hierarchical methods. In the literature, most of the research effort is spent on partitional methods. Roughly, partitions of data can be of two types: crisp and fuzzy (Bezdek 1981; Krishnapuram and Keller 1993). In a streaming environment, fuzzy clustering makes it possible to decrease the effect of data points that do not belong to one single cluster, for example, points located between overlapping clusters or points resulting from noise. These points, by having varying membership values, now have a low influence on the calculation of the cluster centre positions. Hence, with the introduction of the fuzzifier parameter, $m > 1$, the cluster analysis becomes much more robust. The value of the fuzzifier defines the maximum fuzziness or noise in the data. The fuzzy methods adapt to the present amount of noise and avoid erroneous detection of clusters generated by random patterns. Therefore, the challenge consists in determining an appropriate value for the fuzzifier. Thus, despite being less used in data stream analysis, fuzzy clustering offers potential advantages over hard partitioning:

- In the case of non-stationary streams, i.e., in the presence of concept drift, the loss of validity of a clustering model can be better represented by real-valued membership, which changes smoothly, without having to make abrupt changes in the model.
- With Boolean membership, the detection of a concept change requires a sufficiently large dataset to allow reliable estimates, while with fuzzy models, point-wise calculation is possible, allowing much faster detection.
- In the particular case of possibilistic models, the additional robustness property allows point-wise detection of outliers in addition to source change detection.

This motivates the development of an unsupervised evolving intelligent system for data streams, which adapts to any variation in data trends by parameter adaptation as well as structural evolution and discovers the uncertain nature of real-world issues in an online manner.

Most of the fuzzy clustering algorithms are modelled to process data in a batch mode. The sample of observations consists of N, $n-$dimensional feature vectors $X=(x_1,x_2,\ldots x_k,\ldots x_N)\subset R^n$. This dataset has to be partitioned into m clusters with membership levels μ_{jk} for each kth feature vector x_k to j^{th} cluster, $j=1,2,\ldots m$, where the number of clusters m is considered to be a priori unknown. The set $C=(c_1,c_2,\ldots c_k,\ldots c_N)$, with $c\leq N$ and typically $c\ll N$. Each element $c_i\subset R^n$ is called a prototype \ (centroid) and usually is in a one-to-one relationship with the cluster. The inclusion of kth observation in cluster j is measured by the membership μ_{jk}. Prototypes are locations in the data space that minimize their average squared distance to points in their cluster. An alternative representation of a cluster is by means of the point in the cluster that minimizes the average absolute distance from all others. This is termed a medoid. Fuzzy clusters are characterized by real-valued memberships, $\mu_{jk}\epsilon[0,1]\subset R$. Each object can be a member of all the clusters to which it has in general different membership degrees, each less than 1. The probabilistic constraint is:

$$\sum_{j=1}^{m}\mu_{jk}=1,\ \forall k$$

The model is termed as possibilistic (Krishnapuram and Keller 1993) if the above constraint is not enforced. The fuzzy clustering task becomes significantly more complicated if data come sequentially in an online mode as a data stream. This situation is then subject to data stream mining (Aggarwal 2007; Bifet 2010; Gama 2010), and naturally the sample volume N in this case is not fixed, and it keeps on increasing. Let, $O=o_1,o_2,\cdots o_N$ denote a stream of N objects, for example, traffic flow patterns on a street network or web activity profiles on an e-commerce website. In the streaming data model, N is very large (in principle, unbounded) and objects are observed sequentially. Individual observations in the stream are represented by a fixed set of features. The existing works are solely dependent on the user parameters and not data-driven and this concept fails to provide real time analysis. This motivates the data-driven, explainable determination of fuzzifier values and the number of clusters present in the dataset.

The proposed method in its core uses the FCM method with an initial small number of clusters and recursively calculates the entropy of the data. It determines whether the entropy has increased or not with respect to the last instance. The intuition behind the entropy change is that it indicates a change in the distribution of the arrived data points with respect to the existing data. Depending on this detection, the proposed method modifies the number of clusters and assesses whether the cluster validity index is improved or not, progressing accordingly. As the number of clusters does not analyse the clustering procedure alone, i.e., the fuzzifier controls the fuzziness of the environment, the proposed method also alters the fuzzifier depending on the entropy value and identifies the proper fuzzifier value. The major concern of the existing work is the search for proper parameters and the methods performed without holding the incoming data flow. The proposed method determines the parameters with each iteration where it takes new data points as new information too. This mechanism helps the proposed method detect concept drift and evolution with more information and makes the algorithm capable of performing in real time. This real-time analysis without holding incoming data points makes it realistic and more applicable in real life. The experimental study on synthetic datasets and real-life datasets shows the efficacy of the algorithm to determine concept drift and evolution in a streaming environment. The article in the next section discusses the existing works in the corresponding field and their applicability. Then the proposed method is demonstrated in detail and the experimental analysis is presented thereafter.

11.2 RELATED WORK

The work aims to comprehend unsupervised evolving systems for data streams, which respond to changes in data trends through parameter adaptation as well as structural evolution. The methods should be able to detect the uncertain character of real-world issues online and adapt accordingly. Existing studies in this field can be classified into two categories:

11.2.1 Evolving Neural Networks Based Methods

When data is fed sequentially with separable clusters, the self-organizing map by Kohonen (1990) has the ability to adapt successfully to the clustering task. These neural networks have a single layer with lateral connections and learn using the winner-take-all principle. Because of the need to resolve fuzzy clustering tasks in real-time, Tsao, Bezdek, and Pal (1994) developed the self-learning neuro-fuzzy systems, which are hybrid systems composed of the self-organizing map (SOM) and the fuzzy C-means algorithm. Because of the use of particular tuning techniques, these hybrid systems have considerable capability; however, in order to accomplish clustering tasks, these systems must first specify the number of clusters and a preset fuzzifier value.

There are evolving connectionist systems that change over time and can solve clustering with an unknown number of clusters. This technique of evolving clustering algorithms is followed by EFuNN by Kasabov (2001), DENFIS by Kasabov and Song (2002), eTS by Angelov and Filev (2005), ECoS by Kasabov (2007). In supervised learning mode, these systems modify their parameters, and clustering is employed for scatter partitioning of the input space. Due to this supervised learning of parameters, the applicability of these methods decreases when applied in real time, as real-time data annotation is quite expensive.

There are probabilistic fuzzy clustering procedures effective for data stream mining by Bezerra, Costa, Guedes, & Angelov (2016); Deng and Kasabov (2000); Maciel, Ballini, and Gomide (2017); Ravi, Srinivas, and Kasabov (2007) and procedures based on the Gustafson-Kessel algorithm by Dovžan and Škrjanc (2011); Škrjanc and Dovžan (2015), which adapt to a recurrent form for the processing of streaming data. In these methodologies, the number of clusters present in the data is a priori known, and the evolution in learning occurs based on the purposeful change of the threshold parameter, which identifies the quality of an obtained solution during changes to the number of possible clusters. This a priori knowledge of the number of clusters and the data-independent threshold parameter makes the learning of these methods restricted for data stream processing. All the aforementioned methods also lack the intuition to analyse the fuzzifier value during adaptation.

Since it is impossible to choose the fuzzifier value a priori, cluster ensembles are employed to address this task. A pool of similar algorithms differing in fuzzifier values is assessed and based on the quality of the obtained solution, the fuzzifier value is determined. As data streams are generally evolving in nature, there must be a procedure to choose the best neurons of the pool in the cluster ensembles. Cascade Neural Networks by Fahlman and Lebiere (1990) and Hybrid Neural Cascade systems by Bodyanskiy, Tyshchenko, and Kopaliani (2015) implement this intuition where in each cascade the clustering is performed independently. Then every succeeding cascade increases the number of possible clusters by one until the clustering of desired quality is achieved. Thus, the process becomes equivalent to hierarchical divisive clustering by Rodrigues and Gama (2007). Merging the ideas of Online Data Stream Clustering, Evolving Systems, and Hybrid Cascade systems an Evolving Cascade Connectionist System the work by Bodyanskiy, Tyshchenko, and Kopaliani (2017) is introduced which is able to process data sequentially when the number of clusters as well as the fuzzifier value is apriori unknown and may change while the process runs. Due to this computationally heavy mechanism, dependency on the initial parameters of other dependent systems, and large search space, the methodology is restricted for data stream processing.

11.2.2 Clustering Based Methods

The methodology proposed by Zhang, Qin, Wang, Wang, and Xue (2016) to detect concept drift in a streaming environment uses fuzzy membership to detect concept drift. The membership degree of fuzzy clustering is used to calculate the information entropy of the data and according to the entropy, concept drift is detected. The data stream FCM clustering algorithm by Gao, Li, and Meng (2017) inherits the membership degree matrix to adapt to new data points. It detects the amount of change in the membership degree matrix, and depending on that, either it updates with the current matrix or inherits the existing matrix. The proposed model by Al-Khamees,

Al-A'araji, and Al-Shamery (2021) suggests a clustering mechanism that is based on optimizing the e-Cauchy algorithm that computes the density for each data sample. The density determines the cluster membership for the data point. The algorithm determines the number of clusters and converges to an ideal number by implementing evolving mechanisms. The methodology by Sangma, Rani, Pal, Kumar, and Kushwaha (2022) implements a fuzzy hierarchical clustering method for clustering multiple data streams using a feature-based clustering approach. The fuzzy affinity of data streams to different clusters is calculated using normalized behaviour cosine similarity to the cluster centroids. It handles the concept evolution by updating the hierarchical clustering structure by either merging and/or splitting the nodes depending on the extent to which the nodes' entropy changes. The major assumption of the methodology is the independence of features from each other, and this assumption cannot be guaranteed for a data stream.

The discussion on fuzzy clustering by two genres emphasizes the fact that starting models are parameter dependent, and the improvement over manual parameter selection increases the search space and processing cost. The existing approaches' reliance on user-supplied parameters motivates the proposed work to a data-driven determination of both parameters with reduced search space and processing expense.

11.3 PROPOSED METHODOLOGY

The primary goal of the proposed Fuzzy clustering algorithm for the data stream is to determine the fuzzifier parameter and the number of clusters present in the accumulated data in real time. Due to the requirement of real-time analysis, data-driven determination of the parameters is the only feasible way. The intuition behind the proposed methodology is that the entropy of the data is a good indicator of how much fuzziness is present in the data. If the fuzziness in the accumulated data is low, i.e., data points can be clustered with proper separation, then the entropy value will be low. On the other hand, if the fuzziness in the data is high, i.e., the data points tend to form overlapping clusters, the entropy value will be high. Now, in this scenario, if the entropy is low, then the algorithm is on a proper path; otherwise, it is necessary to increase the fuzzifier value and decrease the entropy value accordingly. Now, it is necessary to consider that for the other parameter, the number of clusters, the increase in clusters may form a clustering where a large number of clusters are formed but in an ideal case, the number of clusters is low. To handle this issue, a cluster validity index will be a good indicator of whether the increment in the number of clusters is improving the performance of the algorithm or not. So, the inclusion of performance measures using cluster validity index will bound the search for lower entropy with a proper number of clusters, which is, in reality, unbounded. So, the initial step should be low in the count for the number of clusters and also the fuzzifier value. If entropy suggests an increment of the number of clusters and gets support by the cluster validity indices, then the number of clusters will be increased. The aforementioned condition is also applicable for the decrement of the fuzzifier value and the number of clusters. From this discussion, it is understandable that the proposed algorithm in core depends on the entropy of the data, cluster validity index, and performs the Fuzzy C-means algorithm.

11.3.1 Pre-requisite

11.3.1.1 Fuzzy C-Means Algorithm

Fuzzy C-means clustering is a fuzzy generalization of standard (K-means) clustering (Hartigan 1975). In fuzzy clustering, an object x may belong to different clusters at the same time, and the degree to which it belongs to the ith cluster is expressed in terms of a membership degree μ_i. Consequently, the boundary of single clusters and the transition between different clusters are usually smooth rather than abrupt. The fuzzy C-means minimizes the following objective function:

$$\sum_{i=1}^{n}\sum_{j=1}^{K}\left\|x_i - c_j\right\|^2 \left(\mu_{ij}\right)^m$$

where $\mu_{ij} = \mu_i(x_i)$ is the membership of the ith object x_i in the jth cluster, and c_j is the jth cluster centre. In the probabilistic version of fuzzy C-means, it is assumed that

$$\sum_{j=1}^{K}\mu_{ij} = \sum_{j=1}^{K}\mu_j(x_i) = 1$$

for all x_i (Höppner, Klawonn, Kruse, and Runkler 1999). This constrained optimization problem is solved using Lagrange's Multiplier Method. The possibilistic fuzzy C-means minimizes the Lagrange function:

$$L\left(\mu_{ij}, c_{ij}, \lambda_i\right) = \sum_{i=1}^{n}\sum_{j=1}^{K}\left(\mu_{ij}\right)^m \left\|x_i - c_j\right\|^2 + \sum_{i=1}^{n}\lambda_i\left(\sum_{j=1}^{K}\mu_{ij} - 1\right)$$

where λ_i is an undetermined Lagrange multiplier. The algorithm approximates an optimal (sub-optimal) solution by means of an iterative scheme that alternates between recomputing the optimal centres according to

$$c_j = \frac{\sum_{i=1}^{n}\left(\mu_{ij}\right)^m x_i}{\sum_{i=1}^{n}\left(\mu_{ij}\right)^m}$$

and membership degrees according to

$$\mu_{ij} = \frac{\left\|x_i - c_j\right\|^{\frac{2}{1-m}}}{\sum_{j=1}^{K}\left\|x_i - c_j\right\|^{\frac{2}{1-m}}}$$

The model is termed as possibilistic if the probabilistic constraint is not enforced. The idea is simply to view membership degrees as typicalities. This allows points to have low memberships to all clusters, which could indicate a point being an outlier.

11.3.1.2 Determination of Data Entropy

In information theory, entropy is a measure of the amount of uncertainty or randomness in a system or dataset. It is often denoted by the symbol H_n and can be calculated using the following formula:

$$H_n = -\sum_{i=1}^{n} p_i \ln(p_i)$$

where p_i is the probability of the occurrence of an event i. The equation sums over all possible events in the system or dataset, weighted by their probability and the logarithm of their probability. Entropy can be interpreted as the expected amount of information that is needed to describe the outcome of a random event. In other words, high entropy means that the outcome of an event is highly unpredictable or uncertain, while low entropy means that the outcome is more predictable or certain.

In a streaming environment, when the underlying data distribution changes, the entropy value abruptly increases from the previous one because the clustering algorithm was accustomed to a certain data distribution. However, that sudden change in data distribution made the algorithm less efficient, resulting in a high the entropy value. This indicates that the fuzzifier value should be increased. Also, it is true that with a decrement in the entropy value, the algorithm senses that the fuzzifier value should be decreased.

11.3.1.3 Xie-Beni Cluster Validity Index

The Xie-Beni index (Xie and Beni 1991) is a clustering validity index used to evaluate the quality of a clustering solution. The Xie-Beni index measures the compactness and separation of clusters by comparing the distance between data points and their cluster centres within clusters, and the distance between cluster centers across clusters. A lower Xie-Beni index value indicates better clustering performance, meaning that the clusters are more compact and well-separated. It is evaluated as below:

$$I_{XB} = \frac{\sum_{i=1}^{n}\sum_{j=1}^{K}(\mu_{ij})^m d_{ij}^2}{n\left(\min_{j\neq l}\left\|\bar{v}_j - \bar{v}_l\right\|^2\right)}$$

where $\sum_{i=1}^{n}\sum_{j=1}^{K}(\mu_{ij})^m d_{ij}^2$ is the sum of the squared distances between data points and their cluster centres and $n\left(\min_{j\neq l}\left\|\bar{v}_j - \bar{v}_l\right\|^2\right)$ is the minimum distance between any two data points in the dataset.

11.3.2 Adaptive Determination of the Number of Clusters and Fuzzifier Value

The proposed clustering algorithm for data streams, being adaptive, focuses on changing the number of clusters and fuzzifier value when it senses a concept drift or concept evolution. Depending on the aforementioned discussion, the algorithm determines the entropy of the accumulated data and determines whether it has changed or not with respect to the last instance. The proposed algorithm initially starts with a number of clusters of 2 due to the reason that the entropy will guide it to increase and the number of possible clusters can be high but cannot be lower than 2. At any time instance t, suppose there are N data points, and the entropy of the data is calculated using the following equation:

$$E_n = -\frac{1}{N}\sum_{j=1}^{c}\sum_{i=1}^{N}\mu_{ij}\ln\left(\mu_{ij}\right)$$

If the calculated entropy is changed over the previous entropy calculated at time instance $t-1$ by a certain percentage (η), then the algorithm detects a possible concept drift or concept evolution and tries to re-cluster the data by the following steps:

1. If the optimal number of the cluster at time instance $t-1$ be c, then perform FCM on the accumulated data arrived till t by increasing and decreasing c by 1, i.e $c+1$ and $c-1$.
2. The algorithm then compares the clustering quality of the clusters found with the number of clusters $c-1$, c, $c+1$ using Xie-Beni index and keeps the best clustering result.

Without the incorporation of entropy, there will be a swinging motion where the algorithm always tries to better the performance with $c-1$, c, $c+1$ numbers of clusters. The incorporation of entropy will indicate that there is a change in the dataset but to precisely get the impact, whether the change is in the number of clusters or in fuzzifier value that is very important. If the change is in the number of clusters, then it will be handled accordingly, and an improvement in performance measures will be cited. Otherwise, at any instance of time t, suppose the optimal number of clusters is c and the proposed algorithm calculates a range of fuzzifier values with the following equation (Huang, Xia, Wang, Zeng, and Wang 2012):

$$m \in \left[1+\frac{c-1}{c}\star\frac{2}{\delta}\star|\Delta|,\ \frac{2\log d}{\log\left(\frac{\delta}{1-\delta}\star\frac{1}{c-1}\right)}+1\right]$$

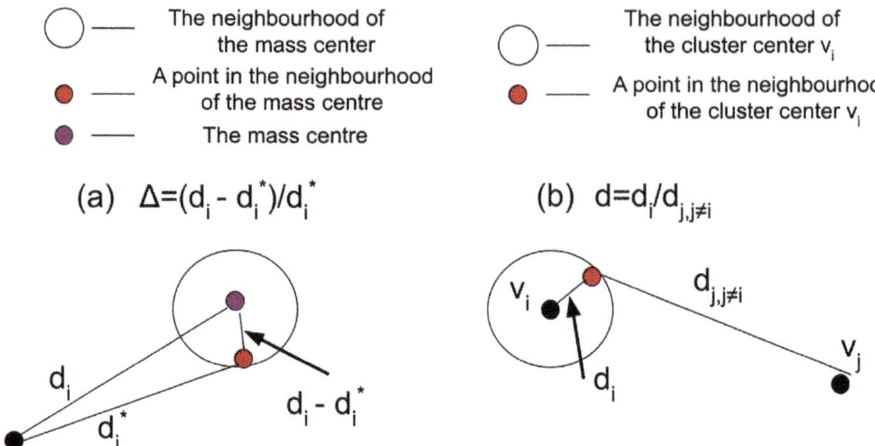

FIGURE 11.1 Determination of dependent variables for fuzzifier range search.

It then iterates over the range with an interval of 0.5 (Torra 2015). For each iteration, it compares the clustering result with others using the Xie-Beni index and thus selects the best fuzzifier value for the data stream at that time instance (Figure 11.1).

11.3.3 Concept Drift and Concept Evolution Analysis

Concept drift is a common phenomenon in the streaming environment, where the underlying data distribution changes over time, leading to a degradation of the performance of machine learning methods developed on earlier data. This can be caused by various factors, such as changes in the data generation process, shifts in user behaviour, or system malfunctions. In data stream mining, concept drift is a significant challenge, as the data is continuous and unbounded, and the model needs to adapt in real time to the changing data distribution. In the case of concept evolution, a new concept emerges, i.e., new clusters form that were absent earlier. The entropy value of the dataset abruptly increases when a concept drift or evolution occurs in a data stream. The proposed algorithm tries to detect that abrupt change in entropy which eventually leads to the detection of the concept drift that happens within the data stream. By the determination of the fuzzifier value, the concept drift will be handled and the determination of a proper number of clusters will address the issue of concept evolution.

11.3.4 Adaptive Fuzzy C-Means Algorithm for Data Stream

Algorithm: Adaptive Fuzzy C-Means Algorithm for Data Stream

INPUT

1. n dimensional data stream X
2. Initial no. of cluster = c,
3. Threshold parameter h

OUTPUT

1. Optimal number of clusters
2. The fuzzy weight matrix
3. Final cluster centres

STEPS

At any time instance t

1. Perform FCM on the data points with number of cluster = c
2. Calculate entropy E_t
3. If $\frac{|E_{t-1} - E_t|}{E_{t-1}} > h$
   a. Perform FCM on the with number of cluster = $c-1, c, c+1$
   b. Compare the results of $c-1, c, c+1$ using the Xie-Beni index and keep the best as an optimum result.
4. Else
   a. Go to 1 when new data points come in the next time instance.

5. Repeat the above steps until the data stream ends.

11.4 EXPERIMENTAL RESULTS AND ANALYSIS

Initially, the description of the datasets is given, and the performance of the proposed algorithm is studied and compared with the state of the algorithms.

11.4.1 Dataset Description

The performance of the proposed algorithm is analysed on both synthetic and real datasets.

11.4.1.1 Synthetic Dataset

Six synthetic datasets have been created incorporating normal distribution to analyse the performance. The description is provided in Table 11.1.

11.4.1.2 Real Dataset

The proposed algorithm was run on six real-life dataset (Dua and Graff 2017). The description of the datasets is as follows (Table 11.2):

11.4.2 Performance Analysis and Comparative Study

The results of the proposed algorithm on the mentioned datasets are as follows:

11.4.2.1 Impact of Entropy Calculation

The performance of the proposed method is studied with and without entropy calculation to establish the effectiveness of entropy calculation. The performance is studied on the synthetic datasets, and the comparative study is presented in Table 11.3.

TABLE 11.1
Description of the Six Synthetic Datasets

Dataset	Cluster	Mean	Covariance	Instances
Syn1	Cluster A	[6,15]	[2,0;0,2]	1000
	Cluster B	[7,26]	[2,0;0,2]	1000
	Cluster C	[15,15]	[2,0;0,2]	1000
Syn2	Cluster A	[11,11]	[2,0;0,2]	1000
	Cluster B	[20,16]	[2,0;0,2]	1000
	Cluster C	[13,21]	[2,0;0,2]	1000
Syn3	Cluster A	[13,13]	[0.5,0;0,0.5]	1200
	Cluster B	[15,17]	[1,0;0,1]	1200
	Cluster C	[20,22]	[2,0;0,2]	600
Syn4	Cluster A	[12,12]	[0.5,0;0,0.5]	1200
	Cluster B	[13,16]	[1,0;0,1]	1200
	Cluster C	[16,21]	[2,0;0,2]	600
Syn5	Cluster A	[10,10]	[2,0;0,2]	1400
	Cluster B	[17,17]	[2,0;0,2]	1000
	Cluster C	[22,25]	[1,0;0,1]	800
	Cluster D	[20,30]	[1,0;0,1]	500
	Cluster E	[22,16]	[2,0;0,2]	300
Syn6	Cluster A	[10,10]	[2,0;0,2]	500
	Cluster B	[15,15]	[1,0;0,1]	500
	Cluster C	[15,25]	[1,0;0,1]	500
	Cluster D	[22,25]	[2,0;0,2]	500
	Cluster E	[17,30]	[1,0;0,1]	500
	Cluster F	[15,35]	[1.5,0;0,1.5]	500
	Cluster G	[30,35]	[2,0;0,2]	500
	Cluster H	[25,7]	[2,0;0,2]	500

TABLE 11.2
Description of the Six Real Life Datasets

Dataset Name	Instances	Features
Gas Sensor	13,910	128
GPS tracking of smart device	153,540	50
Wine	4898	10
Key-stroke	78,095	38
Electricity	45,312	8
NSL-KDD99	150,121	29

It is evident from the study that when the optimization with entropy calculation is performed, the number of iterations to reach the optimal solution becomes considerably smaller compared to that without entropy calculation, thereby reducing the complexity of the proposed algorithm. As it is essential for a fuzzy clustering algorithm developed for data streams to have a low complexity for real-time analysis, the

TABLE 11.3
Comparative Analysis Shows the Efficacy of Entropy Calculation be Means of Reduction in the Number of Iterations

	Without Entropy		With Entropy	
	Number of Iterations	Xie-Beni Index Value	Number of Iterations	Xie-Beni Index Value
Syn1	300	1.02	1	0.07
Syn2	300	1.5	1	0.1
Syn3	300	2.1	1	0.13
Syn4	300	0.88	1	0.27
Syn5	400	0.09	1	0.12
Syn6	400	0.51	4	0.12

algorithm proposed here can be implemented efficiently on data streams due to its low complexity. Also, it can be seen that when using the algorithm without entropy calculation, the number of cluster detections may not be optimal, as the algorithm was randomly increasing or decreasing the number of clusters. However, entropy helps the algorithm to detect concept drift and evolution and change the number of clusters accordingly.

11.4.2.2 Performance Analysis on Synthetic Dataset

The synthetic datasets, being two-dimensional, are visualized to show how the data points arrive with increasing time and how the clusters are formed and reformed by the proposed algorithm (Figures 11.2–11.25).

All the visualizations and performance show the efficacy of the proposed algorithm in the detection of cluster numbers and fuzzifier values. It also shows the efficiency of the algorithm in the detection of concept drift and evolution.

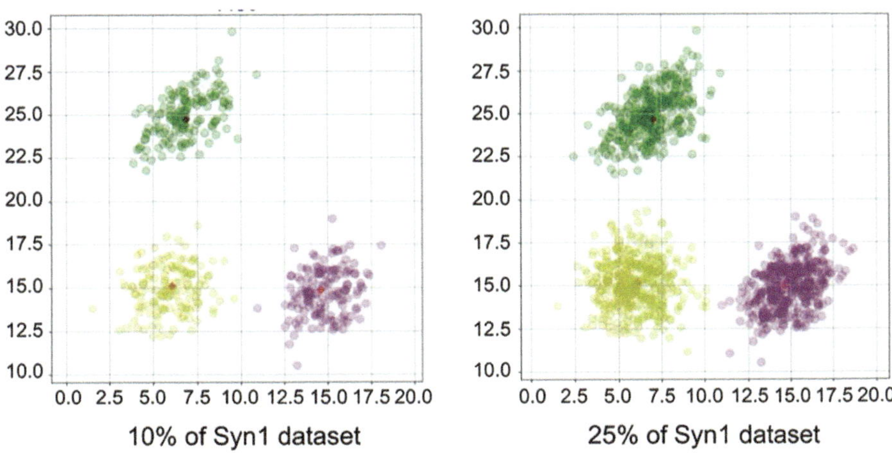

FIGURE 11.2 Initial cluster formation by the proposed algorithm on Syn1.

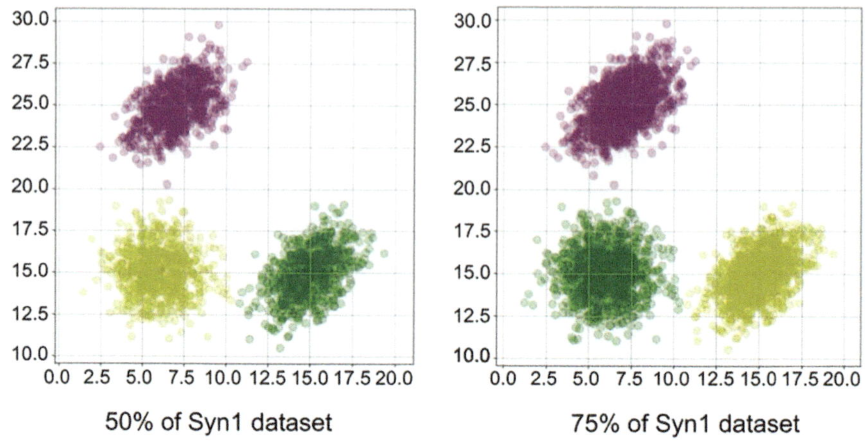

FIGURE 11.3 Cluster adaptation by the proposed algorithm on Syn1.

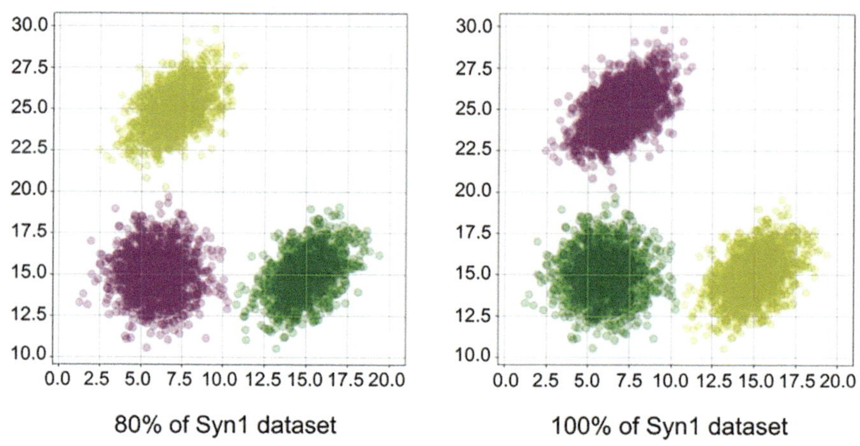

FIGURE 11.4 Final cluster formation by the proposed algorithm on Syn1.

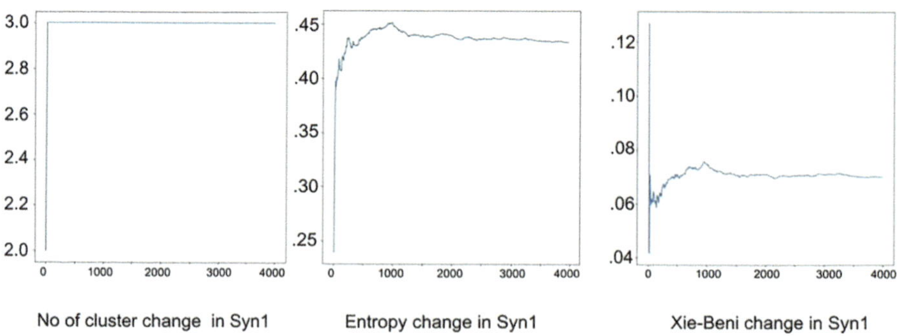

FIGURE 11.5 Change of parameters and performance by the proposed algorithm on Syn1.

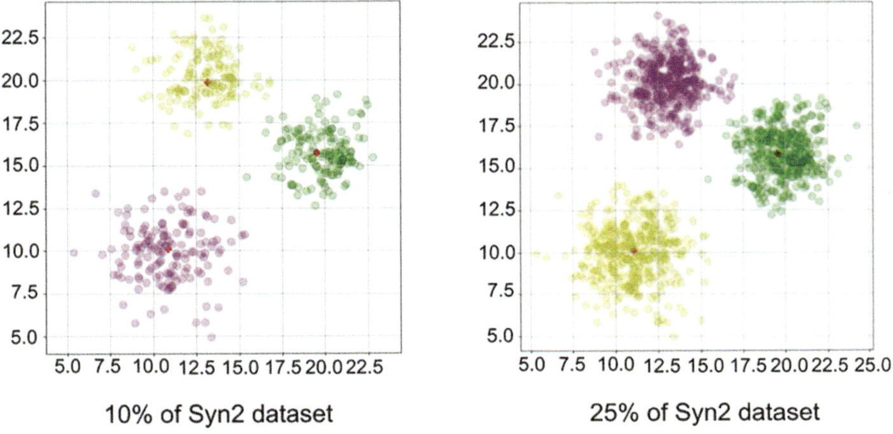

FIGURE 11.6 Initial cluster formation by the proposed algorithm on Syn2.

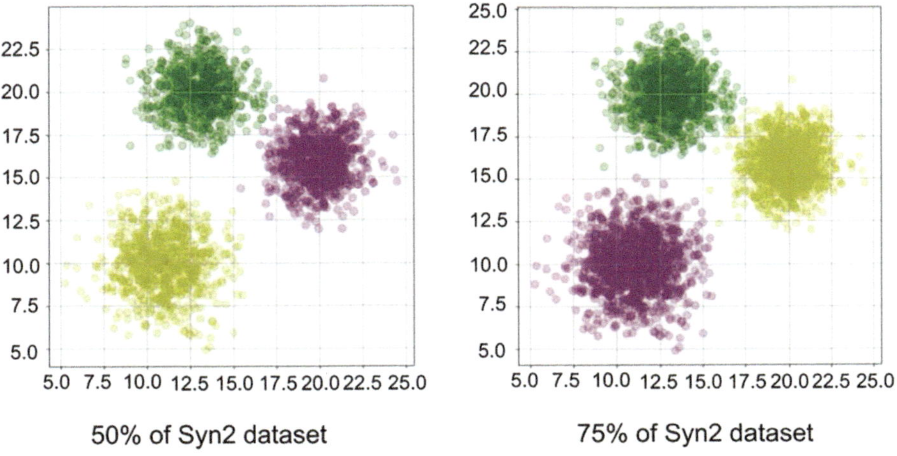

FIGURE 11.7 Cluster adaptation by the proposed algorithm on Syn2.

11.4.2.3 Performance Analysis on Real-Life Dataset

The performance of the proposed algorithm on real-life dataset is analysed (Figures 11.26–11.30).

The graphs show that the proposed algorithm determines the number of clusters in the dataset which get disrupted with the change in entropy score, indicating the awareness the algorithm possesses. The stability in Xie-Beni index value also signifies the steady performance of the proposed algorithm.

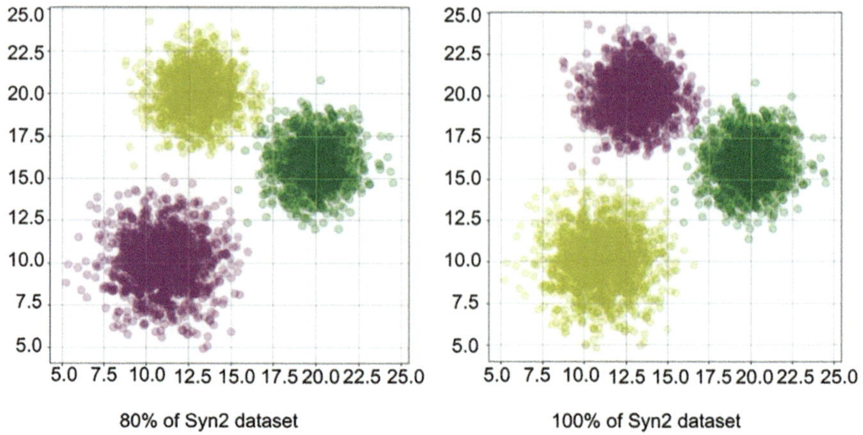

FIGURE 11.8 Final cluster formation by the proposed algorithm on Syn2.

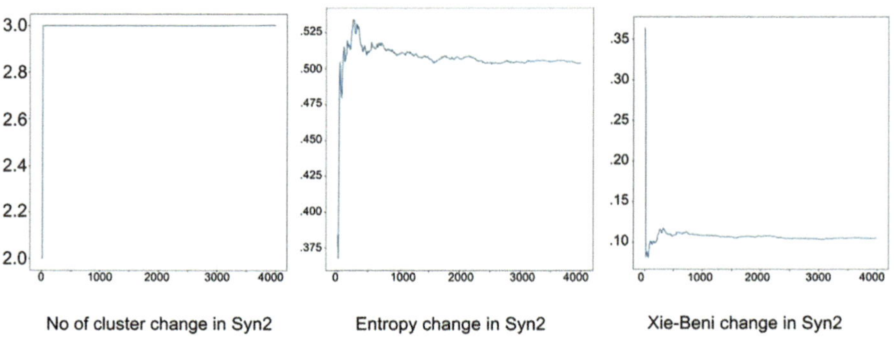

FIGURE 11.9 Change of parameters and performance by the proposed algorithm on Syn2.

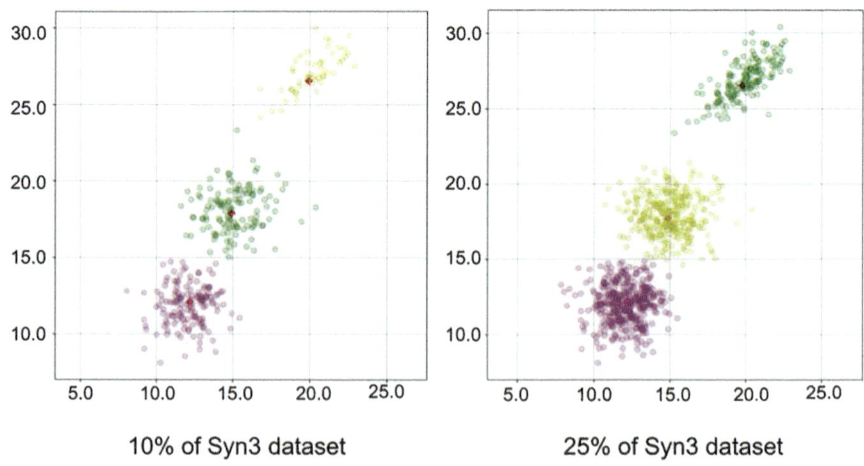

FIGURE 11.10 Initial cluster formation by the proposed algorithm on Syn3.

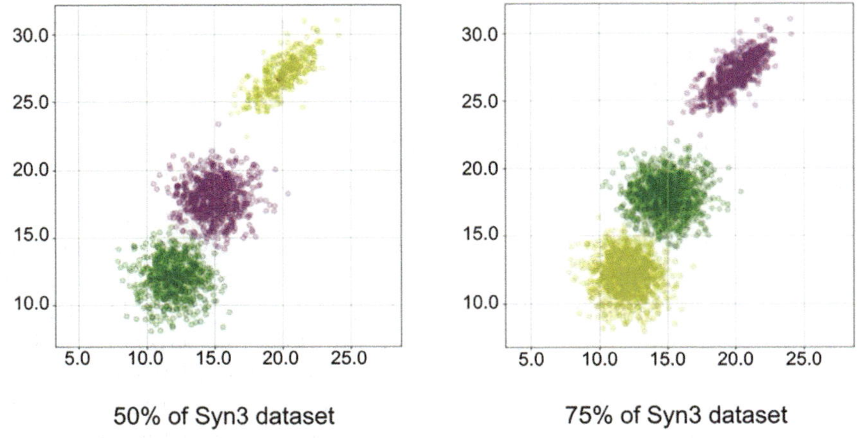

FIGURE 11.11 Cluster adaptation by the proposed algorithm on Syn3.

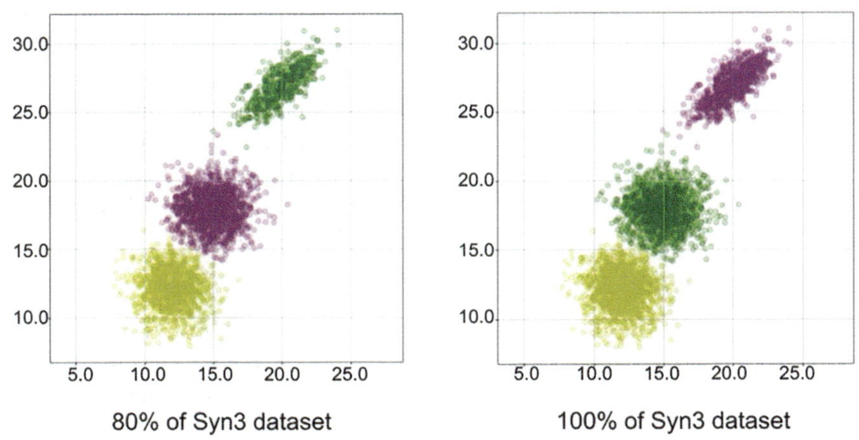

FIGURE 11.12 Final cluster formation by the proposed algorithm on Syn3.

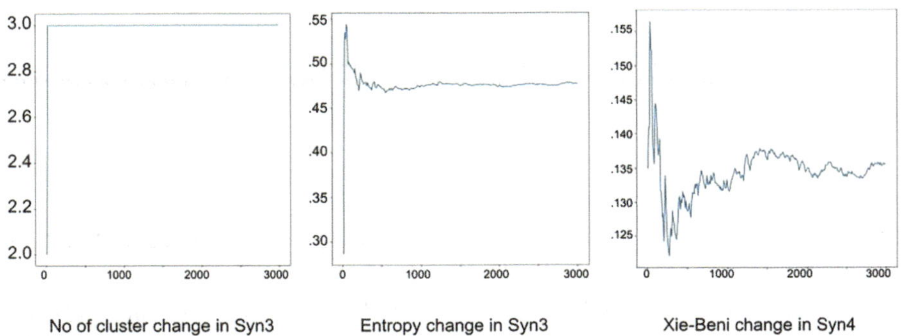

FIGURE 11.13 Change of parameters and performance by the proposed algorithm on Syn3.

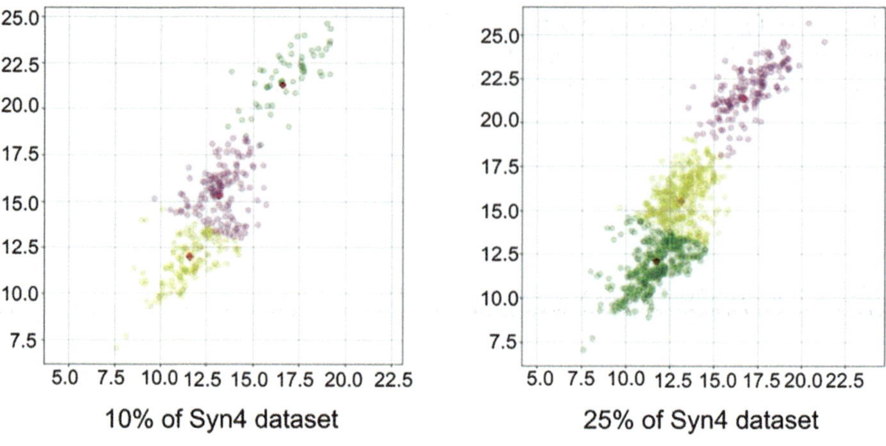

FIGURE 11.14 Initial cluster formation by the proposed algorithm on Syn4.

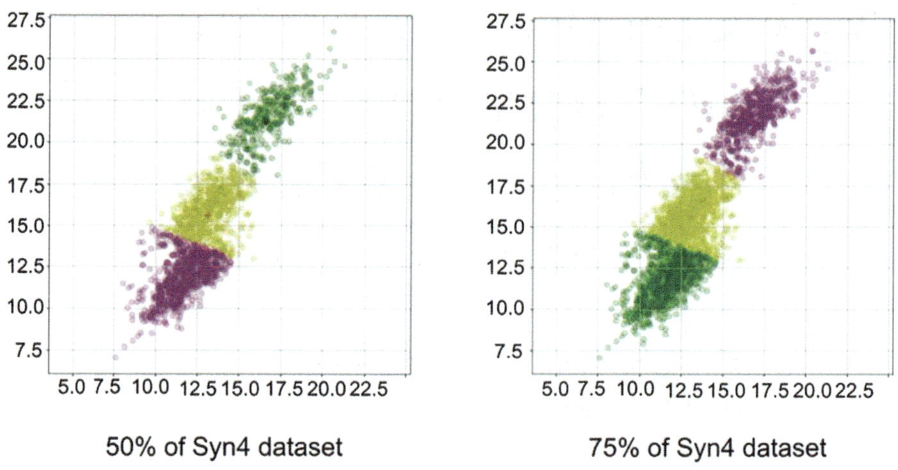

FIGURE 11.15 Cluster adaptation by the proposed algorithm on Syn4.

11.4.2.4 Performance Analysis on Real-Life Dataset and Comparative Study

This section deals with the comparison of the proposed algorithm with the other state-of-the-art algorithms. The algorithms selected for comparison are as follows: OFCM, SPFCM, eGKPCM, GPCMs, and Fuzzy-based Evolving Cauchy Algorithm. All the algorithms are tested on the NSL-KDD dataset (Table 11.4).

It is evident that the proposed algorithm is able to detect the same number of clusters as the other algorithms. The clustering result is then compared with other algorithms using clustering validity indexes like partition coefficient, partition entropy, and Xie-Beni index. It can be observed that these cluster validity indexes of

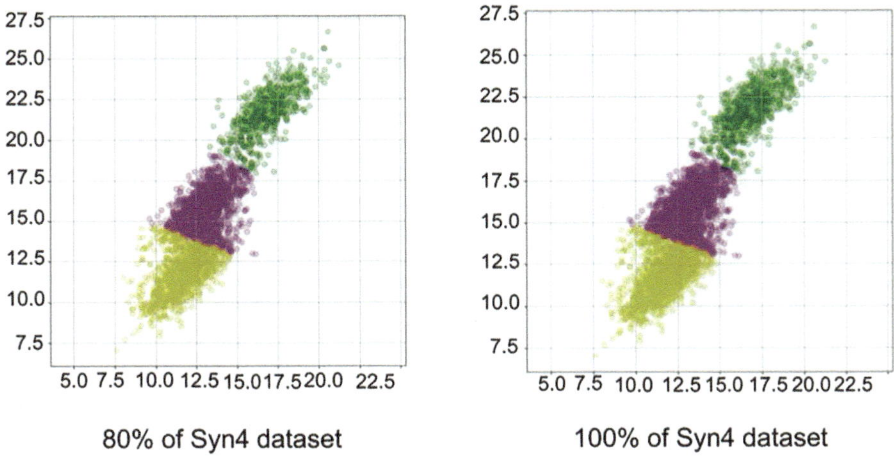

FIGURE 11.16 Final cluster formation by the proposed algorithm on Syn4.

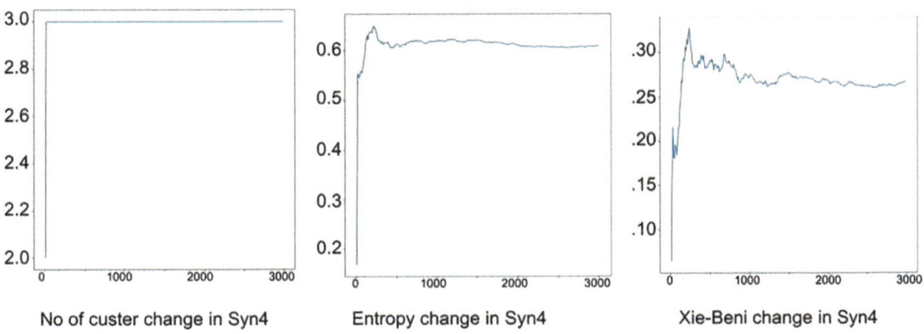

FIGURE 11.17 Change of parameters and performance by the proposed algorithm on Syn4.

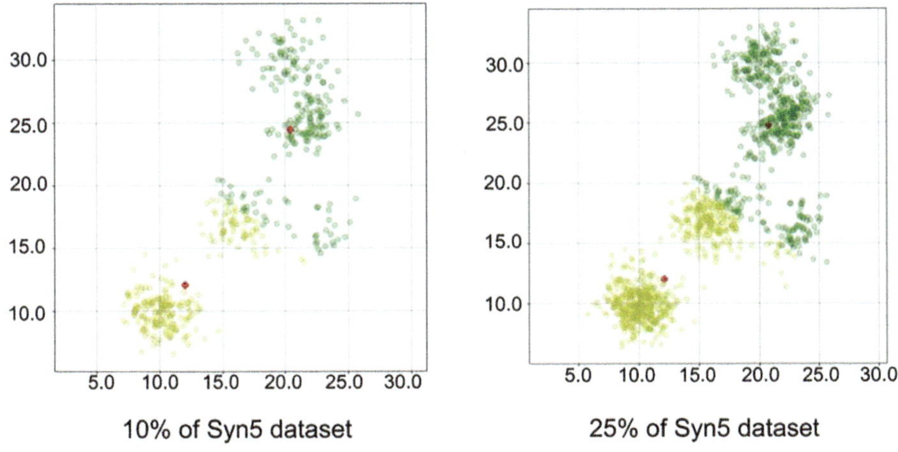

FIGURE 11.18 Initial cluster formation by the proposed algorithm on Syn5.

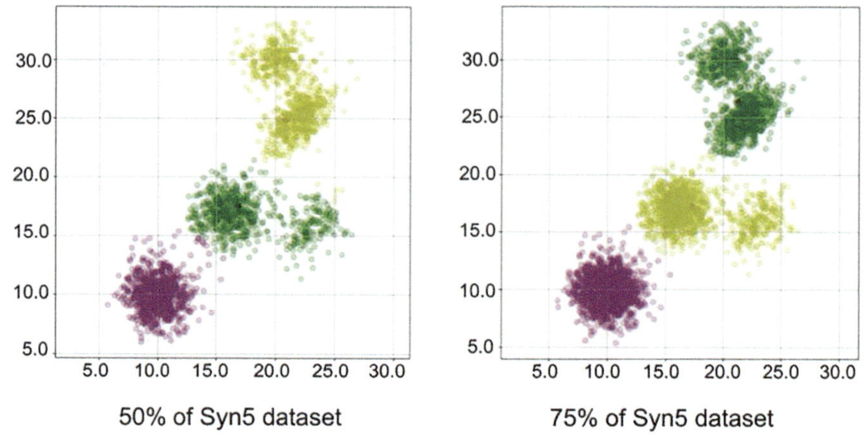

FIGURE 11.19 Cluster adaptation by the proposed algorithm on Syn5.

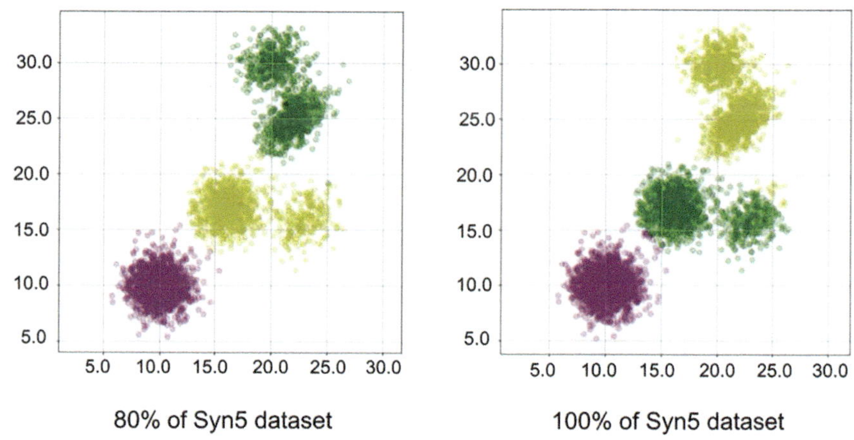

FIGURE 11.20 Final cluster formation by the proposed algorithm on Syn5.

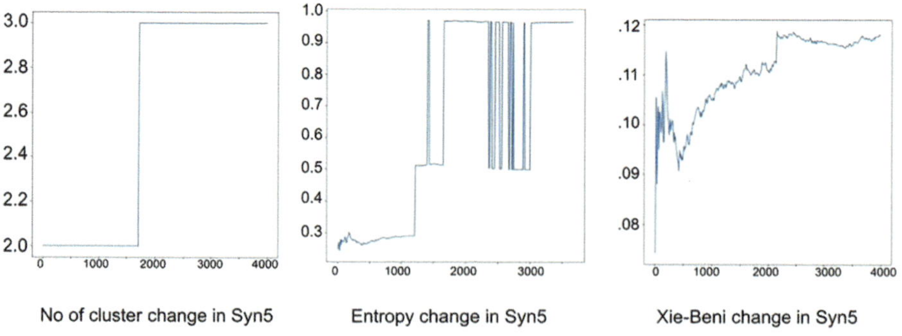

FIGURE 11.21 Change of parameters and performance by the proposed algorithm on Syn5.

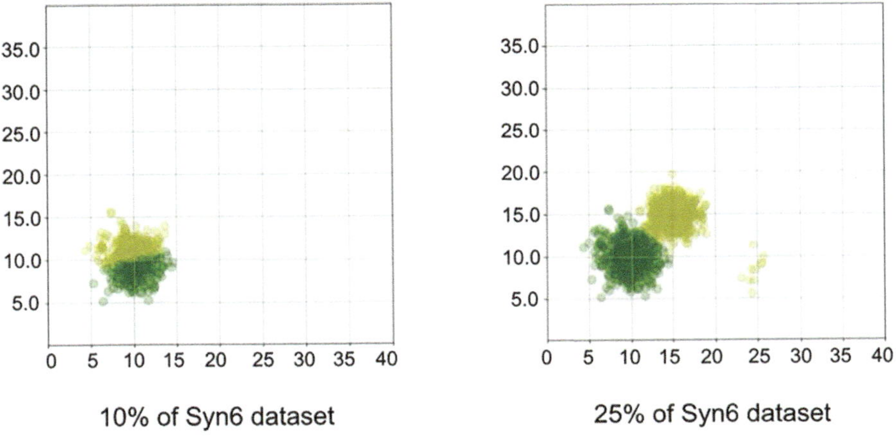

FIGURE 11.22 Initial cluster formation by the proposed algorithm on Syn6.

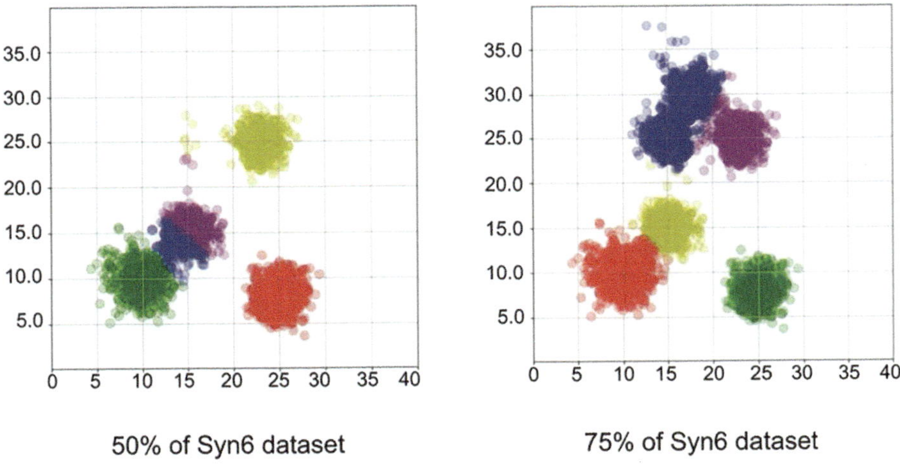

FIGURE 11.23 Cluster adaptation by the proposed algorithm on Syn6.

the proposed algorithm and other state-of-the-art algorithms are almost in the same range, which signifies the competitiveness of the algorithm. The advantage the proposed algorithm provides is the explainability of the algorithm which makes it more real-life application-oriented (Table 11.5).

The proposed algorithm detects the optimal number of clusters as four like another state-of-the-art algorithms. The clustering quality of our proposed algorithm is compared with the method and it is competitive.

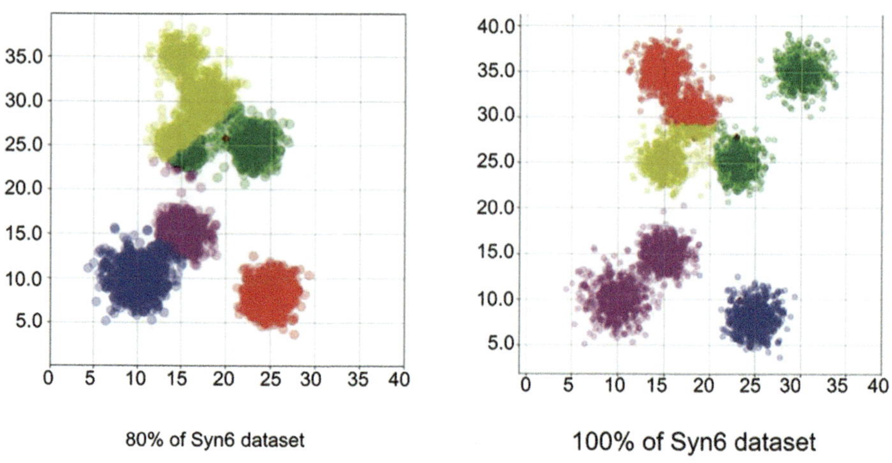

FIGURE 11.24 Final cluster formation by the proposed algorithm on Syn6.

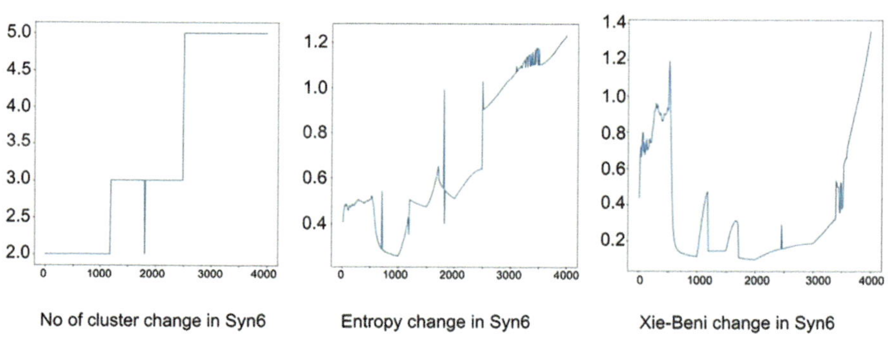

FIGURE 11.25 Change of parameters and performance by the proposed algorithm on Syn6.

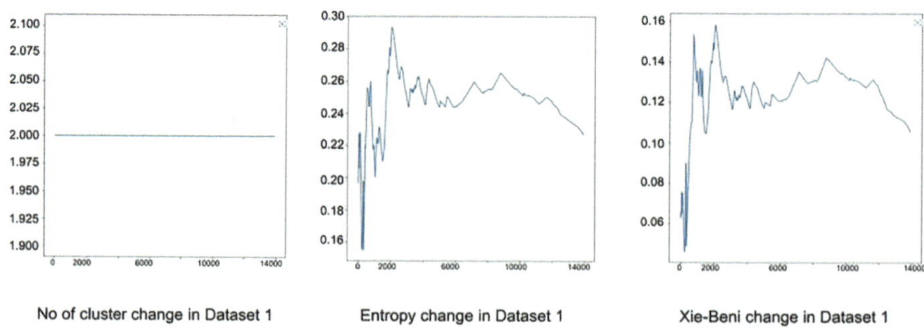

FIGURE 11.26 Change of parameters and performance by the proposed algorithm on Gas Sensor Dataset.

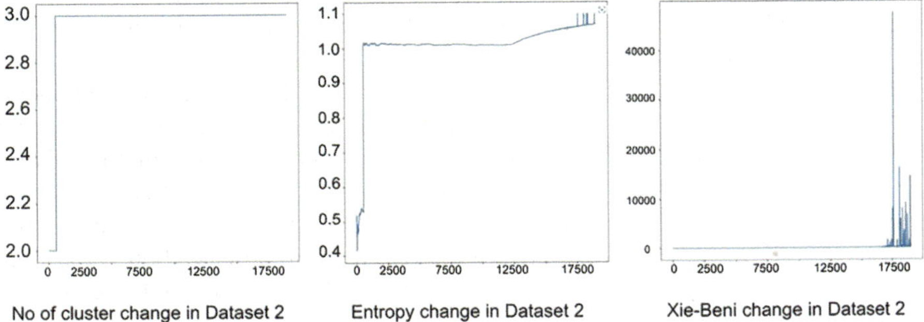

FIGURE 11.27 Change of parameters and performance by the proposed algorithm on GPS tracking of Smart Device Dataset.

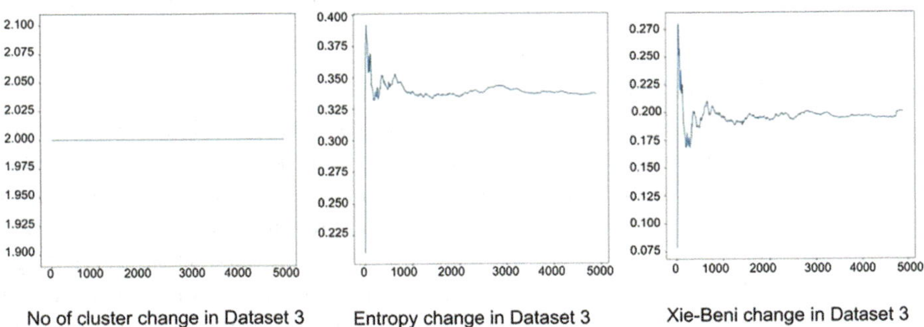

FIGURE 11.28 Change of parameters and performance by the proposed algorithm on Wine Dataset.

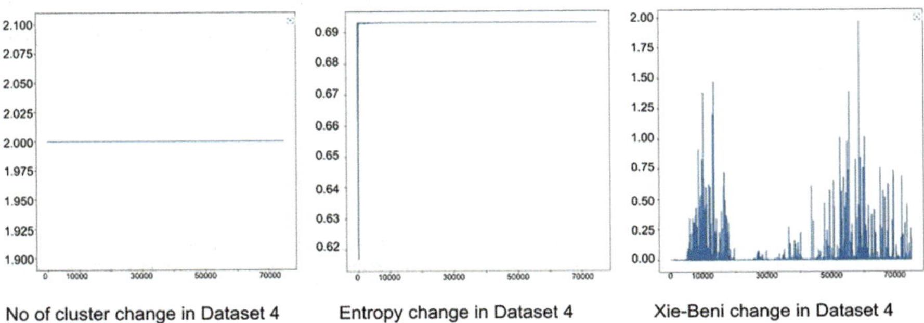

FIGURE 11.29 Change of parameters and performance by the proposed algorithm on Key-stroke Dataset.

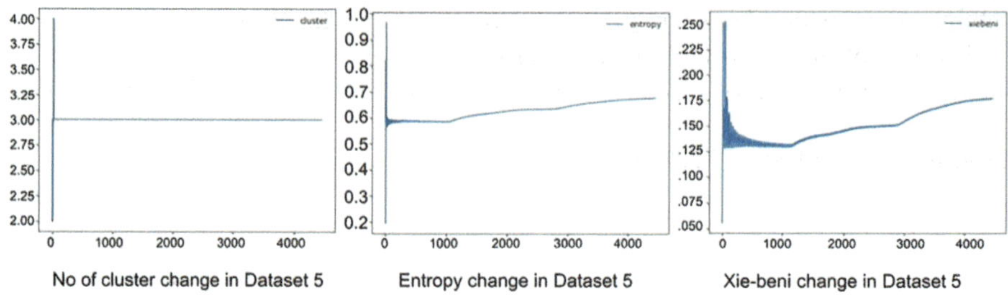

FIGURE 11.30 Change of parameters and performance by the proposed algorithm on Electricity Dataset.

TABLE 11.4
Comparison of the Proposed Algorithm with Other S.O.T.A. Algorithms on NSL-KDD Dataset

	Proposed Method	OFCM	SPFCM	eGKPCM	GPCMs
Number of clusters	4	4	4	4	4
Xie-Beni index value	2.02	2.2	2.3	2.5	0.6
Partition coefficient	0.76	0.78	0.74	1.00	0.84
Partition entropy	0.11	0.4	0.41	0.05	0.17

TABLE 11.5
Comparison of the Proposed Algorithm with Evolving Cauchy Algorithm on NSL-KDD Dataset

	Proposed Method	Evolving Cauchy Algorithm (0.0037)	Evolving Cauchy Algorithm (0.0039)
Number of clusters	4	5	3
Silhouette coefficient	−0.037	0.44	0.20

11.5 CONCLUSION AND FUTURE WORK

The manuscript presents an adaptive algorithm for fuzzy clustering of data streams that effectively detects the number of clusters, concept drift, and concept evolution and adjusts fuzzy membership accordingly. The proposed algorithm uses entropy as a measure of uncertainty to identify changes in the data distribution and dynamically updates the cluster centres to reflect the new data characteristics. The results of the experiments conducted on both synthetic and real-world datasets show that the proposed algorithm outperforms existing fuzzy clustering algorithms in terms of clustering accuracy and adaptivity to changing data streams. Overall, this work contributes to the development of robust and efficient clustering algorithms for real-time analysis of data streams in various application domains such as finance, healthcare, and social media analysis. Further research could focus on extending the proposed algorithm to handle high-dimensional and heterogeneous data streams and exploring its scalability to large-scale datasets.

ACKNOWLEDGEMENT(S)

This work was supported by the Technology Innovation Hub on Data Science, Big Data Analytics, and Data Curation under Grant NMICPS/006/MD/2020–21 dated 16.10.2020 of DST, India.

REFERENCES

Aggarwal, C. C. (2007). *Data Streams: Models and Algorithms* (Vol. 31). Springer Science & Business Media, Berlin.

Al-Khamees, H., Al-A'araji, N., & Al-Shamery, E. (2021). Data stream clustering using fuzzy-based evolving cauchy algorithm. *Int. J. Intell. Eng. Syst.*, 14 (5), 348–358.

Angelov, P., & Filev, D. (2005). Simpl_eTS: a simplified method for learning evolving Takagi-Sugeno fuzzy models. In *The 14th IEEE International Conference on Fuzzy Systems, FUZZ'05, Reno, NV, USA* (pp. 1068–1073). doi: 10.1109/FUZZY.2005.1452543

Barbara, D. (2002). Requirements for clustering data streams. *SIGKDD Explorations*, 3, 23–27.

Bezdek, J. C. (1981). *Pattern Recognition with Fuzzy Objective Function Algorithms*. Advanced Applications in Pattern Recognition. doi: 10.1007/978-1-4757-0450-1.

Bezerra, C. G., Costa, B. S. J., Guedes, L. A., & Angelov, P. P. (2016). A new evolving clustering algorithm for online data streams. In *2016 IEEE Conference on Evolving and Adaptive Intelligent Systems (EAIS), Natal, Brazil* (pp. 162–168). doi: 10.1109/EAIS.2016.7502508

Bifet, A. (2010). *Adaptive Stream Mining: Pattern Learning and Mining from Evolving Data Streams* (Vol. 207). IOS Press, Amsterdam.

Bodyanskiy, Y., Tyshchenko, O., & Kopaliani, D. (2015). A hybrid cascade neural network with an optimized pool in each cascade. *Soft Computing*, 19(12), 3445–3454.

Bodyanskiy, Y. V., Tyshchenko, O. K., & Kopaliani, D. S. (2017). An evolving connectionist system for data stream fuzzy clustering and its online learning. *Neurocomputing*, 262, 41–56.

Deng, D., & Kasabov, N. (2000). ESOM: an algorithm to evolve self-organizing maps from online data streams. In *Proceedings of the IEEE-INNS-ENNS International Joint Conference on Neural Networks, IJCNN 2000*. Neural computing: New challenges and perspectives for the New Millennium, Como, Italy (Vol. 6, pp. 3–8). doi: 10.1109/IJCNN.2000.859364

Dovžan, D., & Škrjanc, I. (2011). Recursive fuzzy c-means clustering for recursive fuzzy identification of time-varying processes. *ISA Transactions*, 50(2), 159–169.

Dua, D., & Graff, C. (2017). UCI machine learning repository. Retrieved from https://archive.ics.uci.edu/ml.

Fahlman, S., & Lebiere, C. (1989). The cascade-correlation learning architecture. In D. Touretzky (Ed.), *Advances in Neural Information Processing Systems* (Vol. 2). Morgan-Kaufmann.

Gama, J. (2010). *Knowledge Discovery from Data Streams*. CRC Press, Boca Raton, FL.

Gao, T., Li, A., & Meng, F. (2017). Research on data stream clustering based on fcm algorithm1. *Procedia Computer Science*, 122, 595–602.

Hartigan, J. A. (1975). *Clustering Algorithms*. John Wiley & Sons, Inc., Hoboken, NJ.

Höppner, F., Klawonn, F., Kruse, R., & Runkler, T. (1999). *Fuzzy Cluster Analysis: Methods for Classification, Data Analysis and Image Recognition*. John Wiley & Sons, Hoboken, NJ.

Huang, M., Xia, Z., Wang, H., Zeng, Q., & Wang, Q. (2012). The range of the value for the fuzzifier of the fuzzy c-means algorithm. *Pattern Recognition Letters*, 33 (16), 2280–2284.

Kasabov, N. (2001). Evolving fuzzy neural networks for supervised/ unsupervised online knowledge-based learning. *IEEE Transactions on Systems, Man, and Cybernetics, Part B (Cybernetics)*, 31 (6), 902–918.

Kasabov, N. (2007). *Evolving Connectionist Systems: The Knowledge Engineering Approach*. Springer, London.

Kasabov, N. K., & Song, Q. (2002). Denfis: dynamic evolving neural-fuzzy inference system and its application for time-series prediction. *IEEE Transactions on Fuzzy Systems*, 10 (2), 144–154.

Kohonen, T. (1990). The self-organizing map. *Proceedings of the IEEE*, 78 (9), 1464–1480.

Krishnapuram, R., & Keller, J. M. (1993). A possibilistic approach to clustering. *IEEE Transactions on Fuzzy Systems*, 1 (2), 98–110.

Lughofer, E. D. (2008). Flexfis: A robust incremental learning approach for evolving takagisugeno fuzzy models. *IEEE Transactions on Fuzzy Systems*, 16 (6), 1393–1410.

Maciel, L., Ballini, R., & Gomide, F. (2017). Evolving possibilistic fuzzy modelling. *Journal of Statistical Computation and Simulation*, 87 (7), 1446–1466.

Ravi, V., Srinivas, E., & Kasabov, N. (2007). On-Line Evolving Fuzzy Clustering. In *International Conference on Computational Intelligence and Multimedia Applications (ICCIMA 2007)*, Sivakasi, India (Vol. 1, pp. 347–351) doi: 10.1109/ICCIMA.2007.111.

Rodrigues, P. P., & Gama, J. (2007). Semi-fuzzy Splitting in Online Divisive -Agglomerative Clustering. In: Neves, J., Santos, M.F., Machado, J.M. (eds) *Progress in Artificial Intelligence*. EPIA 2007. Lecture Notes in Computer Science, vol 4874. Springer, Berlin, Heidelberg. https://doi.org/10.1007/978-3-540-77002-2_12

Sangma, J. W., Rani, Y., Pal, V., Kumar, N., & Kushwaha, R. (2022). FHC-NDS: Fuzzy hierarchical clustering of multiple nominal data streams. *IEEE Transactions on Fuzzy Systems*, vol. 31, no. 3, pp. 786–798, March 2023, doi: 10.1109/TFUZZ.2022.3189083.

Škrjanc, I., & Dovžan, D. (2015). Evolving gustafson-kessel possibilistic c-means clustering. *Procedia Computer Science*, 53, 191–198.

Torra, V. (2015/06). On the selection of m for fuzzy c-means. In Proceedings of the 2015 conference of the international fuzzy systems association and the European society for fuzzy logic and technology (p. 1571-1577). Atlantis Press. Retrieved from https://doi.org/10.2991/ifsa-eusflat-15.2015.224.

Tsao, E. C.-K., Bezdek, J. C., & Pal, N. R. (1994). Fuzzy kohonen clustering networks. *Pattern Recognition*, 27 (5), 757–764.

Xie, X. L., & Beni, G. (1991). A validity measure for fuzzy clustering. *IEEE Transactions on Pattern Analysis & Machine Intelligence*, 13 (8), 841–847.

Zhang, B., Qin, S., Wang, W., Wang, D., & Xue, L. (2016). Data stream clustering based on fuzzy c-mean algorithm and entropy theory. *Signal Processing*, 126, 111–116.

12 Demystifying the Black Box

Unveiling the Decision-Making Process of AI Systems

Anuhya Bhagavatula, Shrusti Ghela, and B. K. Tripathy

12.1 INTRODUCTION

As artificial intelligence (AI) systems are increasingly being used in critical applications, such as healthcare and finance, it is becoming increasingly important to ensure their transparency, trustworthiness, and accessibility. One critical aspect of this is explaining the decisions made by AI systems in a way that is understandable to the end-users. The explanations can help end-users understand how the AI systems arrived at their decisions and can increase their trust in these systems.

Explainable AI (XAI) is a field of study that focuses on developing methods and techniques for generating explanations for AI systems. There are several ways to explain the decisions of AI systems, the most prominent of which are model interpretability, Local Interpretable Model-agnostic Explanations (LIME), SHapley Additive exPlanations (SHAP), as well as many others. These approaches can be used for both global and local explanations and can have different meanings for end-users depending on their perspective.

It is important to carefully choose the appropriate method to ensure that the explanation is accurate, relevant, and understandable to the end-users. The choice of method to explain an AI system's decision depends on several factors such as the type of model used, the end-users, and the context of the decision. Moreover, different end-users may have different requirements for the explanations they receive. For instance, a medical expert may require more detailed explanations than a layperson when interpreting the results of an AI model. Therefore, it is important to tailor the explanations to the end-users and their level of expertise.

This chapter makes a case for why explaining AI decisions is important, provides an overview of common XAI techniques, and discusses how they can be used to generate explanations for AI systems. The chapter also addresses the different end-users or target audiences for XAI decisions and provides references to recent research. In

DOI: 10.1201/9781003442509-12

addition, it covers the limitations of some of the XAI techniques and the factors that determine when to use these decisions.

This chapter aims to provide a comprehensive introduction to the importance of explaining decisions made by AI systems and the different methods available for generating explanations. It highlights the need to tailor the explanations to the end-users and their level of expertise and guides when to use different XAI techniques. By increasing the transparency, trustworthiness, and accessibility of AI systems, we can ensure that these systems are used responsibly and for the benefit of society.

12.2 THE CASE FOR EXPLAINING AI DECISIONS

12.2.1 Significance of Transparent AI Decision-Making

The interpretability of decisions made by AI can be desired due to myriad reasons. Some of these are listed below:

1. **Understandability [1]:** This is referred to as intelligibility [2], and interpretability [3] by some. The broad definition is the ability of humans to understand a system or a model and the function it performs without the need for a detailed explanation of the internal algorithms/structure by which it processes the data.
2. **Transparency [1]:** Providing transparency in the decision-making process is important for ensuring accountability and preventing biases. Transparency helps to identify potential sources of errors or biases, enabling corrective actions to be taken.
3. **Trustworthiness [1]:** Explaining AI decisions increases the trustworthiness of the system by demonstrating that it is making decisions based on clear and well-defined criteria. This increases user confidence in the system and promotes adoption.
4. **Comprehensibility [4]:** Comprehensibility is a way to help users understand the reasons behind their decisions. This improves the user experience and makes it easier to interact with the system.
5. **Controllability [1]:** This is the provision for users to have greater control over the decision-making process, allowing them to adjust the criteria used by the system. This helps to ensure that the system is aligned with user preferences and goals.
6. **Fairness [1]:** Explaining decisions can help ensure that the system is fair by identifying and mitigating biases. This promotes equity and prevents discrimination.
7. **Causality [5]:** Causality enables users to understand the underlying reasons for the decision in terms of finding what variables are related to each other. This helps build trust in the system and promotes confidence in the decision-making process.
8. **Transferability [5]:** Explainability can help ensure that the system is transferable to different contexts by making it clear how the decision-making

process works. This promotes scalability and enables the system to be used in a wider range of applications.

9. **Informativeness [5]:** Explainable ML models provide users with additional information that can be used to improve decision-making and avoid falling into common misconception pitfalls. This can help to identify patterns or trends that would be difficult to discern otherwise.
10. **Confidence [5]:** Explanations can help build confidence in the system by demonstrating that the decisions are based on sound reasoning and data. This promotes trust and adoption.
11. **Accessibility [5]:** Accessibility is a way to ensure that explanations are accessible to all users, including those who are non-experts on the functioning of the system or do not have a technical background when facing algorithms that seem incomprehensible at first sight.
12. **Interactivity [5]:** Interactively explaining AI decisions enables users to engage with the system and provide feedback, promoting a more collaborative decision-making process.
13. **Privacy Awareness [5]:** Privacy-awareness helps protect user privacy by minimizing the amount of personal data that is collected and used by the system.

12.2.2 Decoding the Essence of a "Good Explanation"

There exists no formal definition for a "good explanation" of AI decisions currently. It is currently an open research question that has different criteria in different fields. In reference [6], the authors explore the current state of research on what makes a good explanation for AI systems by exploring the criteria for good explanations in various fields, including cognitive science, computer science, psychology, and philosophy. Miller [7] highlights three key properties of useful explanations in social sciences: *counterfactual* wherein humans understand the cause of certain events better than other events, selective wherein the explanations are focused on a subset of reasons for a decision made by a system, and social wherein the explanation of decisions is interactive focusing on transferring knowledge so as to not overwhelm humans. On the other hand, in the field of computer science, an explanation focuses on the technical workings of the system – what reasoning was applied by the system that led to a particular decision [8]. When there are numerous definitions for a "good explanation", it is hard to pin down an AI system to give an explanation that caters to this vast ambiguity. Thus, it is important to understand the end-users and circumstances in which the decisions are being made and whether the explanation satisfies the definition of a "good explanation" for that circumstance.

12.2.3 Identifying the Intended Audience of AI Decisions

Practical Implementation of Explainability in AI is a major barrier. The inability to explain or fully understand the reasons why state-of-the-art machine learning algorithms work as efficiently as they do is a significant problem that finds its roots in two different causes. The first cause is the gap between the research community and business sectors, particularly in strictly regulated sectors, such as banking, finance, security, and health, among many others. These sectors have traditionally lagged behind in the digital transformation of their processes and have some reluctance to implement techniques that may put their assets at risk. The second axis is that of knowledge. AI has helped researchers across the world with the task of inferring relations that were far beyond human cognitive reach. However, we are entering an era in which results and performance metrics are the only interests shown up in research studies. Although for certain disciplines, this might be the fair case, science and society are far from being concerned just by performance. The search for understanding is what opens the door for further model improvement and its practical utility. Listed below are the different target audiences and the purposes for which XAI is important to them.

Data scientists: Data scientists are among the primary users of XAI systems, as they are responsible for developing and implementing machine learning models. XAI provides data scientists with insights into the decision-making process of the model, helping them improve the accuracy and reliability of their algorithms. Research by Kalia et al. (2021) demonstrated how XAI can help data scientists identify errors and biases in the model's decision-making process, enabling them to refine and optimize the algorithm [9].

Business stakeholders: Business stakeholders include executives and managers. XAI can help these users understand how the model is making decisions and identify areas where the algorithm can be optimized to improve business outcomes. Research by Chang et al. (2020) demonstrated how XAI can help business stakeholders develop more effective marketing strategies by providing insights into the decision-making process of the model [10].

End-users: End-users, including consumers and patients, are an important target audience for XAI decisions. These users rely on the decisions made by machine learning models, and XAI can help them understand how these decisions are reached. Research by Topol (2019) showed how XAI can help patients make more informed decisions about their healthcare by providing them with insights into the decision-making process of the algorithm [11].

Regulatory agencies: Regulatory agencies and policymakers are responsible for ensuring that AI models comply with legal and ethical standards. XAI can help these agencies monitor the performance of machine learning models and identify potential issues related to fairness, accountability, and transparency. Research by Rudin et al. (2019) demonstrated how XAI can help regulatory agencies assess the accuracy and fairness of machine learning models [12].

Legal professionals: Finally, legal professionals, including lawyers and judges, may need to make decisions based on the outputs of machine learning models, and XAI can help them understand how these outputs are generated. Research by Huang et al. (2020) showed how XAI can help legal professionals identify biases and errors in the decision-making process of machine learning models, enabling them to make more informed and equitable decisions [13] (Figure 12.1).

FIGURE 12.1 Goals and target audience of XAI. (Image by the author.)

12.2.4 Determining the Appropriate Context for Utilizing AI Decisions

The decision to use XAI may depend on several factors, including the complexity of the model, the impact of its decisions, and the level of trust required by its users. Local approaches are suitable for explaining specific decisions or predictions, while global approaches are suitable for explaining the overall behavior and performance of the model.

Local approach: In a local approach, XAI is used to explain the decision-making process of a model for a specific instance or observation. This approach is suitable when a user needs to understand how the model made a particular decision or prediction. For example, in medical diagnosis, XAI can be used to explain why a patient was diagnosed with a particular disease based on their symptoms and medical history. It can help doctors understand how an AI model arrived at the diagnosis and identify potential errors or biases in the decision-making process [14]. This approach can also be useful in scenarios where the user needs to take action based on the model's decision. For instance, in autonomous vehicles, XAI can be used to explain why the vehicle made a particular decision, such as braking or changing lanes, allowing the user to take control of the vehicle if necessary [15].

Global approach: In a global approach, XAI is used to explain the decision-making process of a model across all instances or observations. This approach is suitable when the user needs to understand the model's overall behavior and performance. For example, in credit scoring, XAI can be used to explain how the model determines creditworthiness based on various features, such as income, age, and credit history [16]. XAI can help users identify potential biases and discrimination in the model's decision-making process and improve the model's overall fairness and transparency. This approach can also be useful in scenarios where the user needs to compare different models and select the most suitable one.

12.3 COMMON METHODS TO EXPLAIN THE DECISIONS OF SYSTEMS

12.3.1 Model Interpretability

Feature importance, partial dependence plots, and decision trees are widely used Model Interpretability tec hniques in XAI to better understand the decision-making process of machine learning models. In this section, these techniques are explained in detail with examples of how they have been applied in real-world scenarios.

Feature importance is a method to determine which input features are most important in influencing the output of a machine learning model. This technique can help identify the key factors that contribute to the model's predictions and also assist in identifying potential biases in the model. There are various methods to calculate feature importance, such as permutation importance, which involves randomly shuffling a feature and measuring the decrease in the model's accuracy [17]. Feature importance has been used in various applications such as healthcare, finance, and natural language processing (NLP).

Partial dependence plots (PDPs) are another technique in XAI that helps understand the relationship between the input features and model predictions [18]. PDPs visualize the relationship between a model's output and a single input feature while keeping all other features constant. By plotting the partial dependence of the output on each input feature, researchers can gain insights into how each feature influences the model's predictions. PDPs can also help identify interaction effects between input features, which can lead to better model interpretability. This technique has been used in various applications, such as predicting the price of real estate, diagnosing diseases, and analyzing social media data.

Decision trees are a type of model that uses a tree-like structure to represent a series of decisions and their consequences [19]. Each node in the tree represents a decision based on an input feature, and each branch represents the possible outcomes of that decision. Decision trees can be used to visualize the decision-making process of a model, which can help identify important input features and interactions between them. Decision trees have been used in various applications, such as diagnosing diseases, predicting stock prices, and detecting fraudulent transactions.

Feature importance, PDPs, and decision trees are often used together to provide a comprehensive understanding of the model's decision-making process. For example, researchers can first use feature importance to identify the most important input features, then use PDPs to visualize the relationship between those features and the model's output, and finally use decision trees to identify the most important decisions made by the model. This combined approach has been used in various applications, such as predicting customer churn in telecommunications and diagnosing breast cancer.

These XAI techniques have been applied in various domains such as healthcare, finance, and social sciences. For example, feature importance has been used to identify the most important risk factors for cardiovascular disease, cancer, and other diseases. PDPs have been used to study the relationship between weather conditions and traffic accidents and to analyze the impact of environmental factors on crop yields. Decision trees have been used to explain the credit scoring process in finance, the diagnosis process in healthcare, and the voting behavior in social sciences.

In healthcare, XAI techniques such as feature importance, PDPs, and decision trees have been used to diagnose diseases and identify risk factors. In a study on diagnosing heart disease using electronic health records, researchers used feature importance to identify the most important risk factors, such as age and blood pressure. They then used decision trees to visualize the decision-making process of the model and identify the most important decisions made by the model. In another study, researchers used PDPs to analyze the relationship between input features and the risk of stroke.

In finance, XAI techniques such as feature importance, PDPs, and decision trees have been used to predict stock prices and identify credit risk. For example, in a study on predicting stock prices, researchers used feature importance to identify the most important features, such as past stock prices and economic indicators. They then used decision trees to visualize the decision-making process of the model and identify the most important decisions made by the model. In another study, researchers used PDPs to analyze the relationship between credit score and default risk.

Another important aspect of XAI is the ability to identify when a model is making errors. This is particularly important in high-stakes applications such as healthcare, where incorrect predictions can have serious consequences. One approach to identifying errors is to use outlier detection techniques such as Local Outlier Factor (LOF) [20] and Isolation Forest. In a study by Zhang et al., they used LOF to identify abnormal data points in a medical dataset and found that these points were often associated with misclassified cases [21].

When it comes to visualizing the decision-making process of a model, researchers can use various tools such as decision trees, heatmaps, and scatter plots. In a study by Craven et al., they used a heatmap to visualize the contribution of different input features to a deep neural network that was trained to predict speech emotion. The results showed that the model relied heavily on features related to pitch, loudness, and spectral shape.

Although these XAI techniques provide valuable insights into the model behavior, they also have some limitations. Feature importance and PDPs assume that the input features are independent, which might not be the case in practice. Decision trees can become very complex for high-dimensional feature spaces, which can make them difficult to interpret. Moreover, these XAI techniques only provide a local explanation for a particular instance, and they do not necessarily generalize to other instances or the model as a whole.

To overcome these limitations, researchers have proposed various extensions and modifications to these XAI techniques. For example, SHAP values can provide a global explanation of the model behavior by computing the average contribution of each feature across all possible feature combinations. Integrated gradients can provide a more accurate measure of feature importance by taking into account the nonlinearity of the model [22]. Anchors can provide more interpretable explanations by identifying the necessary and sufficient conditions for a certain prediction to hold.

12.3.2 Local Interpretable Model-Agnostic Explanations (LIME)

LIME is a popular interpretability technique that provides explanations for individual predictions made by machine learning models. LIME was introduced in the paper "Why Should I Trust You?": Explaining the Predictions of Any Classifier by Marco Tulio Ribeiro, Sameer Singh, and Carlos Guestrin in 2016 [23].

The basic idea behind LIME is to explain a model's prediction for a particular instance by approximating the model with an interpretable model in the vicinity of the instance. Here is the detailed working of LIME:

Sample instances: LIME generates a set of perturbations of the instance, where each perturbation is a modified version of the original instance that differs only slightly from it. These perturbations are generated randomly or by using an algorithm that ensures the perturbations are meaningful.

Weigh instances: For each perturbation, LIME generates a weight that reflects the degree to which the perturbation influenced the prediction. This weight is calculated based on the distance of the perturbation from the original instance in feature space and a kernel function that assigns more weight to nearby perturbations.

Train interpretable model: LIME then uses these weights to fit an interpretable model that approximates the original model's behavior in the vicinity of the instance. The interpretable model is typically a simple model, such as linear regression, decision trees, or Lasso regression.

Explain prediction: This interpretable model can then be used to generate explanations for the original model's prediction by examining the coefficients of the linear model. For example, the coefficients of the interpretable model are used to identify which features of the instance were most important in determining the prediction (Figure 12.2).

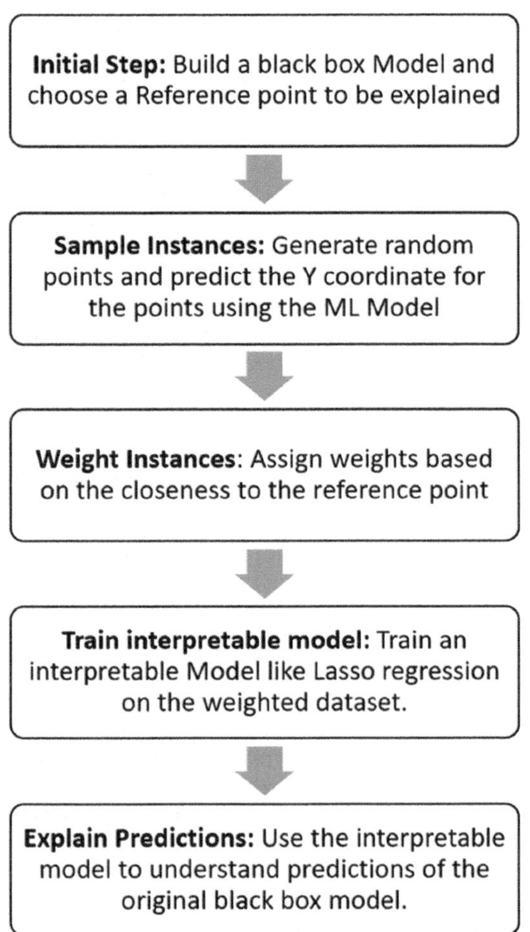

FIGURE 12.2 LIME algorithm step-by-step. (Image by the author.)

LIME has a wide range of applications in XAI, and its ability to provide local, interpretable explanations makes it a valuable tool for understanding the decisions of complex machine learning models. Some of the applications of LIME in XAI include:

1. **Image classification:** An example is using LIME to explain the reasons behind high-level driving behaviors in autonomous vehicles [24].
2. **Text classification:** Using LIME to explain the predictions of neural models for text classification tasks [25].
3. **Fraud detection:** LIME can be used to explain the decisions of fraud detection models and to identify the features that contributed the most to the predictions [26].
4. **Medical diagnosis:** An example would be using LIME to explain the predictions of a machine learning model for predicting the likelihood of heart disease [27].

5. **Recommender systems:** LIME can be used to explain the recommendations made by recommender systems. For example, recommendations made by a movie recommender system [28].

LIME has also been extended to handle more complex models, such as deep neural networks, and to incorporate domain-specific knowledge, such as constraints on feature values [29].

Although LIME is considered an effective technique for explaining the decisions of black-box models, it has some limitations [30,31]:

1. It is unable to capture complex interactions between features in the data. This is because LIME assumes that the relationship between the features and the model's output is linear or locally linear, which may not always be the case. Therefore, the explanations provided by LIME may not be accurate or complete.
2. It may not be able to handle high-dimensional data or data with a large number of features. This is because LIME creates local linear models for each feature, which can be computationally expensive and may not be practical for large datasets. In such cases, other techniques such as SHAP may be more suitable.
3. LIME explanations may not be consistent across different perturbation samples, which can lead to instability in the explanations. This is because LIME selects a random subset of the features for each explanation, which can lead to different explanations for the same instance. This can be problematic in applications where consistent explanations are required.
4. LIME may not be able to handle data with class imbalance, as it may create local models that are biased towards the majority class. This can result in explanations that do not accurately capture the behavior of the model for minority classes.
5. LIME explanations may not be intuitive or understandable for non-experts, as they rely on the concept of linear models and may not be easily interpretable. Therefore, it is important to provide appropriate visualizations and explanations to ensure that the results are understandable and usable for the intended audience.

There are also variants of LIME that have been proposed to address some of its limitations. One variant is Anchors, which is a rule-based approach that generates explanations in the form of if-then statements. Anchors can generate more concise and interpretable explanations than LIME, but may not work well for complex or nonlinear models [32]. Another variant is LIMEtree, which uses decision trees as the interpretable model instead of linear models. LIMEtree can better capture nonlinearities in the data and may be more robust to outliers [33]. LIME also may not be able to provide explanations for non-numeric data, such as text or images. In such cases, other techniques such as LIME-Text or LIME-Image may be more appropriate.

Overall, while LIME is considered a useful technique for XAI, it is important to consider its limitations and potential sources of error when applying it to different datasets and models.

12.3.3 SHapley Additive exPlanations (SHAP)

SHAP is a popular model-agnostic technique that provides a way to attribute the contribution of each feature in a prediction to its final outcome by assigning each feature an importance value based on the Shapley values from cooperative game theory [34]. Here's how SHAP functions in more detail:

Generate sample instances: SHAP generates a set of perturbations of the instance, where each perturbation is a modified version of the original instance that differs only slightly from it.

Compute shapley values: For each perturbation, SHAP computes the Shapley value, which is a measure of the contribution of a feature to the prediction. The Shapley value is calculated based on the idea of "fairness" in game theory and assigns a value to each feature based on its contribution to the prediction while taking into account the interactions between features.

Compute feature importance: SHAP uses the Shapley values to compute feature importance, which is a measure of the importance of each feature in the prediction. Feature importance is calculated as the mean absolute Shapley value across all instances.

Explain prediction: The feature importance scores are then used to generate explanations for the original model's prediction. The feature importance scores identify which features of the instance were most important in determining the prediction (Figure 12.3).

SHAP has been successfully applied to a wide range of machine-learning applications, including image classification, text classification, and time-series forecasting. In image classification, SHAP can identify which pixels are most important in determining the classification. In text classification, SHAP can identify which words or phrases are most important in determining the classification.

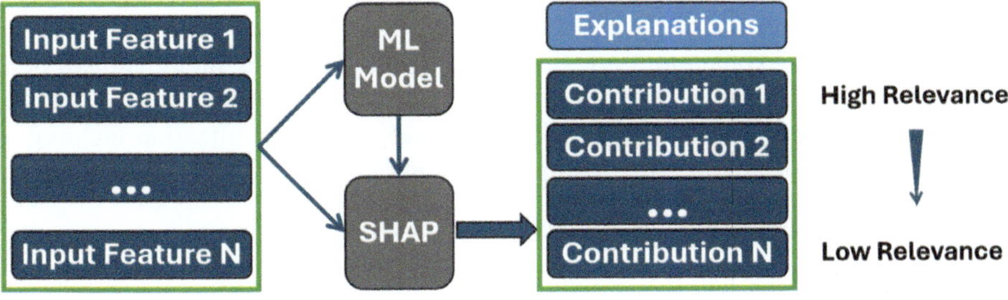

FIGURE 12.3 Overview of SHAP. (Image by the author.)

One of the strengths of SHAP is that it provides not only feature importance scores but also individual instance explanations. This allows users to understand how a model's prediction was made for a specific instance. SHAP also provides a way to explain complex interactions between features, which is often a challenge with other model-agnostic explanation techniques.

Some of the limitations of SHAP are discussed below [30,33].

1. Computational complexity. SHAP involves calculating the Shapley values, which is a computationally expensive process. For large datasets or models with a large number of features, the computation time required to calculate the Shapley values can be significant, making it impractical for some applications.
2. Access to internal structure. One limitation of SHAP is that it requires access to the model's internal structure, which may not always be available. In some cases, such as with proprietary or pre-trained models, the internal structure of the model may not be accessible, limiting the applicability of SHAP.
3. Assumption of additive relationships. SHAP assumes that the relationships between the input features and the output of the model are additive. However, in some cases, the relationships between the features may be more complex, and SHAP may not be able to capture these complex interactions.
4. Sensitivity to baseline choice. SHAP can be sensitive to the choice of the baseline used for calculating the Shapley values. The baseline represents the reference point from which the Shapley values are calculated, and different baselines can lead to different explanations. Therefore, the choice of baseline can significantly affect the interpretation of the results.

Despite its great significance, it is essential to consider these limitations and potential sources of error when applying SHAP to different datasets and models.

12.3.4 Counterfactual Explanations

This method provides an explanation by showing how the outcome would have changed if a particular input feature had a different value [35]. In other words, counterfactual explanations provide a way to explore what-if scenarios and understand the factors that influence a model's predictions.

Counterfactual explanations are generated by modifying the input features of an instance to produce a new instance that is similar to the original instance but with a different prediction. The modifications are made using optimization techniques, such as gradient descent, to find the minimum set of changes needed to produce the desired outcome.

Counterfactual explanations have a wide range of applications in XAI, including:

1. **Debugging models:** Counterfactual explanations can be used to identify and diagnose errors in machine learning models by identifying the features that are responsible for the model's prediction.

2. **Model improvement:** Counterfactual explanations can be used to improve machine learning models by identifying which features are most important for the model's prediction.
3. **Fairness and bias:** Counterfactual explanations can be used to identify and address issues of fairness and bias in machine learning models by exploring what-if scenarios and identifying the factors that influence the model's predictions.
4. **Personalization:** Counterfactual explanations can be used to provide personalized recommendations or treatments by exploring what-if scenarios and identifying the changes needed to achieve a desired outcome.

One example of the application of counterfactual explanations is in healthcare. In a study by Wachter et al. [36], counterfactual explanations were used to generate treatment recommendations for patients with sepsis. The counterfactual explanations were used to explore what-if scenarios, such as "what changes to the patient's condition would result in a different treatment recommendation?". The study found that counterfactual explanations could improve the accuracy of treatment recommendations and help clinicians understand the factors that influence their decisions.

Another example of the application of counterfactual explanations is in image classification. In a study by Kim et al. [37], counterfactual explanations were used to explore the factors that influence the prediction of a deep neural network for classifying images of skin lesions. The study found that counterfactual explanations could help identify and diagnose errors in the model and provide insights into the features that are important for the model's prediction.

Counterfactual explanations are a powerful tool in XAI that can be used to explore what-if scenarios, identify important features, and diagnose errors in machine learning models.

12.3.5 Rule-Based Systems

Rule-based systems (RBS) are a class of XAI techniques that use a set of rules to explain the reasoning behind a decision made by an AI model. In RBS, a set of conditions and actions are defined as rules, which can be used to explain how the model arrived at a particular decision [38]. RBS is based on the idea of creating an explicit set of rules that can be used to explain a model's decision-making process to a human user. The rules are typically represented in the form of if-then statements and can be used to explain both binary and multi-class classification problems.

Rule-based systems consist of two primary components: a knowledge base and an inference engine. The knowledge base contains a set of rules that describe relationships between variables, while the inference engine uses these rules to make inferences and decisions [39]. When new data is presented to the system, the inference engine applies the rules to the data to determine the appropriate response or action.

One of the key advantages of RBS is that it provides a clear and interpretable explanation of the model's decision-making process. This can help users to better understand the model's behavior and improve their trust in the system. Another advantage is that RBS can be used to diagnose errors and inconsistencies in the model's behavior, which can help to improve the model's overall performance [40].

For example, a rule-based system could be used to diagnose medical conditions based on patient symptoms. The knowledge base would contain a set of rules that describe the relationships between symptoms and medical conditions, while the inference engine would apply these rules to the patient's symptoms to arrive at a diagnosis.

Recent research has demonstrated the effectiveness of rule-based systems in a variety of applications. For example, one study used a rule-based system to improve the accuracy of forecasting the power output of wind turbines [41]. The system used a set of rules to adjust the forecasts based on weather conditions, turbine operating characteristics, and other factors. The study found that the rule-based system improved forecasting accuracy by up to 10% compared to other methods.

Another study used a rule-based system to diagnose faults in rotating machinery [42]. The system used a set of rules to identify the causes of faults based on sensor data and other inputs. The study found that the rule-based system was able to diagnose faults with high accuracy and that it outperformed other methods in certain scenarios.

RBS has been used in the context of explainable AI (XAI) to provide interpretable explanations of deep learning models. For example, the RBS-Net model was developed as a way to provide explanations of the decisions made by convolutional neural networks (CNNs) for image classification tasks [43]. The RBS-Net model uses a set of rules to identify the most important features of an input image and provides an explanation of how CNN arrived at its decision.

Some advantages of using RBS are that they are inherently transparent and interpretable because they operate on a set of explicit rules that can be easily understood and verified by human experts [44,45]. This makes them ideal for use in situations where it is necessary to explain the decision-making process to end-users, such as in medical diagnosis or loan approval. RBS can be easily updated or modified by adding or changing rules, which makes them more flexible than other machine learning models that require retraining to incorporate new data. RBS are also generally computationally efficient and require less memory than other machine learning models, which makes them suitable for use in real-time decision-making applications.

However, like all other methods RBS also has some limitations as listed below [45]:

1. **Limited generalization:** RBS are often designed to operate on specific domains or problem spaces and may not generalize well to new or unseen situations.
2. **Rule acquisition:** The process of acquiring and encoding rules into an RBS can be time-consuming and may require significant domain expertise.
3. **Over-reliance on expert knowledge:** RBS rely heavily on expert knowledge to design and encode rules, which may introduce biases or errors into the system.

Recent research has highlighted both the advantages and limitations of RBS in XAI. For example, a study by Hu et al. [46] demonstrated the advantages of RBS in improving the accuracy of wind power forecasting. The study found that the RBS approach outperformed other machine learning models in certain scenarios and was computationally efficient. However, the study also noted the limitations of RBS in terms of limited generalization and rule acquisition.

Similarly, a study by Liu et al. [44] demonstrated the effectiveness of RBS in diagnosing faults in rotating machinery. The study found that RBS was able to achieve high accuracy in fault diagnosis and was computationally efficient. However, the study also noted the challenges of acquiring expert knowledge and designing rules for the system.

Overall, RBS is a powerful and interpretable XAI technique that can be used to provide clear and understandable explanations of AI models. While there are some limitations to RBS, such as the need for domain experts to define the rules and the potential for rules to become too complex, it remains a valuable tool in the XAI toolkit.

12.3.6 Prototype-Based Explanations

This method that illustrates how the prediction was influenced by similar cases or prototypes in the training data. Prototype-based explanations are a class of model-agnostic techniques in Explainable Artificial Intelligence (XAI) that aim to identify and describe representative instances, or prototypes, from a given dataset that explain the behavior of a model. The prototypes can be selected based on their similarity to a given input or based on their coverage of the decision space. These prototypes are then used to generate explanations that highlight the important features or characteristics of the model's behavior. Prototype-based explanations have been used in various domains, including image classification, natural language processing, and fraud detection.

One prototype-based explanation method is "prototype and feature visualization" (PFV), which was proposed by Kim et al. [47] for image classification tasks. PFV identifies a set of representative prototypes from the training data that are similar to the input image and then generates feature maps that visualize the activated features of the prototypes in response to the input. The resulting visualization shows the important image regions and features that the model uses to make its decision.

Another prototype-based explanation method is "prototype selection and distillation" (PSD), which was proposed by Sun et al. [48] for text classification tasks. PSD identifies a set of representative prototypes from the training data that are similar to the input text and then generates a summary of the key features and characteristics of the prototypes that are most relevant to the model's decision. The resulting summary provides an explanation of the model's behavior that is both concise and interpretable.

Prototype-based explanations have also been used in fraud detection systems, where they can help identify the characteristics of fraudulent transactions. For example, Liu et al. [49] proposed a prototype-based method that identifies representative prototypes of fraudulent transactions and generates explanations that highlight the features and characteristics of fraudulent behavior.

Prototype-based explanations have the advantage of being model-agnostic, meaning they can be applied to any machine learning model regardless of its architecture or complexity. They provide more natural and intuitive explanations for users to understand the decision-making process of the model. They can handle high-dimensional data and complex decision boundaries and can also identify similarities and differences among instances, which can help identify potential biases or anomalies in the data.

However, like other XAI techniques, prototype-based explanations also have limitations. For example, they may not be able to capture the full complexity of the model's decision-making process, and the choice of prototypes may affect the resulting explanation as the quality of the explanation depends heavily on the quality and representativeness of the prototypes. The prototypes are often selected based on distance metrics, which may not accurately reflect the true similarity between instances and different metrics may lead to different prototypes and explanations. They are also not suitable for explaining model behavior in regions of the input space where no prototypes exist.

Nevertheless, prototype-based explanations are a promising area of research in XAI, and they have the potential to help users understand and trust the behavior of machine learning models in a wide range of applications.

12.3.7 Saliency Maps

Saliency maps are a type of explanation in XAI that aims to highlight the important regions of an input that contribute most to a model's prediction. Saliency maps are generated by computing the gradient of the model's output with respect to the input features and then visualizing the gradient as a heatmap over the input.

Saliency maps can be used to visualize how a model makes its predictions by highlighting the important regions of the input that contribute to the prediction. They can be used to identify and diagnose errors in machine learning models by pinpointing the features that are responsible for the model's prediction. Saliency maps can improve the interpretability of machine learning models by providing a visual explanation of how the model makes its predictions. They can also be used to detect and diagnose bias in machine learning models by identifying the features that the model relies on to make its predictions.

One example of the application of saliency maps is in object recognition. In a study by Simonyan et al. [50], saliency maps were used to visualize the features that a deep neural network used to classify images of objects. The study found that the saliency maps provided insights into the visual features that were most important for the model's predictions, such as the edges and corners of objects.

Another example of the application of saliency maps is in natural language processing. In a study by Li et al. [51], saliency maps were used to visualize the words in a sentence that were most important for a model's prediction of sentiment. The study found that the saliency maps provided a more interpretable explanation of the model's behavior than traditional feature importance methods.

Saliency maps have limitations that can affect their reliability in XAI. One limitation is the interpretation of the maps. Since saliency maps highlight the pixels or features that the model used to make a decision, it is not always clear what these features represent or how they relate to the model's decision. This can make it difficult to interpret the maps and understand the reasoning behind the model's decision. For example, in image recognition tasks, saliency maps may highlight pixels that represent low-level image features, such as edges or corners, but it may be unclear how these features relate to the semantic content of the image [52].

Another limitation of saliency maps is that they are often sensitive to the choice of model architecture and training data. Different model architectures and training datasets can result in different saliency maps, which can affect the interpretability and robustness of the XAI system. For example, a saliency map generated using a deep neural network trained on a specific dataset may not generalize well to other datasets or architectures [53].

Additionally, saliency maps can be vulnerable to adversarial attacks, where small perturbations to the input can result in significant changes to the saliency map and the model's decision. Adversarial attacks can be used to fool the model into making incorrect or malicious decisions, which can be a significant security risk in XAI systems.

Finally, saliency maps may not always provide a complete explanation of the model's decision. In some cases, the model may rely on more complex reasoning processes that are not captured by saliency maps. In these cases, additional XAI techniques, such as counterfactual explanations or decision trees, may be necessary to provide a more complete understanding of the model's decision.

Overall, saliency maps are a useful tool in XAI that can be used to visualize how machine learning models make their predictions and provide insights into the important features that the model relies on.

12.3.8 Layer-Wise Relevance Propagation

This method assigns a relevance score to each feature by propagating the prediction backward through the network layers. Layer-wise Relevance Propagation (LRP) is an approach for understanding the decision-making process of Deep Neural Networks (DNNs) in XAI. LRP provides a way to decompose the output of a DNN into the input space, assigning a relevance score to each input feature. This approach is based on the idea of propagating the prediction from the output layer of the DNN back through each layer in a way that preserves the proportionality of relevance among the neurons within each layer.

LRP has been applied to various types of DNNs, including Convolutional Neural Networks (CNNs) and Recurrent Neural Networks (RNNs), and has been used in various applications, such as image classification, speech recognition, and natural language processing [54]. LRP is particularly useful in cases where the high-dimensional input data is difficult to interpret, and traditional feature importance methods, such as decision trees and partial dependence plots, are not applicable.

LRP works by assigning relevance scores to the neurons in the output layer of the DNN, which are then propagated backward through the network. In each layer, the relevance scores are assigned to the neurons based on their contribution to the output. The relevance scores are then propagated backward to the previous layer, again assigning scores to the neurons based on their contribution to the output. This process is repeated until the input layer is reached, where the relevance scores are assigned to each input feature.

One of the main advantages of LRP is its ability to provide a clear explanation of how a DNN arrives at its decision, which is crucial for applications such as medical diagnosis and financial risk assessment. LRP has also been used to identify which parts of an input image are most important for a particular classification by generating a heatmap that highlights the relevant areas of the image [55].

LRP has been successfully applied to various types of DNNs, including image recognition models such as Inception and ResNet, and has been used to explain the behavior of models used in medical imaging and natural language processing. For example, LRP has been used to identify which regions of the brain are most important for classifying different types of images in fMRI studies.

One of the limitations of LRP is that it can be computationally expensive, especially for large and complex networks. Several modifications to the LRP algorithm have been proposed to address this issue, such as Randomized LRP and Deep Taylor Decomposition [56]. Another limitation is that LRP does not provide a global explanation of the DNN behavior, as it only considers one input at a time. However, LRP can be used in combination with other XAI methods, such as SHAP and LIME, to provide a more complete picture of the DNN behavior.

LRP is a powerful and flexible method for understanding the decision-making process of DNNs, and it has a wide range of applications in XAI. With the increasing use of DNNs in critical applications, such as healthcare and finance, LRP is becoming an increasingly important tool for ensuring the transparency and accountability of these systems.

12.3.9 Teaching Explanations for Decisions (TED)

TED (Teaching Explanations for Decisions) is introduced in a paper titled TED: Teaching AI to Explain Its Decisions [57], where the authors propose a framework that enables machine learning models to explain their decisions. TED is a modular and extensible framework that can be used with different machine learning models and explanation methods. The framework consists of three components: a representation of the decision space, an explanation generation module, and an explanation evaluation module.

The representation of the decision space captures the relationship between the input features and the model's predictions. The explanation generation module generates natural language explanations for the model's decisions based on the decision space representation. The explanation evaluation module evaluates the quality of the generated explanations by measuring their fidelity to the model's decision-making process and their comprehensibility to humans. The authors demonstrate the effectiveness of TED in several applications, including diagnosing diabetic retinopathy and predicting loan default.

The authors demonstrate the effectiveness of TED in several applications, including diagnosing diabetic retinopathy and predicting loan default. In the diabetic retinopathy diagnosis application, TED helped identify the most important image features used by the model to make its diagnosis. In the loan default prediction application, TED provided explanations that helped lenders understand the factors contributing to the model's predictions.

The authors also conduct a user study to evaluate the effectiveness of the explanations generated by TED. The study participants found the explanations generated by TED to be more helpful than those generated by existing explanation methods. The participants also found the explanations to be more trustworthy and to improve their understanding of the model's decision-making process.

This is a novel approach to addressing the interpretability problem in machine learning models by enabling them to generate natural language explanations for their decisions. The framework is flexible and can be used with different machine learning models and explanation methods, making it applicable to a wide range of applications. The authors show that the explanations generated by TED are effective in enhancing the transparency and trustworthiness of machine learning models, making them more suitable for critical applications. Overall, the paper makes a valuable contribution to the field of explainable artificial intelligence by proposing a framework that can help bridge the gap between machine learning models and human understanding.

12.4 CONCLUSIONS

Artificial intelligence (AI) has been widely adopted in various fields, including healthcare, finance, and cybersecurity, due to its ability to process large amounts of data and make decisions quickly. However, as the complexity of AI systems continues to increase, it becomes more challenging to understand how these systems make decisions. This is where Explainable AI (XAI) comes in, as it enables us to understand and interpret the decision-making processes of AI systems.

XAI is a rapidly growing field that aims to provide insights into the behavior of machine learning models and explain the reasoning behind their decisions. It also aims to increase transparency, trustworthiness, and accountability of machine learning models, especially in critical applications such as healthcare and finance. The techniques discussed in this chapter are just a few examples of the many approaches that are being developed to achieve this goal.

One of the most significant advantages of XAI is its ability to provide a clear explanation of how a machine learning model arrives at its decision. This is crucial for applications such as medical diagnosis and financial risk assessment. XAI techniques, such as layer-wise relevance propagation and the TED framework, have been used to identify which parts of an input image are most important for a particular classification by generating a heatmap that highlights the relevant areas of the image.

Another advantage of XAI is that it can help improve the accuracy of machine learning models. By providing explanations for the decisions made by the model, it is possible to identify errors or biases in the model and make necessary corrections. Moreover, XAI can help make machine learning models more accessible to non-expert users, as it provides a way to explain complex decisions in a language that is easy to understand.

However, XAI also has its limitations. One of the main challenges of XAI is that it can be computationally expensive, especially for large and complex networks. Several modifications to the XAI algorithms have been proposed to address this issue, such as Randomized LRP and Deep Taylor Decomposition. Another limitation is that XAI does not provide a global explanation of the machine learning model's behavior, as it only considers one input at a time. However, XAI can be used in combination with other techniques, such as SHAP and LIME, to provide a more complete picture of the machine learning model's behavior.

12.5 FUTURE SCOPE

XAI is a crucial area of research that plays a significant role in ensuring the transparency and accountability of machine learning models. The techniques discussed in this chapter are just a few examples of the many approaches that are being developed to achieve this goal. While each technique has its own strengths and limitations, they all represent important contributions to the field of XAI and provide valuable tools for improving the transparency and interpretability of AI systems. As AI continues to advance and become more widespread, the need for XAI techniques will only increase, making it an exciting and important area of research.

Moreover, XAI can also have a positive impact on society. The ability to explain AI decisions can help build trust between users and AI systems, which can lead to greater adoption of these systems. This, in turn, can lead to more efficient and effective decision-making, as well as improved outcomes for individuals and society as a whole. Additionally, XAI can also help address issues such as bias and discrimination in AI systems by enabling us to identify and correct these issues.

It is essential to recognize that XAI is not a one-size-fits-all solution. The choice of XAI technique will depend on the context of the decision, the type of model used, and the end-users. As AI systems continue to evolve and become more complex, the development of new XAI techniques will be necessary to keep up with these advancements. Therefore, it is crucial to continue investing in XAI research to ensure that AI systems are transparent, trustworthy, and accountable. The XAI techniques discussed in this chapter represent important contributions to this field, providing valuable tools to achieve these goals.

REFERENCES

1. Laato, S., Tiainen, M., Najmul Islam, A. K. M., & Mäntymäki, M. (2022). How to explain AI systems to end users: A systematic literature review and research agenda. *Internet Research*, 32(7), 1–31.
2. Ehsan, U., Tambwekar, P., Chan, L., Harrison, B., & Riedl, M. O. (2019). Automated rationale generation: A technique for explainable AI and its effects on human perceptions. In *Proceedings of the 24th International Conference on Intelligent User Interfaces (IUI'19)* (pp. 263–274). New York: Association for Computing Machinery.

3. Köhl, M. A., Baum, K., Langer, M., Oster, D., Speith, T., & Bohlender, D. (2019). Explainability as a non-functional requirement. In *2019 IEEE 27th International Requirements Engineering Conference (RE)* (pp. 363–368). Jeju, Korea (South). doi: 10.1109/RE.2019.00046.
4. Oh, C., Song, J., Choi, J., Kim, S., Lee, S., & Suh, B. (2018). I lead, you help but only with enough details: Understanding user experience of co-creation with artificial intelligence. In *Proceedings of the 2018 CHI Conference on Human Factors in Computing Systems (CHI'18)*, Paper 649 (pp. 1–13). New York: Association for Computing Machinery.
5. Arrieta, A. B., Díaz-Rodríguez, N., Del Ser, J., Bennetot, A., Tabik, S., Barbado, A., ... & Herrera, F. (2020). Explainable Artificial Intelligence (XAI): Concepts, taxonomies, opportunities and challenges toward responsible AI. *Information Fusion*, 58, 82–115.
6. Confalonieri, R., Coba, L., Wagner, B., & Besold, T. R. (2021). A historical perspective of explainable Artificial Intelligence. *Wiley Interdisciplinary Reviews: Data Mining and Knowledge Discovery*, 11(1), e1391.
7. Miller, T. (2019). Explanation in artificial intelligence: Insights from the social sciences. *Artificial Intelligence*, 267, 1–38.
8. Guidotti, R., Monreale, A., Ruggieri, S., Turini, F., Giannotti, F., & Pedreschi, D. (2018). A survey of methods for explaining black box models. *ACM Computing Surveys (CSUR)*, 51(5), 1–42.
9. Kalia, A., Goel, A., & Jain, V. (2021). Explainable AI: A review. *Artificial Intelligence Review*, 54(2), 1209–1238.
10. Chang, Y., Jin, G. Z., & Lu, L. (2020). AI and marketing: Implications and challenges for marketing research and practice. *Journal of Marketing Research*, 57(1), 1–19.
11. Topol, E. J. (2019). High-performance medicine: The convergence of human and artificial intelligence. *Nature Medicine*, 25(1), 44–56.
12. Rudin, C., Wang, C., Coker, B., Slavkovic, A., Liu, Y., & Madigan, D. (2019). The human side of machine learning: Strategies for balancing predictive accuracy and fairness. *IBM Journal of Research and Development*, 63(4/5), 1–14.
13. Huang, L., Joseph, K., Nelson, B., Rubinstein, B. I., & Tygar, J. D. (2020). An empirical study of the impact of machine learning models on decision-making. *Proceedings on Privacy Enhancing Technologies*, 2020(1), 41–61.
14. Samek, W., Wiegand, T., & Müller, K. R. (2017). Explainable artificial intelligence: Understanding, visualizing and interpreting deep learning models. *arXiv preprint arXiv:1708.08296.*
15. Omeiza, D., Webb, H., Jirotka, M., & Kunze, L. (2021). Explanations in autonomous driving: A survey. *IEEE Transactions on Intelligent Transportation Systems*, 23(8), 10142–10162.
16. Demajo, L. M., Vella, V., & Dingli, A. (2020). Explainable ai for interpretable credit scoring. *arXiv preprint arXiv:2012.03749.*Chicago.
17. Altmann, A., Toloşi, L., Sander, O., & Lengauer, T. (2010). Permutation importance: A corrected feature importance measure. *Bioinformatics*, 26(10), 1340–1347.
18. Dorie, V., Hill, J., Shalit, U., Scott, M., & Cervone, D. (2019). Automated versus do-it-yourself methods for causal inference: Lessons learned from a data analysis competition. *Statist. Sci.* 34 (1) 43–68.
19. Bragg, J., & Ren, Y. (2019). Explaining recommendations: Satisfaction vs. promotion. In *Proceedings of the 25th ACM SIGKDD International Conference on Knowledge Discovery & Data Mining*, Anchorage AK USA August 4–8 (pp. 1452–1460).
20. Breunig, M. M., Kriegel, H. P., Ng, R. T., & Sander, J. (2000, May). LOF: Identifying density-based local outliers. In *Proceedings of the 2000 ACM SIGMOD International Conference on Management of Data*, Dallas Texas, USA, May 15–18 (pp. 93–104).

21. Zhang, K., Hutter, M., & Jin, H. (2009). A new local distance-based outlier detection approach for scattered real-world data. In *Advances in Knowledge Discovery and Data Mining: 13th Pacific-Asia Conference, PAKDD 2009 Bangkok, Thailand, April 27–30, 2009 Proceedings 13*(pp. 813–822). Berlin: Springer.
22. Sundararajan, M., Taly, A., & Yan, Q. (2017, July). Axiomatic attribution for deep networks. In *International Conference on Machine Learning*, Sydney, Australia, PMLR 70 (pp. 3319–3328). PMLR.
23. Ribeiro, M. T., Singh, S., & Guestrin, C. (2016, August). "Why should i trust you?" Explaining the predictions of any classifier. In *Proceedings of the 22nd ACM SIGKDD International Conference on Knowledge Discovery and Data Mining*, San Francisco California, USA, August 13–17 (pp. 1135–1144).
24. Madhav, A. S., & Tyagi, A. K. (2022, July). Explainable Artificial Intelligence (XAI): connecting artificial decision-making and human trust in autonomous vehicles. In *Proceedings of Third International Conference on Computing, Communications, and Cyber-Security: IC4S 2021* (pp. 123–136). Singapore: Springer Nature Singapore.
25. Zhang, Y., Zhao, T., Ma, L., & Sun, Y. (2018). Interpretation of Neural Models for Text Classification. arXiv preprint arXiv:1805.01086.
26. Wu, T. Y., & Wang, Y. T. (2021, November). Locally interpretable one-class anomaly detection for credit card fraud detection. In *2021 International Conference on Technologies and Applications of Artificial Intelligence (TAAI)* (pp. 25–30). IEEE.
27. Liu, Y., Huang, X., An, B., & Zeng, D. D. (2018). Evaluating machine learning interpretability methods on a medical case. In *Proceedings of the 27th International Conference on Computational Linguistics*, Santa Fe, New Mexico, USA, COLING (pp. 3637–3647).
28. Bragg, J., & Ren, Y. (2019). Explaining recommendations: Satisfaction vs. Promotion. In *Proceedings of the 25th ACM SIGKDD International Conference on Knowledge Discovery & Data Mining,* Anchorage, AK, USA, August 4–8 (pp. 1452–1460).
29. Wang, S., Yao, S., Li, B., Li, M., Zhang, Y., Xu, X., & Xu, Z. (2021). LIME-tree: An interpretable and computationally efficient model-agnostic explanatory method. *IEEE Transactions on Neural Networks and Learning Systems*, 32(5), 2098–2107.
30. Molnar, C. (2020). Interpretable machine learning. Lulu.com.
31. Guidotti, R., Monreale, A., Ruggieri, S., Turini, F., & Giannotti, F. (2018). A survey of methods for explaining black box models. *ACM Computing Surveys*, 51(5), 93.
32. Ribeiro, M. T., & Singh, S. (2020). Anchors: High-precision model-agnostic explanations. In *Proceedings of the AAAI Conference on Artificial Intelligence*, Hilton New York Midtown, New York, New York, USA (Vol. 34, No. 05, pp. 8063–8071).
33. Sokol, K., & Flach, P. (2020). LIMEtree: Interactively customisable explanations based on local surrogate multi-output regression trees. *arXiv preprint arXiv:2005.01427.*
34. Lundberg, S. M., & Lee, S. I. (2017). A unified approach to interpreting model predictions. In *Advances in Neural Information Processing Systems*, NeurIPS proceedings, Long Beach, California, USA, 4–9 December, *30.*
35. Fernández-Loría, C., Provost, F., & Han, X. (2020). Explaining data-driven decisions made by AI systems: the counterfactual approach. *arXiv preprint arXiv:2001.07417.*
36. Wachter, S., Mittelstadt, B., & Russell, C. (2018). Counterfactual explanations without opening the black box: Automated decisions and the GDPR. *Harvard Journal of Law & Technology*, 31(2), 841–887.
37. Kim, Y., Kim, S., Lee, H. W., & Kim, S. J. (2021). Visualizing counterfactual explanations for deep neural networks. *arXiv preprint arXiv:2106.00756.*
38. Liao, T. W. (2019). An overview of rule-based systems for explainable artificial intelligence. *IEEE Access*, 7, 133661–133673.
39. Stanciulescu, A., & Belciug, S. (2021). Rule-based systems for decision-making in healthcare. In *Computational Intelligence Methods for Healthcare IT* (pp. 77–95). Springer. doi: 10.1007/978-981-99-8853-2.

40. Perales, F. J., Ramirez-Gonzalez, G., Prieto-Gonzalez, L., & Gonzalez-Carrasco, I. (2021). Rule-based systems for autonomous driving. *Sensors*, 21(12), 3976.
41. Hu, J., Shi, J., & Sun, J. (2021). Wind turbine power output forecasting based on machine learning and rule-based methods. *Applied Energy*, 291, 116865.
42. Liu, H., Sun, B., & Jiang, C. (2021). Fault diagnosis of rotating machinery based on a hybrid rule-based system and deep belief network. *Mechanical Systems and Signal Processing*, 149, 107196.
43. Zhang, Y., Liu, S., Zhang, Y., & Li, H. (2020). RBS-Net: A rule-based system network for explainable deep learning. *IEEE Transactions on Neural Networks and Learning Systems*, 32(10), 4641–4653.
44. Liu, Y., Sun, X., & Jiang, J. (2021). Fault diagnosis of rotating machinery based on a rule-based system. *Mechanical Systems and Signal Processing*, 147, 107219.
45. Wang, C., Zhou, C., & Leung, H. (2021). Explainable AI: A review of rule-based systems. *IEEE Transactions on Computational Social Systems*, 8(4), 897–911.
46. Hu, J., Shi, X., & Sun, Y. (2021). A rule-based system for wind power forecasting. *Energy*, 222, 119969.
47. Kim, J., Lee, J., Kim, S., & Choi, S. (2020). Prototype and feature visualization for interpretable deep neural networks. *IEEE Transactions on Visualization and Computer Graphics*, 26(1), 471–481.
48. Sun, Z., Feng, S., Li, J., Li, X., & Li, L. (2020). Learning Text Representations via Prototype Selection and Distillation. *arXiv preprint arXiv:2002.06791.*
49. Liu, X., Zhang, H., He, Y., & Yin, G. (2021). Fraud detection for insurance claims via prototype-based explanations. *Knowledge-Based Systems*, 215, 106763.
50. Simonyan, K., Vedaldi, A., & Zisserman, A. (2013). Deep inside convolutional networks: Visualising image classification models and saliency maps. arXiv preprint arXiv:1312.6034.
51. Li, J., Monroe, W., Ritter, A., Galley, M., Gao, J., & Jurafsky, D. (2016). Understanding neural networks through representation erasure. arXiv preprint arXiv:1612.08220.
52. Samek, W., Binder, A., Montavon, G., Lapuschkin, S., & Müller, K. R. (2019). Evaluating the visualization of what a deep neural network has learned. *IEEE Transactions on Neural Networks and Learning Systems*, 30(11), 3369–3383.
53. Nguyen, A., Yosinski, J., & Clune, J. (2016). Multifaceted feature visualization: Uncovering the different types of features learned by each neuron in deep neural networks. arXiv preprint arXiv:1602.03616.
54. Bach, S., Binder, A., Montavon, G., Klauschen, F., Müller, K. R., & Samek, W. (2015). On pixel-wise explanations for non-linear classifier decisions by layer-wise relevance propagation. *PLoS One*, 10(7), e0130140.
55. Samek, W., Montavon, G., Binder, A., Lapuschkin, S., & Müller, K. R. (2017). Evaluating the visualization of what a deep neural network has learned. *IEEE Transactions on Neural Networks and Learning Systems*, 28(11), 2660–2673.
56. Lapuschkin, S., Binder, A., Montavon, G., Muller, K. R., & Samek, W. (2016). Analyzing classifiers: Fisher vectors and deep neural networks. In *Proceedings of the IEEE Conference on Computer Vision and Pattern Recognition*, 27–30 June, Las Vegas, NV, USA (pp. 2912–2920).
57. Hind, M., Wei, D., Campbell, M., Codella, N. C., Dhurandhar, A., Mojsilović, A., … & Varshney, K. R. (2019). TED: Teaching AI to explain its decisions. In *Proceedings of the 2019 AAAI/ACM Conference on AI, Ethics, and Society*, Honolulu, HI, USA, January 27–28 (pp. 123–129).

13 Explainable Deep Learning Architectures for Product Recommendations

Boominathan Perumal, Swathi Jamjala Narayanan, and Sangeetha Saman

13.1 INTRODUCTION

Online shopping provides a convinient way of purchasing products that anyone can desire anytime and under any category. Since platforms tend to sell a wide variety of products, it is necessary to incorporate a good navigation or recommendation system within the platform. Hence, there must be a system that provides navigation and product recommendations to the customers. Product recommendation is a method of filtering products and showing them to the user with the intention that the user might find the product interesting and buy or search for more such products.

According to the findings of Chai et al. (2002), many e-commerce websites use a menu-based system for recommending and searching for products. Not only is it inconvenient for the customer to use, but it is also very tiresome and overwhelming to interact with the system. Another very common method of recommending/navigating products through the website is by using keyword-based searches.

Due to the growth in popularity of these websites, the number of customers using the market has also increased. The customers can be of two types: either they can be first-time users or they can be regular users of the platform. Both demographics have their way of searching for and buying the products. The proposed system can recommend products to first-timers that they might find interesting or may just observe the activity of the first-time customers. Eventually, when those first-time customers become regular, the system will start recommending products based on any previous purchases, such as books, gifts, accessories for already purchased items, or subscriptions, or according to their address (according to the items that other people in the locality or the city might have bought).

A product recommendation system delivers information, items, or services that are customized to the user's specific tastes and preferences (Shahbazi and Byun, 2018). The purpose of the recommended system is to provide customers with accurate and

DOI: 10.1201/9781003442509-13

diversified recommendations while also increasing user satisfaction with the recommendation algorithm. Based on how the recommendation is produced, recommendation systems are grouped into three types: content-based, collaborative filtering, and hybrid techniques. The collaborative method suggests the item to the specific user depending on the rating of the other users in the system. Content-based techniques predict items relying on their previous history. A hybrid approach combines content-based and collaborative filtering techniques in several ways. These approaches provide the necessary explanatory reasoning and can be utilized to produce explanations. Several deep learning (DL), machine learning (ML), and black-box models have been presented in recent years to handle many types of problems. The forms of interpretability of ML algorithms can differ significantly. Some approaches are naturally transparent and comprehensible, while others are too difficult to understand and thus require specialized approaches to obtain interpretation. Before discussing the various methods defined in the explainability of machine learning techniques, it is necessary to establish the interpretation and explanation of the term to distinguish between these two levels of observation. The question of "how" an algorithm makes a decision is answered by interpretation, whereas explanation aims to discover "why" it made such a conclusion.

Artificial intelligence (AI)-based techniques are becoming more frequently used in ordinary life. Although their superior performance, deep learning algorithms are inexplicable to humans due to their complex structure. Consumers want to understand a model prior to making a decision, thus they might have to go with more widespread former methods that are relatively simpler to interpret. Because suitable model explanations may increase user acceptability, facilitate explanatory problem-solving, and encourage understanding, the demand for explained black-box modelling is growing rapidly in a variety of domains. It is difficult to construct explanations using deep learning-based recommendation systems because the interactions that take place among the different model components are not explicitly stated. As a result, we describe a personalized recommendation system based on deep learning which offers distinct detailed explanations and consequences. A successful recommendation system must not only correctly recognize the demands of users but also analyze their psychology and provide relevant recommendations in a direction that users can accept. The goal of explainable recommendations is to give customers recommendations as well as a justification for the recommendation, hence increasing user trust and acceptance of the recommendation system (RS). In aspects of interoperability and fidelity, then we developed that the techniques presented in the present study are superior to those suggested in earlier studies. Several explainability methods have developed in recent years, each with its own set of benefits and drawbacks. They are classified into various kinds, the most prevalent of which are model-agnostic or post hoc methods and model-intrinsic methods.

The remainder of this chapter is organized as follows. In Section 13.2, related research on machine learning, deep learning, and explainable artificial intelligence (XAI) for product recommendation is discussed. Then, in Section 13.3, we outline different ways for presenting recommendations. In Section 13.4, we present commonly used machine and deep learning approaches used for recommendation. In Sections 13.5 and 13.6, we explain the two types of explainable product recommendation systems and present case studies based on Amazon and Netflix recommendations. Finally, in Section 13.7, we discuss our findings and provide recommendations for further study.

13.2 RELATED WORKS

Machine learning for recommendation systems is highly connected to model-based explainable recommendation research. We will first provide an overall review of machine learning for customized recommendations in this section.

Deep learning models have also been extended to Sentiment Analysis of customer reviews for better product recommendations. Sentiment Analysis is understanding the emotions behind the human author of a text by a computer.

Besides these machine learning-based approaches to Product Recommendation, there is a new approach that has recently been on the rise, explainable machine learning. Explainable machine learning offers the potential to provide stakeholders with insights into model behavior by using various methods such as feature importance scores, counterfactual explanations, or influential training data. Some work related to explainable machine learning algorithms has been highlighted below.

The research paper published by Bhatt et al. (2020) discusses how organizations view and use explainability for stakeholder consumption. This study synthesizes the limitations of current explainability techniques that hamper their use for end users. Although explainability approaches are increasingly being used by ML engineers as sanity checks throughout the development process, the authors discovered that there are still considerable restrictions present in techniques that inhibit their usage to directly address end users. These drawbacks include having to look for explanations to be evaluated by domain experts, the possibility of inaccurate correlations being represented in model explanations, the absence of causal intuition, and the delay in calculating and displaying explanations in real time.

Explainable Machine Learning in Credit Risk Management is a topic covered in another work on explainable machine learning algorithms, which was offered by Bussmann et al. (2021). The authors' introduction of an agnostic, post-processing technique based on correlation network models was driven by the need to make use of the high predicted accuracy provided by advanced machine learning models while also making them interpretable. TreeSHAP is employed, which uses polynomial time instead of the conventional exponential runtime to compute the SHapley Additive exPlanation (SHAP) for trees. This study has significant policy ramifications for legislators and regulators working to safeguard users of artificial intelligence services. While artificial intelligence significantly increases the accessibility and simplicity of financial services, it also introduces new hazards. According to the research, network-based explainable AI models can significantly increase our understanding of the factors that influence financial risks, particularly credit risks.

Explainable machine learning models are widely applied in the medical field. The study by Scott Lundberg et al. (2018) discusses explainable machine-learning predictions for the avoidance of hypoxemia during surgery. In this work, the authors describe the creation and evaluation of a machine-learning-based system that, while a patient is under general anaesthesia, predicts the likelihood of developing hypoxemia and explains the risk factors. When anaesthesiologists were provided with interpretable hypoxemia risks and contributing variables, the system performed better because it was trained using minute-by-minute data from the electronic medical documentation of over 50,000 surgical procedures. The results indicate that if anaesthesiologists anticipate 15% of hypoxemia events now, many would predict 30% among them with this approach, and the majority of these occurrences might benefit from early detection since they are connected to modifiable factors.

A significant research question in recommender systems is the prediction of individual ratings. Despite the high accuracy of the latent factor model's (e.g., matrix factorization) rating prediction, they have several faults, including lack of transparency, cold-start, and unsatisfactory outcomes with particular user-item groupings. Another study by Cheng et al. (2018) uses ratings in addition to textual evaluations and item photos to overcome these drawbacks. To represent users' preferences and objects' attributes from many angles, the authors first propose an aspect-aware multi-modal topic model (MATM) for product images and text reviews, as well as determine the importance of an item's features to the user. A novel aspect-aware latent factor model (ALFM) is then introduced, which discovers the latent characteristics of users and items based on user ratings. This approach might reduce the issue of data sparsity and improve suggestion interpretability. The Amazon product dataset and the Yelp 2017 Challenge dataset have both been the subject of extensive experimental study. Results demonstrate that their method significantly outperforms robust standard approaches, especially for customers with limited ratings; product visual features provide a detailed, explicit interpretation of the predicted results.

A framework for interactive and understandable machine learning is proposed in a new study by Spinner et al. (2019) that facilitates consumers to (i) perceive the models developed by machine learning, (ii) identify the framework deficiencies using various XAI methods, and (iii) enhance and optimize the models. Their architecture combines an iterative XAI process with eight global surveillance and steering methods, including quality tracking, provenance monitoring, comparison of models, and developing trust. To operationalize the framework, ExplAIner, an automated visual predictive analytics system for highly interactive and explicable learning processes, generates each stage of the recommended pipeline within the well-known Tensor Board setting.

Building the trust of doctors is necessary for successfully transferring machine learning models to clinical practice in the healthcare field. Explainability, or the capacity of a machine learning model to defend its conclusions and aid doctors in justifying model predictions, has usually been seen as essential to building confidence. However, the absence of precise definitions for useful justifications in many contexts hurts the discipline. Therefore, a study by Sana Tonekaboni et al. (2019) focuses on finding certain explainability components that can encourage the development of ML model trust. Medical practitioners in two distinct acute care disciplines were polled. (Intensive Care Unit). They analyzed their input to define the circumstances under which explainability increases physicians' confidence in machine learning models. Additionally, the authors defined the categories of explanations that doctors had determined to be most relevant and essential for successful translation to practice in hospitals. Finally, the authors provide specific measures for an in-depth examination of clinically explained approaches. Researchers are interested in promoting the endorsement, wider acceptance, and continued usage of ML systems in healthcare by incorporating clinicians' and ML researchers' perspectives on explainability.

Another study by Reis et al. (2019) looks into fake news detection by making use of explainable machine learning models. They were able to create several models that provide very accurate judgments, even though the great majority of models are useless and fail to distinguish between real news and fake news. They specifically concentrated on models with a greater than 0.85 probability of ranking a randomly selected false news article higher than a randomly selected fact. They discovered a strong correlation between features and model predictions for these models, demonstrating that some features are tailored for identifying particular kinds of fake news and demonstrating that different combinations of features each cover a particular area of the fake news space. Finally, they explain the variables that went into the model's judgments, fostering civic thinking by enhancing their capacity to assess digital information and draw justifiable conclusions.

Deep learning models have also gained a lot of fascination in the field of recommendation research. Deep models work when there is adequate data for training. Many deep models for explainable suggestions have been developed, which will be discussed further below. Kotouza et al. (2020) developed one method for product recommendation, specifically for clothing items. In addition to gathering data using natural language processing, which involved data collection from two sources, the data were clustered and sent as input to a machine-learning algorithm to recommend products. The system utilizes two components, online and offline. The offline component is used for data collection and processing, while the online component serves as the User Interface for customers to interact with. Once the data was collected, they would be clustered using several different algorithms and then sent to a Deep Reinforcement Learning Algorithm. The customers then had the choice to either like the recommended product or not, and this information would also be considered as a data point in the form of user feedback for future processing.

Other approaches in the area of Deep Learning have also been explored as follows. According to Shankar et al. (2017), Tuinhof et al. (2018), and Jo et al. (2020), proposed models that utilize CNN for the recommendation of fashion-based products. The main idea behind it is to feed CNN with visual data such as images. The results of which can be used to recommend the products. The models have been tested against state-of-the-art datasets such as *Exact Street2Shop* and have been deployed on big-name e-commerce websites such as Flipkart.

According to Lee et al. (2020), another approach involves using Recurrent Neural Networks (RNN) for recommending products, which helps in reducing waste, shopping time, and effort. Lee, et al. (2020) developed an LSTM-based multi-period product recommender system. The principle behind the model is that it learns the patterns of the customers' purchases and recommends products accordingly. The RNN model identifies any repetitions in the pattern to learn it. Unlike the usual Collaborative Filtering (CF) Technique, the model will take time-stamped data as input. The LSTM-based model outperformed the CF-based model by providing more accuracy and diversity. An interesting observation from the experimentation was that the accuracy of the model remained consistent for different sizes of the dataset. The proposed model also stands out with its multi-period perspective taken into consideration.

According to Ying et al. (2018), another efficient approach towards Deep Learning for product recommendation is by using Graph Convolution Neural Networks (GCNN). Ying et al. (2018) introduced a deep recommendation engine called PinSage. This model is capable of being deployed on a large scale, learning embeddings for the nodes on a web-scale graph that contains billions of objects. In this model, random walks are combined with graph convolutions to generate nodes that represent the items, thus incorporating graph structure and node features. This model is successful to the extent of being deployed on a large scale at Pinterest. Apart from the training strategy, this model also utilizes highly efficient random walks for structuring the convolutions, which distinguishes it from other GCNN models. The MapReduce Pipeline is used for efficiently generating the embeddings. The key concept of the model's operation lies in using localized graph convolutions. This can be seen when the modules are created. It starts with generating the embeddings for a node, by applying multiple convolution modules that aggregate information, mostly the visual and textual features, from the local graph neighborhood of the node. This process is repeated for all the modules, after which the modules are stacked to gather the information about the entire local network topology. All the nodes share the parameters of these localized convolutional modules, which gives the model another advantage of having the same parameter complexity regardless of the size of the input. Another technique used for this model is importance pooling and curriculum training, which was able to improve the embedding performance. The deployed model at Pinterest has been trained with over 7.5 billion examples, with a graph containing over 3 billion nodes and 18 billion edges. The nodes represent the pins and boards of the website.

According to Chen, Xu, et al. (2019), most existing recommendation models consider user preference as invariant and generate static explanations. However, rarely that is the case, as customer preferences are dynamic, meaning that the preferences towards a product may change based on different states of mind or circumstances. Due to this phenomenon, customer experience with the existing models may tarnished, leading to degraded customer satisfaction. Hence, Chen, Xu, et al. (2019) came up with a model for accurate user modelling and explanation called Dynamic Explainable Recommender (DER). The main concept behind the model is to combine a time-stamped recurrent unit with a Convolutional Network that reads the review sentences of a particular item. The model learns by aggregating the user's historical data to the Gated Recurrent Unit (GRU), where the last step of the GRU unit encodes the current preference of the user. The preferred item is then profiled against the review sentences for the item and the sentence reviews are projected by the CNN onto a latent vector. Finally, the different sentence reviews are merged with the current preference to determine the final rating. After learning, attention weights generate the importance of each review sentence, aiding the prediction process for the recommendation. The model's evaluation occurred through institutive analysis of the review information and through crowd-sourcing, demonstrating its superiority compared to other state-of-the-art models in the market and verifying its accuracy. As a result, there is a discrepancy between explainability in reality and the idea of transparency, as explanations benefit internal stakeholders rather than external ones.

13.3 APPROACHES TO PRODUCT RECOMMENDATIONS

We outline a description of the many forms of recommendation explanations in this section to provide readers with a realistic perspective on how an explanation might appear in reality.

13.3.1 Item-Based or User-Based Explanations

The explanations can be presented as either a histogram of all neighbors' ratings or as individual evaluations from each neighbor. The method through which this item was chosen might then be included as part of the explanation. The system may also display multiple objects at the same time. A user's interests are likely to be varied in a large domain like news. The system may also display multiple things on the same theme. When a user expresses a preference for one or more things, the recommender system can suggest alternatives. The similarity among items is more consistent than the similarity among users since the science book will always be a science book, whereas the user can change their opinion.

13.3.2 Opinion-Based Explanations

It can also be considered that there are some characteristics of an item that a user appreciates more than others. In that situation, the user should be able to express this in a wider context like an opinion. A feedback option could allow consumers to define how they connect to various aspects of an item. Furthermore, it assists in the generation of extremely fine and more accurate explanations, allowing customers to make more informed conclusions.

13.3.3 Feature-Based Explanations

In a feature-based explanation, users are shown features relevant to their profiles. Numerous studies have shown that incorporating feature-based explanations can assist in enhancing recommendation effectiveness. Recent research looked into demographic information such as age, gender, and residential place in a social media setting for product recommendation and analyzed the results by weighting the attributes according to the ranking function.

13.3.4 Visual-Based Explanations

It has been attempted to use item images for explaining recommendations based on the perception of visual images by researchers. This is extremely noteworthy because online purchasing is mostly a visual experience, and many goods and services are rated based on appearance. With supervised learning, you select an image of a product and request the system to provide n of the most similar items that may also be of interest to you.

13.3.5 Sentence-Level Explanations

The sentence-level method uses templates to provide users with explanation sentences. Researchers are also using sentiment analysis to convert textual comments into explicit ratings that may be processed by the Collaborative filtering algorithm. In recommendation, the sentiment analysis technique is used not only to detect sentences with distinct sentiment values but also to assess the degree of strength or intensity of such sentences.

13.4 METHODOLOGIES ADOPTED IN PRODUCT RECOMMENDATION SYSTEMS

This section describes existing research for the Product Recommendation System, including machine and deep learning models.

13.4.1 Deep Learning-Based Methodologies

Generally, Deep Learning models (Shankar et al., 2017) employed for the Product Recommendation System, work on the concept of taking images of the products as inputs and looking for similarities with other images of other products. This technique is most commonly seen for fashion products like clothing, jewelry, and accessories. However, this is not the only way of getting recommendations for products. There are other methodologies such as image-based, vector-based, and sketch-based models used for product recommendation based on deep learning as explained in the subsection.

13.4.1.1 Image-Based

As mentioned above, this methodology (Tuinhof et al., 2018) tends to take images as input to find similarities among the products that can be utilized for product recommendation. These models tend to use CNN for this purpose. The idea behind them is to find similarities among the images, based on which embeddings for the CNN are generated. These embeddings play the role of visual descriptors, by capturing the different patterns and colors in the image. The dissimilarities in the embeddings are learned by utilizing the Euclidean distance between the embeddings, while the similarities can be found using the k-Nearest-Neighbor (kNN) technique. Since products are of different types and can differ from each other, a good strategy is by using CNNs for the classification of the image (product) based on visual and textual features and then perform feature extraction on the images. The extracted features can be plotted to perform k-NN to find similarities. The first phase involves training different CNNs to predict the category and texture type. Each CNN is expressed as follows:

$$N_i\left(w_i,.\right) \sim S_i\left(V_i, F_i\left(U_i,.\right)\right) \text{ for } i = 1,2,3,\ldots \tag{13.1}$$

where w_i represents the weighted vectors, S_i (V_i,.) represents the fully connected softmax layers, and F_i represents the CNNs without the last layer. The actual classification is done in the softmax layer, and the whole classification from the embedding layer to prediction (see Figure 13.1). Afterward, the softmax output S_i is removed, and the CNN is concatenated to extract the feature vector $F(X)$. This vector is used to carry out the k-NN algorithm.

The most common CNNs used for classification are:

AlexNet: AlexNet is a benchmark CNN network used for image classification. It is the most common model that can be used for any type of image classification and is a popular choice for product recommendation. It consists of 8 layers and takes a 224×224-sized image. The reason for the popularity of this CNN network is due to the non-saturating rectified linear function, ReLU(x), and the presence of a normalization layer after the ReLU activation. The normalization layer plays an important role in improving the generalization ability of the network significantly.

BN-Inception: Batch-Normalized (BN) Inception is also a benchmark network and a popular choice for image classification. It is the extension of the GoogleNet architecture and consists of 34 layers and also takes an input image of 224×224 size. This network allows parallel layers to work by mapping the output to many layers at the same time, which are later concatenated. This network shows the regularization effect and updates the weighted vectors after every training iteration. This addresses the issue of latent input distribution of the hidden layers.

VGG network: This network has 16 layers; hence it is called the 16-layered VGG network. This is a dense network with small receptive fields, thus finding the detailed patterns better than AlexNet, which is a much sparser network. This gives the VGG network the advantage of using the details for finding the similarities in the embeddings rather than only using the

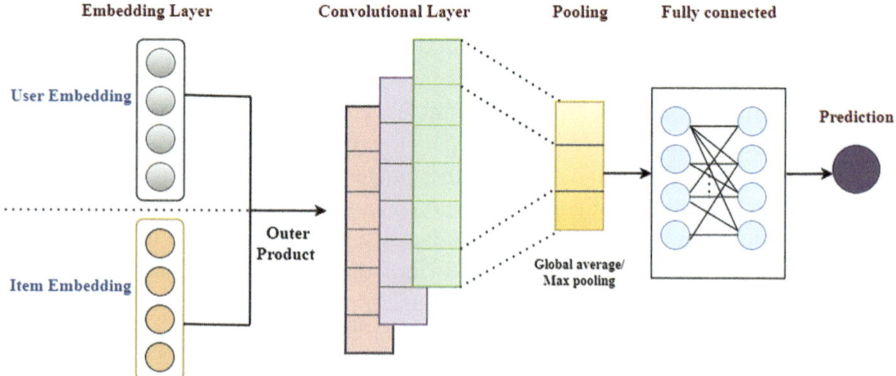

FIGURE 13.1 CNN-based product recommendation. (Photograph by the author.)

network for classification. For training, a triplet, <q, *p*, *n*>, is required as input. The contents of the triplet are a query image (*q*), a positive image (*p*), and a negative image (*n*). The training input also requires labeling. But the labeling is relative. This gives the advantage of not setting up a threshold for the criteria of finding a match or not.

13.4.1.2 Vector-Based

A feature extraction approach indicated Image2Vec may transform an image into a real number vector in a specific dimension. Other images that resemble the original image will be located close to the mapped vector when the real number acquired inside that dimension, a vector space map of the image is developed. Hence, it becomes easy to find images that are similar to the original image, which aids the process of product recommendation.

Vector-based image processing (Jo et al., 2020) requires the image to undergo the process of feature extraction. This is the basis under which the vectors are generated and mapped in the vector space. The vector-based recommendation model is based on the deep neural networks (DNN) recommendation model. The learning process of which can be divided into two stages. The first stage is to get default weights to the data points, that is, the fashion products, by collecting data that has been reviewed by humans. Second, the pre-weighted data points are taken as the default and are updated as the training process goes on.

Using the Image2Vec model, the system shifts the image's top and bottom into vectors. These vectors are later concatenated. During this, there will be some pre-trained data that is given default weights and passed through the model for weight training. The entire Recommendation System will learn two models through the course of the training.

Style recommender: The system recommends a product based on the user and their preference, that is, the system will recommend the top and bottom vectors based on the user profile.

Apparel matching: When the user presents with a top, the system will recommend a bottom that will be best suited for it, and when the user presents a bottom, the system recommends a top relevant to it.

The image feature extraction responsible for generating the top and bottom vectors can be represented as:

$$v(\text{top}) = \text{ImageFeatureExtractor}(t)\, v(\text{bottom}) = \text{ImageFeatureExtractor}$$

$$(b)\, x = v(\text{top})\, \text{concat}\; v(\text{bottom}).$$

13.4.1.3 Sketch-Based

The sketch-based recommendation system also utilizes the Image2Vec model and is based on extracting the features of the image. It transforms images of fashion products using a feature extractor for search. Images have been generated using sketch-up sample models together with that inquiry. These are converted into vectors and are mapped in the vector space and measured for similarities with the other vectors. For this type of recommendation, the Generative Adversarial Network (GAN) model is often used (Jo et al., 2020), which trains two neural networks used for image processing due to their superior performance. According to Goodfellow et al. (2014), GANs are made up of a Discriminator and a Generator neural network. As it tries to identify between fake and real samples, the Generator generates a fake random number of data, including images, text, and other types of data. Each neural model competes with the other during training. Both models repeatedly go through the steps required to achieve in their respective fields. As a min-max game, the GANs have been created. More specifically, they want to increase the probability that the discriminator will make a mistake by increasing the probability of the discriminator's reward. The Discriminator calculates the probability while attempting to reduce its loss by taking into account the fact that the sample it got came from the probing set rather than the produced one.

Moreover, an ANN usually comprises many levels, including an input layer, a hidden layer or layers, and an output layer, with a collection of neurons in each layer. An overview of the other deep neural network architectures and applications (Bhattacharyya et al., 2020) that are closely connected to the product recommendation models is provided in this subsection.

13.4.1.4 Recurrent Neural Networks (RNN)

One of the limitations of traditional neural networks is their inability to store prior knowledge for decisions and forthcoming events. By using memory, RNNs address this problem while preserving the information. RNNs are very effective and suited when applied to data that is in sequence since they can recall and distinguish the patterns found across periods. RNNs are made up of input, output, and hidden units, which are required for storing data. Compared to Multilayer perceptrons (MLPs), RNNs feature a directed loop that saves the prior data and transfers it to the output. By getting a predicted score closer to the actual score, it aims to reduce error and provide reliable predictions. While other RNN variations may be applied based on the specific requirements of the application, the most popular ones are long short-term memory (LSTM) and gated recurrent unit (GRU). When predicting significant occurrences with long intervals, LSTM is a good approach. When trying to determine whether or not the information contained in a "cell" is useful, LSTM increases it with the distinction between short- and long-term memory. A cell memory, for instance, has three gates: an input, a forget, and an output gate. Only informative data gets stored during training; otherwise, it is forgotten via the forget gate. The cell state sends related information to the next level. The GRUs, in contrast, use the hidden state for transferring information rather than the cell state. Additionally, the reset and update gates have been added as two additional gates.

13.4.1.5 Deep Reinforcement Learning (DRL)

In a DRL method, an agent acts in a given environment at a given time step t. The environment responds to the agent in two different ways: (i) with a reward that offers quantitative feedback on the action the agent made at a specific time set t, and (ii) with a state in which the environment has changed as a result of the agent's activity. All these steps are repeated until it reaches a final condition. Deep learning (DL) and reinforcement learning (RL) models are integrated with DRL to address increasingly challenging NLP and product recommendation issues. It uses gradient descent operation to approximate values while taking into account a researcher's preference dynamics.

13.4.1.6 Variational Autoencoder (VAE)

Variational autoencoders are a generative model that combines the following two major concepts: machine learning using Bayes and deep learning. In neural net terminology, a variational autoencoder is made up of an encoder, a decoder, and a loss function. A neural network serves as the encoder, which has weights and biases, produces a hidden representation as its output, and a data point as its input. The encoder aims to provide effective data compression for this lower-dimensional region. Another neural net serves as the decoder, which also contains weights and biases and takes a representation as input. It outputs parameters to the probability distribution of the data. A Bernoulli distribution may be employed to represent the probability distribution of a single pixel. The latent representation of data is sent into the decoder, which then outputs one Bernoulli parameter for each pixel in the image.

13.4.2 Machine Learning-Based Methodologies

Typically, machine learning models make use of content-based filtering and collaborative filtering and proceed as follows: Based on the user's prior clicks and activity, the recommendation system offers content-based filtering recommendations for goods. A machine learning classifier (XGBoost and Random Forest are the most prominent ones) is used to predict the model for the recommendation based on this dataset. Similar to other systems, such as knowledge-based systems, the recommender system is classified into two categories: collaborative filtering and content-based filtering. The Product recommendation is a system created to provide customers with information and views on various products to ease the shopping experience. User-product relationships, user-user relationships, and product-product relationships are the three techniques to build a product recommendation system.

Content-based filtering (CBF): CBF (Shahbazi et al., 2019) collects user metrics including user clicks, purchases, pages viewed, amount of time spent on a website, product categories, etc. Based on this data, a customer profile is created, and this data is then utilized to suggest products in this category.

Collaborative filtering (CF): CF (Thakkar et al., 2019) extracts data based on user behavior and prioritization and estimates how closely related users are to one another. For instance, if user one orders strawberries and user two also orders strawberries, the algorithm will know that both users have the same options and will suggest a few goods that are comparable to them.

Besides Collaborative and Content Filtering, standard machine learning procedures are also followed, starting with Data Analysis. Usually, data mining and web scraping are performed to obtain data about online shopping, followed by some steps to make the data easier to process:

- Collection of data
- Data cleaning
- Management of missing values
- Extraction of relevant data
- Acquiring features

13.4.2.1 Data Pre-processing, Analysis, and Visualization

Data preprocessing is essential when online product data has been collected to exclude all irrelevant data. To accomplish this, raw data is transformed into a form that can be used to perform prediction and data analysis using data preprocessing methods. To make the dataset easier to understand, duplicate files are first eliminated. Second, records for users that don't include any purchases are deleted. Third, only the data required for the recommendation system method is retained.

After acquiring the newly preprocessed data, data analysis and data visualization can be performed. Following are some commonly adopted ways to perform the same:

Time series analysis: Time series analysis can be used to generate unique and important information for recommending products. To conduct the time series analysis, data duration is collected.The collected information is then analyzed on a monthly and daily basis to generate insights into product shopping frequency.

User interest analysis: Each user who visits online shopping websites has preferences for shopping. In this analysis, a list of the most purchased products is extracted based on users' shopping records. The most purchased item shows the user's preferences and interests and importantly, the quality of the product.

Access page analysis: Access page analysis shows the records of user-clicked web pages. The most visited pages are one of the options for recommending the product provided on that access page. User clicks and reviews products and increase the weight of that product for recommending neighborhood products.

Purchased product frequency: The process of product purchasing is divided into three layers. The input layer contains the extracted valuable features out of raw data. Following encoding, the retrieved features are sent to the analysis layer where analysis methods are used on the bought product dataset. Finally, the system visualizes and classifies online shopping products by their purchase frequency.

13.4.2.2 Machine Learning Algorithms

Once data analysis and data visualization are completed and weights are assigned to the different parameters of the dataset based on the results from data analysis and visualization, we can use machine learning algorithms to make the final prediction. For this, the following algorithms are popularly used:

XG boost: The strength of this potent technique resides in its scalability, which promotes highly optimized memory utilization and parallel and distributed computing for quick learning. XGBoost (Shahbazi et al., 2020) is a technique that falls under ensemble learning. It might not always be enough to rely solely on a single machine learning model's output. A methodical approach to combining the prediction capacity of various learners is provided by ensemble learning. A single model that provides the combined output from many models is the outcome. The foundation learners, sometimes referred to as the ensemble's models may come from the same learning algorithm or other learning algorithms. Ensemble learners that are often employed include bagging and boosting. Although these two methods can be used in a variety of statistical models, decision trees have historically been the most typical approach.

Random forest: Random Forest (Khanvilkar and Vora, 2018) and other supervised machine learning algorithms are commonly used in classification and regression problems. It builds decision trees from different samples, using their average for categorization and majority vote for regression. One of the most important features of the Random Forest Algorithm is its capacity to handle data sets including both continuous variables, as in regression, and categorical variables, as in classification. The ensemble uses two kinds of techniques:

1. **Bagging:** A distinct training subset is formed by applying replacement from sample training data, and the final decision is made on a majority vote. Think about Random Forest.
2. **Boosting:** This method transforms weak learners into strong learners by creating sequential models with a high level of accuracy. XGBoost and AdaBoost, for instance (Shahbazi et al., 2020). The operation of Random Forest depends on the Bagging principle, as was just before mentioned.

Support vector machine: SVM, also known as support vector regressor (SVR), is a machine learning technique that can also be referred to as a regression model (Bhatt et al., 2020). SVR is a different classification method from the others described, yet it has some similarities. It has gained popularity among machine learning algorithms as a rapidly evolving approach. A regression model works as a building block of SVR. For the system's prediction requirements, it executes the support vector machine. SVR is intended to obtain the best line within the defined error values, thus is one of the reasons it differs from other machine learning algorithms.

Linear regression: According to Kotouza et al. (2020), a linear regression model is a technique for establishing a relationship between target variables by modeling data into linear equations. The remainder is referred to as the dependent variable, and each variable is considered an ex-positive variable. The first kind of regression model, known as linear regression, has mostly been employed in practical applications. In general, there are two forms of models: one that is typically linearly connected to its parameters and is simpler to fit than the other, which is non-linearly dependent. The estimators for the first model may also be found with ease. Thus, linear regression is more beneficial.

Finally, after applying these machine learning algorithms, we can generate a list of top "x" (top 5, top 10, top 50, etc.) product recommendations for a given user.

13.5 EXPLAINABLE RECOMMENDATION SYSTEMS

Recently, it has become more common to use machine learning and deep learning to solve real-world marketing issues on EC (e-commerce) platforms. Among these, machine and deep learning-based recommender systems have proven to be quite effective in driving sales on EC sites. Though a wide range of effective machine learning and deep learning algorithms used for recommender systems fail in interpretability in terms of the reasoning behind every recommendation. In other words, the model's process of learning operates on its own (a "black box"), and the explanations behind the recommended modifications performed are not mentioned. A large number of explanations and recommendation models (Zhang and Chen, 2020) are still largely black boxes, and we do not completely understand why one item would be recommended over another. The reason is mostly that most deep neural networks' hidden layers contain meanings that are not transparent. As a consequence, explaining the deep models is a significant effort. An explainable recommendation system encourages the design of methods that generate not only accurate recommendations but also appropriate explanations, including why specific items are recommended to a provided target user, resulting in clearer, persuasive, efficient, reliable, and user-friendly outcomes.

Explainable recommendations employ a range of additional data to clarify why a certain product was suggested to a particular user. On EC sites, you may find information about sellers, price ranges, brands, types of items, data on user-bookmarked brands, and the age and gender of users. The development of explainable recommendations has focused on two approaches: intrinsic models, which describe the recommendation procedure, and post-hoc models (see Figure 13.2), which explain the recommendation outcome.

13.5.1 Model-Intrinsic Explanation Approaches

An AI system that explains its predictions using inference techniques is supported by model-intrinsic AI. Using this technique, explanations can be directly derived from a challenging recommendation model. Based on the methodological framework, explainable recommendation models can be classified into the following categories: Matrix Factorization, Topic modeling, Knowledge-Graph based methods, and Deep learning-based explanation models.

13.5.1.1 Matrix /Tensor Factorization Based Explainable Methods

Based on the input data provided, the recommender system approach may be classified into two categories: single feedback (explicit or implicit) and multiple feedback.

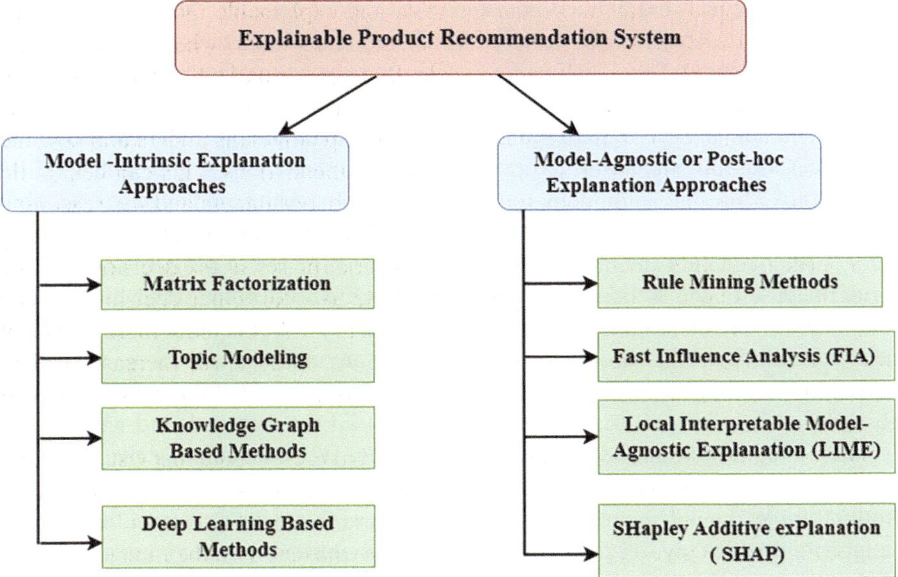

FIGURE 13.2 Categorization of explainable product recommendation systems. (Photograph by the author.)

Matrix Factorization (MF), Probabilistic Matrix Factorization, Non-negative Matrix Factorization, and Deep Matrix Factorization are some of the systems that use explicit feedback. When no explicit feedback is available, the recommender system relies on implicit feedback. Some works dealing with implicit feedback include using weighted matrix factorization to convert real-value implicit feedback into a level of confidence for TV recommendation.

Matrix factorization (MF): MF is a significant method in matrix theory. Considering that user purchase behavior in MF (Pujahari and Sisodia, 2020) is influenced by numerous factors known as latent factors. Each user and item is modelled as a vector of latent factors under this assumption. As a result, the recommendation list can be constructed by sorting the predicted ratings in descending order. The MF algorithm's main purpose is to accurately forecast unknown ratings. It cannot, however, generate novel and diversified item recommendations. The model's capacity to recommend unfamiliar and unpopular long-tail items is reflected in the model's novelty of recommendations. Furthermore, variety represents the similarity of recommended goods, denoting that the system will show different goods to target users, not just comparable ones.

Automated collaborative filtering systems (ACF): An Automated collaborative system's (ACF) operation is similar to human word-of-mouth recommendation. Users of an ACF system are provided with the three-step conceptual model following for how an ACF system operates: the user enters a rating profile. The ACF system (Herlocker et al., 2000) finds people with similar profiles (neighbors) and combines their evaluations to produce suggestions. In the first step, we can gather a massive quantity of implicit and explicit preference data from the user, including page visits and numerical ratings. To define the different types of preference information applied in a particular explanation, an explanation can be necessary. The next step is when ACF systems distinguish themselves from conventional human word-of-mouth recommendations. ACF systems can evaluate thousands of applicants for soulmates and select the individual who closest matches the others. One component vital for collaborative filtering technology to work well is the method employed to find other people with similar profiles. The ideal recommendations will be made if the neighbors of the model are the most accurate classifier of the user's demand for presence data. Because of the fact consumers typically seek out information when estimates and shortcuts are employed, this is essential in sectors with greater levels of risk. The selections made profiles are frequently those that are most comparable to those generated by sampling the high-speed memory's available profiles, rather not necessarily the "best" neighbors. Making a description of all the data gathered when trying to find a neighbor might help explain a prediction. When calculating the similarity between two profiles, most ACF systems will give equal weight to all items present in both profiles. The user may be given the option to evaluate the ratings of the selected neighbors, and if they discover an inappropriate neighbor, they may ignore the prediction or manually exclude that neighbor from evaluation.

In the last step, describe the data and the method used for aggregating neighbors' ratings into a final prediction. Many of the symptoms of poor predictions can be detected and explained at this level. The data are essential in describing this phase. Users may benefit significantly by knowing the way every one of their neighbors rated the item that is recommended, or simply assuming there are many ratings, the distribution of ratings. The user may also be interested in how neighboring ratings are used to create a prediction. The prediction usually represents only an average weight of the neighbor's ratings.

ACF systems are based on complex mathematical models, making it difficult to determine what to explain and how to explain it. Rating histograms appear to be the most convincing way of explaining the data underlying a forecast. A high recommendation may also be justified by signs of prior performance, similarities to other items with a comparable rating, and specific domain content features.

13.5.1.2 Topic Modelling

Existing feature extraction-based granulation methods generally rely on the extraction of latent features and are incapable of directly exposing the contents of each level or explaining their relationship with each other. The interpretability of recommendation results in online recommendations can help the platform summarize user preferences while also improving user beliefs about recommendation results. Thus, developing a method of granulation that collects interpretable granular features is intuitive and meaningful. Collaborative topic regression (CTR) is used to extract the latent features and items based on the user feedback information to attain an interpretable structure. CTR is a tightly coupled recommendation method that incorporates latent Dirichlet allocation with probabilistic matrix factorization (PMF). CTR can also extract an item's topic distribution from its context information. A CTR-based cost-sensitive consecutive three-way recommendation is developed to solve the problem of decision cost inequality and the issue of conflict among decision and time costs.

LDA Topic Model (Latent Dirichlet Allocation): LDA is a hierarchical probabilistic model (Wang et al., 2022) that can categorize the combination of documents into several subjects. Based on the distribution of document topics and the topic assignments for document words, LDA suggests that a collection of document G is created from specified subjects.

The probabilistic graphical model can be used to describe the LDA topic model's generation process.

a. Draw T topics from a Dirichlet distribution: $\varphi_t \sim \text{Dir}(\beta), t \in \{1,\ldots,T\}$;
b. For the l_{th} document in $G\left(l \in \{1,\ldots P\}\right)$:

1. Using a symmetric Dirichlet distribution, draw a topic proportion: $\theta_l \sim \text{Dir}(\alpha)$;
2. For the q_{th} word of the l_{th} document $q \in \{1,\ldots Q\}$;

i. Using multinomial distribution, draw a topic assignment: $z_{l,q} \sim \text{Mult}\left(\theta_l\right)$, $z_{l,q} \in \left\{1,2,\ldots,T\right\}$
ii. Draw a word $w_{l,q}$ from the multinomial distribution: $w_{l,q} \sim \text{Mult}\left(\varphi_{z_{l,q}}\right)$

As a result of LDA topic modelling, Equation (13.2) gives the joint probability distribution of the documents in the corpus. There are four elements in Equation (13.3), the posterior distribution θ, z, φ, and w indicates the proportion of words in documents assigned to specific topics for D, and the words in D as a whole, respectively.

$$p\left(\theta, z, \varphi, w \middle| \alpha, \beta\right) = \prod_{l=1}^{P} p\left(\theta_d \middle| \alpha\right) \prod_{t=1}^{T} p\left(\varphi_t \middle| \beta\right) \prod_{q=1}^{Q} \times p\left(z_{l,q} \middle| \theta_l\right) p\left(w_{l,q} \middle| z_{l,q}, \varphi\right) \tag{13.2}$$

$$p\left(\theta, z, \varphi \middle| w, \alpha, \beta\right) = \frac{p\left(\theta, z, w, \varphi \middle| w, \alpha, \beta\right)}{p(w)} \tag{13.3}$$

To estimate the hidden parameters θ, φ and z, the LDA topic model must approximate the posterior distribution $p\left(\theta, z, \varphi \mid w, \alpha, \beta\right)$. The variational Bayesian approach and the Gibbs sampling method are two general methods for approximating the posterior distribution of LDA.

13.5.1.3 Knowledge Graph-Based Methods

Knowledge graphs have been attracting public interest in recent years as an idea for cutting-edge information generation and recommendations. Knowledge graphs (KG) may explain the recommended items and offer more accurate recommendations, which enables monitoring of the recommended items. They can also help alleviate issues caused by sparse data and cold starts. The knowledge graph (Khan et al., 2022) is employed to connect three categories of media, namely users, products, and features, to choose the most effective way to provide explainable recommendations to users. According to recent findings, the two primary ways for introducing knowledge graphs to recommendation systems are (i) the embedding-based method and (ii) the path-based method. Feature mining impacts performance, as items in content-based RSs are recommended based on how similarly their attributes match the interests of customers. These systems endure difficulties due to cold start limitations and a large amount of direct feature extraction. Furthermore, aggregation of interactions and content-level similarities occurs when collaborative and content-based RSs are incorporated as hybrid RSs. Although aggregation of RSs may address the constraints outlined above, it leads to a trade-off between complexity and performance. This is referred to as KG-based recommendation systems.

Knowledge graph reasoning-based explainable product: The definition of a knowledge graph is a set of triplets $G = \left\{(a_h, m, a_t) | a_h, a_t \in \varepsilon, m \in M\right\}$ that entities are represented by $\in$ and relations are represented by M connecting two entities. Each triplet characterizes a relation $m \in M$ between a head entity $a_h \in \varepsilon$ and a tail entity. We assume a knowledge graph that includes at least two different kinds of entities: a collection $U \subset \varepsilon$ of users and a set $X \subset \varepsilon$ of products. User feedback is modelled by the relationship $m_f \in M$ between a user and a product, which is domain-specific. Reasoning over knowledge graphs G between two entities, termed k-hop reasoning (Balloccu et al., 2023), can be defined as follows.

K-Hop Reasoning: An entity-to-entity k-hop path denoted as $a_0, a_k \in \varepsilon$ is made up of $k+1$ entities joined together by k relations.

$$l^k_{a_0,a_k} = \left\{a_0 \xleftrightarrow{m_1} a_1 \xleftrightarrow{m_2} \ldots \xleftrightarrow{m_k} a_k\right\} \tag{13.4}$$

where $a_{i-1} \xleftrightarrow{m_i} a_i$ with $0 < i \le k$ is thought to represent as $(a_{i-1}, m_i, a_i) \in G$ or its inverse $(a_i, m_i^{-1}, a_{i-1}) \in G$. We represent as L^k_u the collection of all feasible k-hop paths that connect any product to a user u and. The type of reasoning path for the given k-hop reasoning path is the last relation $m_k \in M$. As a result, the example path is "directed".

If user u interacted with product x, then user-product feedbaack $F(u,x) = 1$ where $F = M^{|U|*|X|}$ otherwise $F(u,x) = 0$. The conventional recommendation model that does not explicitly use knowledge graphs strives to calculate the relevances $F(u,x) \in [0,1]$ of unseen items in F and use those estimates to rank products. The abstraction of this operation is learning a model $\theta : (U, X) \to M$. By decreasing user relevance, the top-n $(n \in N)$ products X^n_u are recommended after they have been ranked.

Given a fixed k $\in N$, our goal is to recommend to the user u a beneficial collection of products X^n_u, where each product $x \in X^n_u$ is linked to a reasoning path $l^k_{u,x} \in L^k_u$ that will be used as input for textual explanation creation. The model chooses this path as the most indicative of the recommended product x from all the predicted k-hop paths $L^k_{u,x}$ between user u and product x.

13.5.1.4 Deep Learning Based Method

Long Short-Term Memory (LSTM) for Explainable Product Recommendation: An architecture based on deep learning neural networks and LSTM aimed at providing consumers with not only accurate suggestions but also the reasoning behind recommendations. This kind of architecture would provide recommender system users with a visual explanation technique to explain every recommended item as they receive suggestions from other users in the system.

The LSTM deep learning framework for explainable recommendations (Zarzour et al., 2020) is divided into three major components: First, the ratings and reviews data input gate; second, the long short-term memory model; and third, the resulting explainable recommendations. The first part discusses our model's input, whereas the second part describes its output, which will be shown to customers. The LSTM framework model's input data consists of two vectors: ratings and associated reviews. The user's V rating score for a particular item P is expressed as $R_{i,j}\left(V_i, P_j\right)$, where $R_{i,j}$ is the rating, followed by V_i for the ith user, P_j for the jth item. ID_v, and ID_p of the user and an item can be used to represent them. To develop automatic reasons about why a specific item P is recommended by an active user, ratings and review vectors are used. Deep neural networks encounter several difficulties during the training process, namely vanishing gradients and expanding gradients (Seo et al., 2017). To maintain information about the temporal state, LSTM uses memory cells with self-connections and input and output gates to control information flow. The model's memory cell block is made up of these gates and memory cells. The following implementation takes an input $X = \left(x_1, x_2, \ldots, x_n\right)$ and returns it as a result $Y = \left(y_1, y_2, \ldots, y_m\right)$ by calculating $W_{xi}\ X_t$ network activation (Equation 13.7):

$$p_t = \sigma\left(W_{xp}\, X_t + W_{hp}\, h_{t-1} + W_{sp}\, s_{t-1} + b_p\right) \tag{13.5}$$

$$s_t = f_t\, s_{t-1} + p_t \tanh\left(W_{xs} X_t + W_{hs} h_{t-1} + b_s\right) \tag{13.6}$$

$$f_t = \sigma\left(W_{xf} X_t + W_{hf} h_{t-1} + W_{sf} c_{t-1} + b_f\right) \tag{13.7}$$

$$h_t = o_t \tanh\left(s_t\right) \tag{13.8}$$

$$o_t = \sigma\left(W_{xo} X_t + W_{ho} h_{t-1} + W_{so} c_t + b_o\right) \tag{13.9}$$

where X_t represents the present sample and W_{xp} represents the weight matrix. p_t represents the input gate (Equation 13.5), s_t represents the cell state (Equation 13.6), h_t represents the hidden layers (Equation 13.8), and o_t represents the output gate (Equation 13.9).

The model's output is the explainable recommendation, and the justifications for figuring out the recommendations are tailored based on a few prior reviews of the target object. Usually, the automatically developed explainable recommendation is a series of a word $Z = \left(z_1, z_2, \ldots, z_m\right)$, which is understandable, readable, and consistent with the rating words.

A dual-attention-based, interpretable convolutional neural network (D-Attn): CNNs can derive high-level contextual information from text, and recommender systems have effectively used these properties with lower prediction error compared to collaborative filtering/matrix factorization.

D-Attn (Seo et al., 2017) predicts product ratings by combining review text and ratings. Utilizing aggregated reviews, it learns the latent representation of users and items and enables interaction among user and item models, similar to matrix factorization. Because D-Attn uses both local and global attention layers in its evaluation and rating of data, as well as combining these two in one network training, it is inherently more robust to noise and discrepancies as a result of its use of both collaborative filtering and local attention layers $c(i) = g\left(X_{l\text{-att},i} * W^1_{l\text{-att}} + b^1_{l\text{-att}}\right),\ i \in [1,T]$, and topic-based techniques. The embedding layer uses word embeddings for input review documents defined in Equation (13.10). It has a weight associated with the embedding layer i.e., $W_a \in \mathbb{R}^{l \times |M|}$

$$x_t = W_a a_t \tag{13.10}$$

where l is a collection of words and $|M|$ is vocabulary size. Using the L-Attn algorithm, which determines which words are more informative in a local window of words. The purpose of the L-Attn algorithm is to determine, given a limited window of words, which words contain the most relevant information. After entering the parameter matrix $W^i_{l\text{-att}} \in \mathbb{R}^{w \times l}$ and a bias $b^i_{l\text{-att}}$, the weighted scores for each word are computed (Equations 13.11 and 13.12) as follows:

$$X_{l\text{-att},i} = \left(x_{i+\frac{-w+1}{2}}, x_{i+\frac{-w+3}{2}}, \ldots, x_i, \ldots, x_{i+\frac{w-1}{2}}\right)^T \tag{13.11}$$

$$\mathrm{c}(i) = \left(X_{l\text{-att},i} * W^1_{l\text{-att}} + b^1_{l\text{-att}}\right),\ i \in [1,T] \tag{13.12}$$

$C(i)$ is a notation that stands for the attention scores that are taken as a weight for the i-th word embedding. As an outcome of this, if one word gets a lower score in comparison to another, we can infer that the word with the lower score (Equation 13.13) is less relevant.

$$\hat{x}^L_t = c(t) x_t \tag{13.13}$$

The weighted series of word embeddings is processed using a convolutional layer with a kernel size of one, which reduces overfitting. The sequence is then max-pooled to produce a fixed-length summary vector. As a result, following convolution with a matrix yields a local attention representation (Equation 13.14) of a given text $W^2_{l\text{-att}} \in \mathbb{R}^{l \times n_{l\text{-att}}}$ and a bias $b^2_{l\text{-att}} \in \mathbb{R}^{n_{l\text{-att}}}$ and max pooling (Equation 13.15):

$$Z_{l\text{-att}}(t,i) = g\left(\hat{x}^L_t * w^2_{l\text{-att}}(:,i) + b^2_{l\text{-att}}(i)\right),\ i \in \left[1, n_{l\text{-att}}\right] \tag{13.14}$$

$$z_{l\text{-att}}(i) = \text{MAX}\left(Z_{l\text{-att}}(:,i)\right) \tag{13.15}$$

where $n_{l\text{-att}}$ represents the total number of filters and $g(.)$ denotes a tan$h$ function. The module, G-Attn, receives the words from embeddings in their original order as input. Global attention scores are multiplied by one another to pass through the global attention layer. While the next CNN layer more accurately captures the entire semantic meaning, the G-Attn layer decreases the impacts of words that aren't very informative. The filter for the convolution layer functions on w_f words. When there are $n_{g\text{-att}}$ filters, the convolution filters $w_{g\text{-att}} \in \mathbb{R}^{w_f \times l \times n_{g\text{-att}}}$ are employed for a series of w_f weighted embeddings of words $\hat{X}_{g\text{-att},i} \in \mathbb{R}^{w_f \times l}$ (Equation 13.16), and output features $Z_{g\text{-att}} \in \mathbb{R}^{(T-w_f+1)\times n_{g\text{-att}}}$ defined in Equation (13.17).

$$\hat{X}_{g\text{-att},i} = \left(\hat{x}_i^G, \hat{x}_{i+1}^G, \ldots \hat{x}_{i+w_f-1}^G\right)^T \tag{13.16}$$

$$Z_{g\text{-att}}(i,j) = g\left(\hat{X}_{g\text{-att},i} * W_{g\text{-att}}(:,:,j) + b_{g\text{-att}}(j)\right), \quad i \in [1, T - w_f + 1], \qquad j \in [1, n_{g\text{-att}}], \tag{13.17}$$

Concatenation of outputs from the local and global attention modules (Equation 13.18) occurs through several fully connected layers W_{FC}^1 and W_{FC}^2 (Equation 13.19):

$$z_{\text{out}}^1 = z_{l\text{-att}} \oplus z_{g\text{-att}}^1 \oplus \cdots \oplus z_{g\text{-att}}^{n_w} \tag{13.18}$$

$$z_{\text{out}}^2 = g\left(w_{FC}^1 \cdot z_{\text{out}}^1 + b_{FC}^1\right) \tag{13.19}$$

$$\gamma_u = g\left(w_{FC}^2 \cdot z_{\text{out}}^2 + b_{FC}^2\right). \tag{13.20}$$

The inner product (Equation 13.21) of these two latent representations, L-Attn and G-Attn, can be used to estimate ratings of attention-based semantic representations.

$$\hat{r}_{u,i} = {\gamma_u}^T \gamma_i \tag{13.21}$$

As a result, this explainable CNN-based method opens up a new path for using representation learning in recommendation systems.

13.5.2 Model-Agnostic or Post-Hoc Explanation Approaches

The recommendation method can be difficult to define at times. In such cases, Model-agnostic explainability is interested in discovering model outputs without knowing the model's internal mechanism. Model-agnostic methods, in particular, enable the decision procedure to remain a black box. After the recommendation model has been developed, an explanation model provides explanations for the recommendations (thus the term "post-hoc").

13.5.2.1 Rule-Based Methods

The application of rule-based explainability improves the transparency and interpretability of the recommendation output. In particular, the explainable module creates rules that determine whether a user is likely to be interested in the products they have already rated. A rule is of the form $X \rightarrow Y$, where X and Y are product sets, denoting a relationship between a collection of transactions, each involving a set of products. According to this logic, the existence of X in a transaction automatically implies the inclusion of Y. Earlier, association rules were created to extensively use the rated items. This strategy aims to discover all significant associations between elements. The purpose of association rule mining is to identify all association rules that have more than the user-specified minimum support or confidence.

Neuro-Fuzzy Approach for Explainable Product Recommender System: The neuro-fuzzy approach (Rutkowski et al., 2018) is a unique method that translates nominal attribute values into numerical values in a neuro-fuzzy system employing the Mendel-Wang algorithm for rule construction. It significantly enhances recommendation efficiency by analyzing the context and importance of elements that demonstrate the circumstances of a person involved in the recommendation. We must select which item location of the attribute value will be assessed and how important that position will be for nominal attributes. The next step is to calculate the occurrences of each characteristic's values according to their position, individually for both positive and negative samples.

The first stage in the computations is described in the following Equation (13.22), which allows you to produce x_i, which is the normalized form of weight for attribute values.

$$x_{i,(n;t)} = \sum_{j=1}^{n} t_j * a_j \tag{13.22}$$

where x_i is computed independently for negative and positive samples, the index i indicates the subsequent attribute values, and index j represents the important position index. The next step is to calculate normalized weights w_i for positive and negative samples using Equation (13.23), where x_i is computed by Equation (13.22) and k_i represents all attribute value occurrences.

$$w_{i,(n;t)} = \frac{x_{i,(n;t)}}{\text{COUNT}_i} * \frac{k_{i,(n;t)}}{\text{ALL}} \tag{13.23}$$

where k_n and k_t are the sums of all negative and positive attribute value occurrences on each position. *COUNT* equals the sum of k_n and k_t. Equation (13.22) is used to calculate the columns x_n and x_t. The following formula (Equation 13.24) is proposed to generate a final value of attribute value weight W_i.

$$W_i = \frac{1 - \frac{wn_i + wt_i}{\text{MAX}}}{2} \tag{13.24}$$

A higher weight indicates a positive recommendation, whereas a lower weight indicates a negative recommendation. Utilizing the attributes analysis technique and computing composite attribute values, we can examine the components of numerous attribute values and reduce them to a single numerical value (Equation 13.25).

$$X_k = \frac{\sum_{j=1}^{n} t_{k,j} * W_{k,z}}{\sum_{j=1}^{n} p_{k,j}} \tag{13.25}$$

As shown in Equation 13.25, index k is the number of the selected attribute k and index z is the value of that selected attribute. We can design rules in a neuro-fuzzy system because we can transform all numerical values. This study used fuzzy rules generated by Wang and Mendel (1992) and a backpropagation algorithm to tune the simulations.

13.5.2.2 Fast Influence Analysis (FIA)

Although the fundamental influence computation with the complexity of $O(np)$ is substantially more efficient than explicitly computing Hessian and its inverse $H_{\hat{\theta}}^{-1}$, it nevertheless incurs a high computation cost when applied to real-world datasets.

When we use the aforementioned computing approach on the Movielens 1M dataset (Harper and Konstan, 2015) it can take up to an hour to evaluate the influence of training points on a single test case, which is unacceptable in today's recommender systems. An FIA technique is employed for recommendation (Cheng et al., 2018) to speed up the influence analysis process based on the following two essential observations.

Only a small proportion of MF parameters contribute to the prediction of y_t a given test case (v_t, r_t). In particular, the prediction of y_t is calculated by $\theta_t = \{m_{vt}, q_{rt}\}$, where m_{vt} and q_{rt} are latent vectors for v_t and m_t, respectively. As a result, only changes in the parameters in θ_t affect the forecast y_t.

We simply need to quantify the influence of training points R_t on θ_t now that we have focused on the analysis of the parameters θ_t. Other training points do not produce gradients θ_t and can therefore be discarded during influence computation. Based on the findings above, we can calculate the change in MF's prediction when a training point z is removed by using Equation (13.26):

$$\Delta g\left(v_t, r_t, \hat{\theta}_{-z}\right) = \frac{1}{n'} \nabla_{\theta_t} g\left(v_t, r_t, \hat{\theta}\right)^T H_{\hat{\theta}}^{-1} \nabla_{\theta_t} L(z, \hat{\theta}) \tag{13.26}$$

To summarize, the overall complexity of FIA for MF is $O(n'K)$, which is significantly less than the $O(np)$ cost of the original process. K is the dimension latent vectors. It is also worth noting that n' and K are often tiny and irrespective of the training set's scale. On the other hand, the value of n and p are proportional to the number of training examples. Thus, FIA allows us to undertake efficient MF influence analysis for explainable recommendations even on large datasets.

13.5.2.3 LIME (Local Interpretable Model-agnostic Explanation)

LIME is a method developed by Ribeiro et al. (2016) that estimates a complicated (non-linear) classifier using a sample using sparse linear models. The linear model explains which feature(s) of the dataset contributed directly to the predicted label. In several studies, LIME has often been employed to provide recommendations. With this approach, the local structure of a portion of a complex decision-making framework's output is discovered ex post facto using a simple and very interpretable model.

LIME (Shimizu et al., 2022) has the benefit of being simple to explain and adaptable to any complex model. Explanations are defined formally as a model $m \in M$, in which M is a class of interpretable models, such as linear models, decision trees, or falling rule lists, such that $m \in M$ the user can readily experience an explanation via textual artifacts. In other words, m is $\{0,1\}^{d'}$, i.e. g acts over whether the interpretable components are absent or present. The explanation $m \in M$ may not be interpretable on its own – so we used $\Omega(m)$ to be an explanation $m \in M$'s complexity (as opposed to its interpretability).

In the decision, trees $\Omega(m)$ may be the depth, while in linear models, $\Omega(m)$ may be the number of non–zero weights. Let's consider the model being explained $f : R^d \rightarrow R$, and $f(x)$ is the probability of which x belongs to a particular class in classification. Using $\prod_x(z)$ as a proximity measure between an instance u and x, we can define the locality around x. As a final step, let $L(f, m, \prod_x)$ be the amount of unfaithfulness displayed by m in approximating f in the locality defined by $\prod_x$. Interpretability and local fidelity can only be ensured if we minimize $L(f, m, \prod_x)$ while making $\Omega(m)$ be low adequacy to be able to be interpreted by humans. The following Equation (13.27) provides the explanation given by LIME.

$$\xi(x) = \underset{m \in M}{\arg\min}\, L(f, m, \textstyle\prod_x) + \Omega(m) \tag{13.27}$$

With various explanation families M, complexity measures Ω, and fidelity functions L, the above method is accessible. Here, the attention is on applying perturbations to conduct the search and sparse linear models to provide the explanations.

13.5.2.4 SHAP (SHapley Additive exPlanation)

SHAP is a unified technique for interpreting model prediction that evolved from the SHapley value (Zhong and Negre, 2022) in game theory. Based on game theory, this strategy can provide global and local interpretations. Each attribute is seen as a "player", and the "game" itself is the work of prediction. As a result, calculating the contribution of each "player" becomes an issue. SHAP computes the average of the marginal contributions for each feature across all permutations. SHAP values provide a measure of additive feature importance and simplify inputs by using conditional expectations. SHAP explanation model with simplified input mapping is defined in Equation (13.28). As a result of this definition of SHAP values $h_x(u') = u_s$, features not in the set S will have missing values in u_s. Most models are incapable of handling arbitrary patterns of missing input values, so we approximate $f(u_s)$ with $E\left[f(u)|u_s\right]$. It is possible to simplify the computation of the expected values by using a feature independence assumption (Equation 13.30) and a linear model assumption (Equation 13.31) when using these novel SHAP methods (Max SHAP, Deep SHAP).

$$f(h_x(u')) = E\left[f(u)|u_s\right] \text{ SHAP with simplified input mapping} \quad (13.28)$$

$$= E_{u_{\bar{S}}|u_s}\left[f(u)\right] \quad (13.29)$$

$$\approx E_{u_{\bar{S}}}\left[f(u)\right] \text{ Feature independence assumption} \quad (13.30)$$

$$\approx f\left(\left[u_s, E\left[u_{\bar{s}}\right]\right]\right) \text{ Linear model assumption} \quad (13.31)$$

where the set of non-zero indices in u' is called s. This SHAP value description is intended to closely correlate with the Shapley regression, Shapley sampling, and quantitative input influence feature attributions.

The core research focus of this chapter

Our goal for the future generation of a recommendation system is similar to that of a best friend — one who understands a user's interests, dislikes, priorities, and worries while also being aware of the goods to be recommended. The recommendations of such a system are not simply words or phrases, but rather definitive recommendations that would have been provided by a best friend who understands the user. We call them Explainable Recommendations. Those recommendations might also: (i) include real details about the product that the user will find helpful. (ii) be able to explain not only what features of the thing a user likes or dislikes, but also why. (iii) be expressed in natural language. (iv) be able to explain not only what features of the thing a user likes or dislikes but also why.

13.6 CASE STUDY

Systems for recommending products or rating items are known as recommendation systems. Businesses like Netflix (Töscher et al., 2009), Spotify, and YouTube mostly employ them to create playlists for viewers. To recommend items to its customers, Amazon employs recommender systems. Most recommender systems employ user history to learn more about the user. Two main methods are used by recommender systems. These filters are either collaborative or content-based. Content-based filtering methods are beneficial in situations where information about the item but not the user is available. It performs a user-specific classification function. It simulates a classifier that models the user's preferences regarding the characteristics of an item. However, today's recommender systems are invisible black boxes that don't reveal how a recommendation works. Explanations provide this transparency by disclosing the rationale and supporting evidence for a recommendation. Explanations are an essential component of a personalization system. A person will be able to make informed decisions about the path of a campaign or the structure of a product page if they have access to accurate and timely reporting.

Amazon uses association rule mining for product recommendation by analyzing patterns of transactional co-occurrence, employing knowledge-based methodologies for recommendations based on user behavior. In systems based on deep learning, RNNs are used to examine purchase statistics over time, while CNNs are utilized to evaluate product ratings. To perform recommendations, opinion-based systems verify the opinions of individuals with similar interests. Amazon's product recommendations re generated using a relational analysis of item similarity and user preferences by fuzzy rule-based systems that are explainable.

Netflix employs association rule mining to recommend playlists or films to viewers based on movie review data from users. Utilizing knowledge-based methodologies to generate recommendations based on the expressed preferences of the user, Netflix generates playlist recommendations based on positive and negative user preferences using imprecise rule-based systems that are explainable. To achieve recommendations, opinion-based systems verify the opinions of individuals who enjoy the same shows/movies. In systems based on deep learning, RNNs are employed to examine subscription statistics based on past music purchasing and review site analysis.

13.7 CONCLUSION & FUTURE WORK

We presented a brief outline of techniques used in explainable recommendation research in this chapter. Explainable recommendations are often employed for decision-making activities, where the results and processes must be comprehensible so that customers of the system may understand why a desired outcome is generated and how to apply the result to make decisions. We discussed several categories of explanations, such as machine learning-based, knowledge graph-based, matrix factorization methods, rule-based approaches, topic modelling, deep learning-based, and so on.

As a look ahead, we described many possible future views on explainable recommendation research. Knowledge graphs, user behavior analysis, deep learning, topic modelling, and cognitive foundations are all predicted to assist in the establishment of explainable recommendations.

- In a general context, AI organizations have recognized the need for Explainable AI techniques, which aim to address a broad variety of AI explainability challenges related to deep learning and computer vision product recommendation activities.
- Currently, however, novel deep networks such as few-shot learning, transformers, and adversarial training have limited applications in these product recommendation activities.

- Adaptive recommendation explanations and user visual preference are becoming increasingly popular trends to enhance recommendation performance.
- When user ratings are extremely inadequate, the recommendation performance often suffers greatly. In addition, they cannot be applied to propose new products that have not yet received any user ratings, which is known as the cold-start problem. A future direction for improving automated machine-generated reviews would be to provide more precise recommendations and personalized explanations to overcome these limitations such as sparsity, cold start problems, etc.

REFERENCES

G. Balloccu, L. Boratto, G. Fenu, and M. Marras, "Reinforcement recommendation reasoning through knowledge graphs for explanation path quality," *Knowledge-Based Systems*, 260, 110098, 2023.

U. Bhatt, A. Xiang, S. Sharma, A. Weller, A. Taly, Y. Jia, and P. Eckersley, "Explainable machine learning in deployment," In *Proceedings of the 2020 Conference on Fairness, Accountability, and Transparency*, Barcelona, Spain, pp. 648–657, January, 2020.

S. Bhattacharyya, V. Snasel, A. E. Hassanien, S. Saha, and B. K. Tripathy, *Deep Learning: Research and Applications*, Vol. 7, Walter de Gruyter GmbH and Co KG, Berlin, 2020.

N. Bussmann, P. Giudici, D. Marinelli, and J. Papenbrock, "Explainable machine learning in credit risk management," *Computational Economics*, 57(1), 203–216, 2021.

J. Chai, V. Horvath, N. Nicolov, M. Stys, N. Kambhatla, W. Zadrozny, and P. Melville, "Natural language assistant: A dialog system for online product recommendation," *AI Magazine*, 23(2), 63–63, 2002.

X. Chen, Y. Zhang, and Z. Qin, 2019, "Dynamic explainable recommendation based on neural attentive models," In *Proceedings of the AAAI Conference on Artificial Intelligence*, Vol. 33, No. 01, pp. 53–60, July.

W. Cheng, Y. Shen, Y. Zhu, and L. Huang, "Explaining latent factor models for recommendation with influence functions," arXiv preprint arXiv:1811.08120, 2018.

I. Goodfellow, J. Pouget-Abadie, M. Mirza, B. Xu, D. Warde-Farley, S. Ozair, and Y. Bengio, "Generative adversarial nets," *Advances in Neural Information Processing Systems*, 27, 2014.

F. M. Harper, and J. A. Konstan, "The movielens datasets: History and context," *ACM Transactions on Interactive Intelligent Systems (TIIS)*, 5(4), 1–19, 2015.

J. L. Herlocker, J. A. Konstan, and J. Riedl, "Explaining collaborative filtering recommendations," In *Proceedings of the 2000 ACM Conference on Computer Supported Cooperative Work*, Philadelphia, Pennsylvania, USA, pp. 241–250, December, 2000.

J. Jo, S. Lee, C. Lee, D. Lee, and H. Lim, "Development of fashion product retrieval and recommendations model based on deep learning," *Electronics*, *9*(3), 508, 2020.

N. Khan, Z. Ma, A. Ullah, and K. Polat, K, "Categorization of knowledge graph based recommendation methods and benchmark datasets from the perspectives of application scenarios: A comprehensive survey," *Expert Systems with Applications*, 206, 117737, 2022.

G. Khanvilkar, and D. Vora, "Sentiment analysis for product recommendation using random forest," *International Journal of Engineering & Technology*, 7(3), 87–89, 2018.

M. T. Kotouza, S.F Tsarouchis, A.C Kyprianidis, A. C. Chrysopoulos, and P. A. Mitkas, "Towards fashion recommendation: an AI system for clothing data retrieval and analysis," In *IFIP International Conference on Artificial Intelligence Applications and Innovations*, pp. 433–444, Springer, Cham, June 2020.

H. I. Lee, I. Y. Choi, H. S. Moon, and J. K. Kim, "A multi-period product recommender system in online food market based on recurrent neural networks," *Sustainability*, 12(3), 969, 2020.

S. M. Lundberg, B. Nair, M. S. Vavilala, M. Horibe, M. J. Eisses, T. Adams, and S. L. Lee, "Explainable machine-learning predictions for the prevention of hypoxaemia during surgery," *Nature Biomedical Engineering*, 2(10), 749–760, 2018.

A. Pujahari, and D. S. Sisodia, "Pair-wise preference relation based probabilistic matrix factorization for collaborative filtering in recommender system," *Knowledge-Based Systems*, 196, 105798, 2020.

J. C. Reis, A. Correia, F. Murai, A.Veloso, and F. Benevenuto, "Supervised learning for fake news detection," *IEEE Intelligent Systems*, *34*(2), 76–81, 2019.

M. T. Ribeiro, S. Singh, and C. Guestrin, "Why should i trust you?" Explaining the predictions of any classifier," In *Proceedings of the 22nd ACM SIGKDD international conference on knowledge discovery and data mining*, San Francisco, CA, USA, pp. 1135–1144, August, 2016.

T. Rutkowski, J. Romanowski, P. Woldan, P. Staszewski, and R. Nielek, "Towards interpretability of the movie recommender based on a neuro-fuzzy approach," In *International Conference on Artificial Intelligence and Soft Computing*, pp. 752–762, Springer, Cham, June, 2018.

S. Seo, J. Huang, H. Yang, and Y. Liu, "Interpretable convolutional neural networks with dual local and global attention for review rating prediction," In *Proceedings of the Eleventh ACM Conference on Recommender Systems*, Como, Italy, pp. 297–305, August, 2017.

Z. Shahbazi, and Y. C. Byun, "Product recommendation based on content-based filtering using XGBoost classifier," *International Journal of Advanced Science and Technology*, 29, 6979–6988, 2018.

Z. Shahbazi, and Y. C. Byun, "Product recommendation based on content-based filtering using XGBoost classifier," *Int. J. Adv. Sci. Technol*, 29, 6979–6988, 2019.

Z. Shahbazi, D. Hazra, S. Park, and Y. C. Byun, "Toward improving the prediction accuracy of product recommendation system using extreme gradient boosting and encoding approaches," Symmetry, 12(9), 1566, 2020.

D. Shankar, S. Narumanchi, H. A. Ananya, P. Kompalli, and K. Chaudhury, "Deep learning based large scale visual recommendation and search for e-commerce,". arXiv preprint arXiv:*1703.02344*, 2017.

R. Shimizu, M. Matsutani, and M. Goto, "An explainable recommendation framework based on an improved knowledge graph attention network with massive volumes of side information," *Knowledge-Based Systems*, 239, 107970, 2022.

T. Spinner, U. Schlegel, H. Schäfer, and M. El-Assady, "explAIner: A visual analytics framework for interactive and explainable machine learning," *IEEE Transactions on Visualization and Computer Graphics*, 26(1), 1064–1074, 2019.

P. Thakkar, K. Varma, V. Ukani, S. Mankad, and S. Tanwar, "Combining user-based and item-based collaborative filtering using machine learning," *In Information and Communication Technology for Intelligent Systems: Proceedings of ICTIS*, Vol. 2, pp. 173–180, Springer, Singapore, 2019.

S. Tonekaboni, S. Joshi, M. D. McCradden, and A. Goldenberg, "What clinicians want: contextualizing explainable machine learning for clinical end use," In *Machine Learning for Healthcare Conference*, Ann Arbor, Michigan, pp. 359–380. PMLR, October, 2019.

A. Töscher, M. Jahrer, and R. M. Bell, "The BigChaos solution to the netflix grand prize," Netflix prize documentation, 1–52, 2009.

H. Tuinhof, C. Pirker, and M. Haltmeier, "Image-based fashion product recommendation with deep learning," In *International Conference on Machine Learning, Optimization, and Data Science*, pp. 472–481, Springer, Cham, September, 2018.

L.-X. Wang, and M. Mendel Jerry, "Generating fuzzy rules by learning from examples," *IEEE Transactions on Systems, Man, and Cybernetics*, 22(6), 1414–1427, 1992.

H. Wang, Z. Hong, and M. Hong, "Research on product recommendation based on matrix factorization models fusing user reviews," *Applied Soft Computing*, 123, 108971, 2022.

R. Ying, R. He, K. Chen, P. Eksombatchai, W. L. Hamilton, and J. Leskovec, "Graph convolutional neural networks for web-scale recommender systems," In *Proceedings of the 24th ACM SIGKDD International Conference on Knowledge Discovery & Data Mining*, London, United Kingdom, pp. 974–983, July, 2018.

H. Zarzour, Y. Jararweh, M. M. Hammad, and M. Al-Smadi, M, "A long short-term memory deep learning framework for explainable recommendation," In *2020 11th International Conference on Information and Communication Systems (ICICS)*, Irbid, Jordan, pp. 233–237, IEEE, April, 2020.

Y. Zhang, and X. Chen, X, "Explainable recommendation: A survey and new perspectives," *Foundations and Trends(r) in Information Retrieval*, 14(1), 1–101, 2020.

J. Zhong, and E. Negre, "Shap-enhanced counterfactual explanations for recommendations," In *Proceedings of the 37th ACM/SIGAPP Symposium on Applied Computing*, Virtual event, pp. 1365–1372, April, 2022.

14 Metamorphic Testing for Trustworthy AI

Srinivas Padmanabhuni and Neelima Vobugari

14.1 INTRODUCTION

In the recent past, the topic of Artificial Intelligence (AI) has grabbed the attention of human beings, owing to the rapid strides this technology is making in terms of handling diverse human tasks. At a very basic level, the definition of AI can be as follows:

> ability of machines to exhibit human-like intelligence

This term was coined by John McCarthy in 1956 [1] by a small group of scientists who gathered in Dartmouth for the Summer Research Project on learning and related aspects. This event led to the birth of this field called AI. While AI has several manifestations, it is primarily developed and programmed by software developers in software or hardware embodied with intelligent software. Such software or hardware systems embodied with AI are termed *intelligent agents* [2]. The exact constituents of AI are a set of capabilities commonly termed AI components, which form the basis of calling an agent intelligent.

AI has become an integral part of our daily lives, being used across various industries to create intelligent agents that can perform tasks with increased speed and accuracy, reducing human effort. In other cases, they have been used in several scenarios where it is impossible for humans to act, or where the performance of these agents has surpassed the results of humans in the same tasks.

Due to the success of AI, there is now widespread adoption of AI across several use cases across diverse verticals and human work areas. Today, chatbots are handling customer service functions of all kinds of businesses, be it banks, financial services, retail sites, healthcare providers, or even government services. Likewise, automation powered by AI is very common today in manufacturing-focused industries like auto and aero and high-tech industries. In related applications, AI is now being used to increase human workers' efficiency in radiology, healthcare, clinical diagnosis, machine service, etc., to name a few. In some scenarios, the application is now reaching high utility value like in the case of automated driverless cars, automated financial product planning, etc., replacing human effort.

A closer look at the root cause of the success of AI reveals the coincidence of several factors at this time, all of which have come together to give rise to this phenomenal success. The primary factors essential for this AI success have been manifold. To discuss, let us explore the primary factors:

 DOI: 10.1201/9781003442509-14

a. **Big data:** Today there is a huge increase in data of different varieties being generated in several sectors, be it from social media, sensors, or from different internet sources like videos, etc. The availability of such large data of different varieties is a primary enabler of making practical AI possible, which can work on large numbers and provide interesting insights. Big data is usually characterized by large volumes of data, with high velocity and a large variety of formats [3].
b. **Cheap compute:** Today, the availability of complex computational hardware with high capacity is accessible to every person for use through the explosive growth of computational capacity, courtesy of Moore's Law [4]
c. **Scalable algorithms:** Today, the algorithms in AI based on modern deep learning and neural networks are effective for simulating human-like behavior, provided they are trained on large data.

14.2 IT IS NOT THAT ROSY?

While AI has made significant strides in overall applicability for improving and automating several human tasks, it is also to be noted that there is a significant downside to AI, leading experts, researchers, and governments to rethink the AI revolution in terms of potential incidents and accidents involving AI.

One angle of these visionaries and naysayers is based on the prophecy of AGI (Artificial General Intelligence) [5]. The idea is that over the due course of time, the evolution of AI algorithms will enable an intelligent agent to equal the capability of a human in all kinds of tasks. This scenario is often portrayed as a warning symbol for us to reconsider at the AI progress and maybe impose a moratorium on the same.

Another angle is the series of different accidents and incidents that lead to unacceptable behavior of the AI systems. A systematic effort [6] chronicles such incidents related to AI systems, giving AI systems a bad name.

14.2.1 AI Incidents

Let us look at some of the popular incidents that lend AI a bad name due to their implications.

1. In 2018, an Uber self-driving car hit and killed a pedestrian in Arizona [6]. This serious accident involving AI led to the loss of human lives.
2. Tesla's Autopilot system, designed to assist drivers with tasks such as steering, has been involved in several accidents, including a fatal crash where a Tesla on Autopilot hit a semi-truck.
3. Microsoft's chatbot Tay was launched on Twitter in 2016 to learn from user interactions, but it began posting offensive comments within hours, leading to its prompt shutdown.
4. Amazon abandoned an AI-driven recruitment tool in 2018 due to concerns that it exhibited bias against female candidates

5. A fraudulent attempt to scam victims of the collapse of the FTX exchange was made using a visual and audio deepfake of former CEO Sam Bankman-Fried. The deepfake was posted on Twitter and urged people to transfer funds into an anonymous cryptocurrency wallet.

This represents a sample of the several reported or unreported accidents of AI systems in reality. This raises a serious concern among AI developers: namely, how do we develop trustworthy AI systems?

14.3 TRUSTWORTHY AI SYSTEMS

As we have observed from the several reported incidents and accidents involving AI systems, there is a significant responsibility on the community of AI developers to address these concerns. Given this, there is a serious effort within the AI community of experts to develop frameworks, guidelines, and compliance laws.

While some of these, like the EU AI Act [7], are steps by governments to handle these in a manner to provide for safe and trustworthy AI, there are several standards bodies like ISO attempting to standardize the best practices such as the ISO/IEC 23053:2022 [8]. In addition to the initiatives from standards bodies and actions by governments, there is a popular set of notions coming from AI practitioners and companies toward trustworthy AI. One such popular notion being promoted by enterprises in the AI domain is the notion of Responsible AI [9]. Responsible AI pertains to utilizing and creating artificial intelligence (AI) systems that are ethical, transparent, and accountable.

Overall, we can refer to these initiatives as Trustworthy AI Initiatives. Trustworthy AI also requires assessing the possible societal impacts of AI systems and taking measures to mitigate any adverse effects. This includes tackling issues related to bias, discrimination, and unintended consequences, as well as ensuring that AI systems conform to broader societal objectives and values.

To achieve Trustworthy AI, various stakeholders such as AI developers, policymakers, civil society organizations, and impacted communities need to participate. By doing so, AI can be developed and utilized in ways that benefit everyone while minimizing the associated risks and potential harms that come with this rapidly evolving technology.

14.4 TESTING AS THE PATH TO TRUSTWORTHY AI

A key process in software development is the notion of Software Testing. The notion of testing in software is to precisely address the concerns raised earlier in trustworthy AI, namely identifying risks and reducing the related risks of the related software. As per The International Software Testing Qualifications Board (ISTQB), software testing is defined as below:

> The process consisting of all life cycle activities, both static and dynamic, concerned with planning, preparation, and evaluation of software products and related work products to determine that they satisfy specified requirements, to demonstrate that they are fit for purpose and to detect defects.

So, the basic goal of software testing is primarily to ascertain that the system under test (SUT) is performing as per expectations. In this context, a key notion becomes very important, namely, the notion of a test oracle [9]. Software testing techniques check the behavior of software products against the test oracle, which is a mechanism to determine whether the software behaves correctly.

14.5 TEST ORACLE IN SOFTWARE TESTING

The Test Oracle is an external mechanism to a program that checks the accuracy of the program's output for specific test cases. To test a program, the Test Oracle and the program under test are provided with the same test cases, and their outputs are compared to determine if the program has performed correctly. This comparison process is illustrated in Figure 14.1.

A Test Oracle is used to determine whether a test has passed or failed. It is a vital tool used to verify all of the software's functionalities during the testing process. If a Test Oracle is not available, it can make it difficult to carry out software testing. It serves as a reference point or a mechanism for verifying the accuracy of a program's output for specific test cases.

Very often in modern test automation scenarios, test oracles play a crucial role in creating scripts in the form of programs acting like test oracles to pass on the results of the automation to the developer. This allows the developer to get an idea of the corresponding failing tests, which can guide them to correct the programs.

14.6 WHY TEST ORACLE IS NOT POSSIBLE IN AI?

14.6.1 AI Is Based on ML

Modern AI systems are primarily powered by Machine Learning (ML). Machine learning can be defined in different dimensions. However, at the base of all these definitions is the notion of inductive reasoning, where based on a set of ground data and examples, a generic pattern or logical rule is discovered.

More precisely from [10], machine learning helps in developing methods that can automatically detect patterns in data and then use the uncovered patterns to predict the future. This has become the primary area of AI which has over the past few years given rise to several practical applications of AI. In all ML systems, there is a notion of "Data in, Program Out". Hence due to this, there is a strong dependence of the

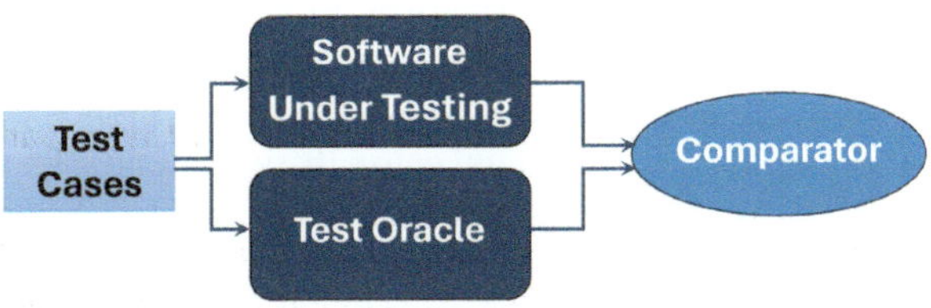

FIGURE 14.1 Test oracle.

output pattern/model on the input data, and all bugs and errors observed in the ML system, or discovered during the ML testing, can be attributed to the training data.

Overall, ML is broadly divided into three categories of popular problems:

a. **Supervised:** This kind of machine learning takes two kinds of data, namely input and output data. Supervised machine learning attempts to find the relationship or pattern of the input data influencing the output data.
b. **Unsupervised:** In this kind of machine learning, there is no notion of any output data. The given data (input) is analyzed for patterns.
c. **Reinforcement:** In this kind of ML, there is no historical data, and the system only learns from experience.

Most of the current AI work involving ML focuses on the idea of supervised ML, where a lot of effort is spent on ensuring that high-quality and correct output data is mapped for the corresponding input data. Further analysis reveals that supervised ML can be divided into two categories:

Classification: In this case, the output variable is discrete and can take only a few values called classes. For example, an incoming mail can be classified as either ham or spam, representing two classes.

Regression: This type of supervised ML requires a continuous value as the output variable. An example is stock price prediction based on previous values and news.

14.7 TEST ORACLE IMPOSSIBILITY FOR ML

ML exhibits probabilistic behavior due to the possibility of several different models for the same input data, which often raises concerns about the repeatability of the model for the same data. It is this probabilistic behavior which causes concerns concerning testing ML systems.

It is very difficult to provide a strictly defined test oracle for an ML system, due to the following reasons:

a. Defining the behavior of a program or model from input data is fraught with the complex task of covering all the input scenarios, which can be quite daunting.
b. The non-determinism of ML systems where there is no guarantee that the same model will be output for the same data every time.
c. As newer data appears, the model might drift, and it is extremely difficult to specify the changes in the test oracle over time.
d. Specification of the comprehensive behavior can be extremely challenging.
e. Issues such as overfitting, where there is a difference in the behavior of the model on training data and test data, cause difficulty in the specification.
f. High dimensionality of the input space makes the process of test generation a complex task.
g. Complex models with limited explainability make it impossible to specify the test oracle.

Given these specifics related to the underlying ML as the engine of AI programs, a test oracle approach to testing ML becomes highly complex and often impossible. There are several approaches to overcome this drawback which may include:

a. **Metamorphic testing:** Instead of a complete test oracle use a pseudo-oracle, which we will study in the next section.
b. **Human oracle:** In many scenarios, humans are the best judge of the correct functioning of the model. It is extremely difficult to perform test automation in such a scenario.
c. **Synthetic data generation:** It is somewhat difficult but often a good idea to try generating all possible input scenarios and simulate such scenarios.

14.8 METAMORPHIC TESTING IS THE ANSWER

Several studies [11–13] have suggested a variation of the notion of a "test oracle" to handle those scenarios where a test oracle is difficult to specify. Such variations are meant to broadly create artificial oracles, or pseudo-oracles, not exact oracles.

The basics of metamorphic testing (MT) [12] are explained in the following paragraphs.

The notion of MT applies to input test generation followed by test result verification.

Metamorphic testing is a technique that makes use of some necessary properties of the software under test, termed metamorphic relations.

Metamorphic relations are expected properties of the functionality of the software, specified without necessarily specifying all expected outputs for individual inputs, as done in a typical oracle. Hence, metamorphic testing based on the idea of metamorphic relations (MRs), is called a pseudo-oracle.

As reference [12] describes several MRs across different test scenarios, it is wise to note that MRs have been applied in complex scenarios where complete Oracle specification is impossible. Some areas where they have been applied are in GUI systems, Real-Time systems, AI, and ML systems, etc., to name a few.

14.8.1 MR Examples

Any MR specification is usually specified in terms of expected behavior for both original inputs and original outputs along with new modified inputs and expected newer outputs. Hence, it does not exactly match the expected output with the original output. Thus, this flexibility gives a lease to testing for complex systems like device drivers, graphics programs, real-time systems, AI, and Machine Learning systems, etc. Let us illustrate a couple of MRs below.

MR1: Let us consider the function sum(L) which returns the sum of all elements of a list L. Now, an MR like MR1 can be as follows:

If we permute the elements of L, giving a new list L1, then we expect sum(L) = sum(L)

This MR1 is fairly easy to understand, where we are specifying that on the permutation of the elements in a list, the sum remains unchanged.

Let us extend this to another MR, say MR2.

Metamorphic test cases for a function

Prod (num1, num2):

1) Prod (num1, num2) = Prod (num2, num1)
2) Prod (num, 0) = 0
3) Prod (num, 1) = num
4) Prod (positive_num1, positive_num2) > 0
5) Prod (positive_num1, negative_num2) < 0
6) Prod (negative_num1, positive_num2) < 0
7) Prod (negative_num1, negative_num2) > 0

FIGURE 14.2 MRs for product.

MR2: If we multiply all elements of a list L by m giving list L1, then sum(L1) = k * sum(L)

This is again fairly simple to examine that this property verifies the sum function without specifying the complete oracle.

A more comprehensive example is shown in Figure 14.2.

An example of extended use of MT is reported in [13] for imaging graphics use cases. As is clear, in graphic processing, it is impossible to specify the exact visual output. Often, MT provides the relevant expertise to test graphics drivers and systems.

14.9 METAMORPHIC TESTING FOR MACHINE LEARNING

As we have already shown, conventional test oracles are ineffective for testing ML applications. To overcome the lack of test oracles, the requisite technique that comes to our midst to help test ML applications is MT.

Several researchers in ML testing have used and proposed several different varieties of MRs to help test a diverse variety of ML applications. Works from [14–20] have given a sufficiently large number of studies that have been applied to a variety of ML applications with a wide variety of data types. In what follows we will illustrate a sample use case of an ML real-life problem where MRs are defined and bugs in ML programs discovered.

Consider the problem of image classification. Let us assume we have developed a complex CNN with different convolution, max pool, and FCN layers to recognize, let us say, masked versus unmasked faces. Let us assume that we propose different MRs as broadly as the accuracy level of the mask classifier will not vary for different transformations of the faces in the input data. For each transformation, we can have an MR. Let us list some MRs in this scenario as illustrated in Figure 14.3:

MR1: For rotation of a masked image up to 30°, the accuracy of the mask classifier does not change significantly (more than 1%).

MR2: For zoom up to 1.5× the accuracy of the mask classifier does not change significantly (more than 1%)

MRs broadly state that the accuracy level of the mask classifier does not vary significantly for image transformations. How does it help in finding corner cases/bugs?

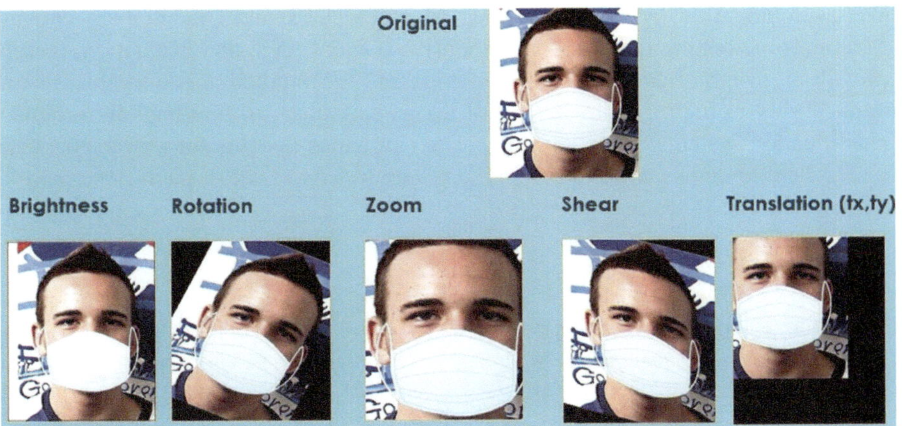

FIGURE 14.3 Different MRs for mask transformations.

In the real-life execution of the model on different variations of the transformed images, it will be observed that certain variations cause the accuracy to change significantly. Those transformations which alter the accuracy levels indicate potential corner cases/bugs. These cases indicate less presence of those variations in the training data, leading to such model performance. A resulting action would be to provide remedial training incorporating these corner cases to help make the model more robust.

14.10 EXTENSIONS

While we have illustrated generic MRs and specific MRs for image classification, across works in [14–20] several transformations of the input scenarios have been tried out. To illustrate, some of these are as follows:

1. Interchanging of class labels;
2. Permutation of an attribute;
3. Adding uninformative attribute;
4. Adding informative attributes
5. Multiplying numerical values by k;

These represent the typical transformations performed on the data and classes, and the resultant changes in accuracy observed. This would give a hint as to the underrepresented classes in the input scenario.

14.11 CONCLUSION

Trustworthy AI is a real need for today's scenario given the multitude of AI accidents, coupled with global AI compliance needs. We highlighted the need for a comprehensive test strategy for AI systems to be able to offer guarantees of trustworthiness. However, AI testing suffers from a crucial bottleneck of not having clear test oracle definitions.

We overcome this by exploring how AI researchers and practitioners can use Metamorphic testing to offer comprehensive test coverage and evaluation, thereby providing insight into potential corner cases. We studied metamorphic testing applicability to highlight the testing of image classifiers, to find bugs in the image classification systems. We hope this chapter provides insights for practitioners and researchers to devise ways to comprehensively test AI systems, paving the path for trustworthy AI.

REFERENCES

1. McCarthy, J., Minsky, M., Rochester, N., Shannon, C. (1955), *A Proposal for the Dartmouth Summer Research Project on Artificial Intelligence.* https://web.archive.org/web/20070826230310/https:/www-formal.stanford.edu/jmc/history/dartmouth/dartmouth.html.
2. Priya Pedamkar. (2023). Intelligent Agent in AI, EDUCBA Blog, last updated: 28 February 2023, https://www.educba.com/intelligent-agent-in-ai/
3. Laney, D. (2001). *3D Data Management: Controlling Data Volume, Velocity, and Variety.* META Group, Rome.
4. G. E. Moore. (1998). "Cramming More Components Onto Integrated Circuits," in *Proceedings of the IEEE*, vol. 86, no. 1, pp. 82–85, January 1998, DOI: 10.1109/JPROC.1998.658762.
5. "Impressed by artificial intelligence? Experts say AGI is coming next, and it has 'existential' risks". *ABC News*, 23 March 2023. https://www.abc.net.au/news/2023-03-24/what-is-agi-artificial-general-intelligence-ai-experts-risks/102035132.
6. Uber's self-driving operator charged over fatal crash - *BBC News*. https://www.bbc.com/news/technology-54175359.
7. The Artificial Intelligence Act. https://artificialintelligenceact.eu/.
8. ISO/IEC 23053:2022- Framework for Artificial Intelligence (AI) Systems Using Machine Learning (ML).
9. Howden, W. E. (1978). "Theoretical and empirical studies of program testing," *IEEE Trans. Softw. Eng.*, vol. 4, no. 4, pp. 293–298.
10. Murphy, K. P. (ed.) (2012). *Machine Learning: A Probabilistic Perspective.* MIT Press, Cambridge, MA.
11. Riccio, V., Jahangirova, G., Stocco, A., et al. (2020). "Testing machine learning based systems: A systematic mapping," *Empir Softw. Eng.*, vol. 25, pp. 5193–5254. https://doi.org/10.1007/s10664-020-09881-0.
12. Chen, T. Y., Kuo, F.-C., Liu, H., Poon, P.-L., and Zhou, Z. Q. (2018). "Metamorphic testing: A review of challenges and opportunities," *ACM Comput. Surv.*, vol. 51, no. 1, pp. 4:1–4:27.
13. Groce, A., Kulesza, T., Zhang, C., Shamasunder, S., and McIntosh, K. (2014). "You are the only possible oracle: Effective test selection for end users of interactive machine learning systems," *IEEE Trans. Software Eng.*, vol. 40, no. 3, pp. 307–323.
14. Guderlei, R., and Mayer, J. (2007). "Towards automatic testing of imaging software using random and metamorphic testing," *Int. J. Softw. Eng. Knowl. Eng.*, vol. 17, no. 6, pp. 757–781.
15. Xie, X., Ho, J.W. K., Murphy, C., Kaiser, G., Xu, B., and Chen, T. Y. (2011). "Testing and validating machine learning classifiers by metamorphic testing," *J. Syst. Softw.*, vol. 84, no. 4, pp. 544–558.
16. Murphy, C., Kaiser, G. E., and Hu, L., "Properties of machine learning applications for use in metamorphic testing," Tech. Rep., June 2008.

17. Murphy, C., and Kaiser, G. E., "Automatic detection of defects in applications without test oracles," Tech. Rep., June 2010.
18. Nakajima, S., and Chen, T. Y. (2019). "Generating biased dataset for metamorphic testing of machine learning programs," In Gaston, C., Kosmatov, N., Le Gall, P. (eds.) *Testing Software and Systems*, pp. 56–64. Springer International Publishing, Cham.
19. Herbold, S., and Haar, T., "Smoke testing for machine learning: Simple tests to discover severe defects," arXiv:2009.01521, 2020.
20. Xie, X., Ho, J., Murphy, C., Kaiser, G., Xu, B., and Chen, T. Y. (2009). "Application of metamorphic testing to supervised classifiers," In *2009 Ninth International Conference on Quality Software*, Jeju, Korea, 24–25 August, pp. 135–144.

15 Software for Explainable AI

Manthan Sanghavi

15.1 INTRODUCTION

Deep neural networks have made significant progress in recent years, leading to their wide-scale adoption in different domains and industries. However, one of the biggest challenges faced is the need for more trustworthiness and interpretability of AI models, as it becomes difficult to understand the predictions, identify biases, detect potential errors, etc. Several algorithms and techniques have been developed to address these challenges. Explainable AI as a research area is rapidly growing and focuses on providing explanations to aid decision-making better. The aim is to make AI models so transparent and interpretable that it is easy for humans to understand what led the AI model to arrive at a prediction (Gunning and Aha 2019; Doshi and Kim 2017; Rudin 2018).

In recent years, several open-source software have been developed to implement different explainable AI systems. These tools aid developers in better interpreting the predictions from AI models, identifying biases, and detecting errors, empowering the developers to monitor better forecasts coming from the AI models while ensuring transparency and accountability (Guidotti et al. 2018; Lipton 2016).

The chapter also provides examples of explainable AI techniques using software packages like Local Interpretable Model-agnostic Explanations (LIME), SHapley Additive exPlanations (SHAP), iNNvestigate, etc. It discusses the features, capabilities, and limitations of these software packages. In conclusion, this chapter provides a guide to software developers to develop better AI models that are transparent, trustworthy, and interpretable (Ribeiro, Singh, and Guestrin 2016; Lundberg and Lee 2017; Molnar 2020; Alber, Lapuschkin, and Seegerer 2019).

This chapter introduces modern software tools and frameworks for building explanatory AI systems. It covers various techniques and approaches used to make AI models explicable, such as model-agnostic methods, custom backpropagation, gradient-based methods, etc. It addresses the challenges of implementing explainable AI systems.

By the end of the chapter, readers will gain a deeper understanding of how to design and develop AI models, and we discuss this by showing step-by-step how explanatory methods can be applied effectively and highlighting basic design patterns.

 DOI: 10.1201/9781003442509-15

15.2 IMPLEMENTING EXPLANATION ALGORITHMS

Implementing explanation algorithms with neural networks can be daunting due to the increased complexity and potential for error. This section introduces critical patterns of explanation algorithms that lend themselves to efficient and structured implementation. It also describes how to approach designing interfaces and tuning parameters.

To make the code more useful, we will use a real-life example by implementing the popular Wide and Deep (Cheng et al. 2016) recommender model with a "movie lens" dataset (Harper, n.d.) for movie recommendations using the PyTorch framework (Paszke et al. 2019). Then, we will provide a step-by-step guide to implementing different explainable AI algorithms using LIME, SHAP, and iNNvestigate.

The model architecture consisted of two parts: wide and deep. The wide part is used to learn interactions between input features using cross-product transformations. In contrast, the deep part is used to understand complex feature representations using stacks of fully connected layers.

The Wide and Deep architecture is prevalent in recommender systems because the model can capture low-order and high-order feature interactions. Low-order interactions refer to simple co-occurrences of input features, such as the user's liking for specific genres of movies. On the other hand, higher-order interactions refer to more complex relationships between input features, such as, a user preferring movies released in a particular year within a specific genre. We used a combination of mean squared error (MSE) loss and binary cross entropy (BCE) loss to train the model. MSE loss was used to predict a user's rating for a particular movie, and BCE loss was used to indicate whether a user would like a specific film.

The training loop consisted of iterating over the training set and computing the loss for each batch of data. We then used the optimizer to update the model parameters based on the gradients calculated by backpropagation. After each epoch, we evaluated the model on the test set, as shown in the code depicted in Figure 15.1.

15.2.1 Model Agnostic Methods

Model-agnostic strategies are a category of XAI procedures that can be connected to any machine learning model, regardless of its design or complexity. This can be contrasted with model-specific strategies, which are designed to work only with a specific sort of model, such as choice trees or neural systems. The critical advantage of model-agnostic strategies is their flexibility and adaptability, which permits them to be utilized across various applications and domains (Lundberg and Lee 2017; Ribeiro, Singh, and Guestrin 2016; Doshi and Kim 2017; Samek, Wiegand, and Müller 2017; Guidotti et al. 2018).

Model-agnostic strategies provide clarifications at a higher level of reflection, as they are not tied to the inner workings of the demonstration. Instead, they treat the demonstration as a black box and point to get its behaviour based on its input-output relationship. This makes model-agnostic strategies especially valuable for complex models that are difficult to interpret, such as deep neural networks.

```
import pandas as pd
import torch
import torch.nn as nn
import torch.optim as optim
from sklearn.model_selection import train_test_split
from sklearn.preprocessing import LabelEncoder, MinMaxScaler

# Load the MovieLens 100K dataset
ratings_df=pd.read_csv("http://files.grouplens.org/datasets/movielens/ml-100k/u.data",
sep="\t", header=None, names=["user_id", "item_id", "rating", "timestamp"])

items_df =pd.read_csv("http://files.grouplens.org/datasets/movielens/ml-100k/u.item", sep="|",
encoding="latin-1",    header=None,    names=["item_id",    "title",    "release_date",
"video_release_date",  "imdb_url",  "unknown",  "Action",  "Adventure",  "Animation",
"Children's", "Comedy", "Crime", "Documentary", "Drama", "Fantasy", "Film-Noir", "Horror",
"Musical", "Mystery", "Romance", "Sci-Fi", "Thriller", "War", "Western"])

# Encode categorical features
user_encoder = LabelEncoder()
item_encoder = LabelEncoder()
genre_encoder = LabelEncoder()

ratings_df["user_id"] = user_encoder.fit_transform(ratings_df["user_id"])
ratings_df["item_id"] = item_encoder.fit_transform(ratings_df["item_id"])
items_df["item_id"] = item_encoder.transform(items_df["item_id"])
items_df["genres"] = items_df.iloc[:, 6:].idxmax(axis=1)
items_df["genres"] = genre_encoder.fit_transform(items_df["genres"])

# Normalize numerical features
scaler = MinMaxScaler()
ratings_df["rating"] = scaler.fit_transform(ratings_df[["rating"]])

# Split the dataset into training and testing sets
train_df, test_df = train_test_split(ratings_df, test_size=0.2)

# Define the wide and deep model
class WideAndDeep(nn.Module):
  def __init__(self, num_users, num_items, num_genres, embed_dim):
    super().__init__()
    self.embed_dim = embed_dim
    self.user_embed = nn.Embedding(num_users, embed_dim)
    self.item_embed = nn.Embedding(num_items, embed_dim)
    self.genre_embed = nn.Embedding(num_genres, embed_dim)
    self.fc_layers = nn.Sequential(
      nn.Linear(embed_dim * 3, 128),
      nn.ReLU(),
      nn.Dropout(0.5),
      nn.Linear(128, 64),
      nn.ReLU(),
      nn.Dropout(0.5),
      nn.Linear(64, 1)
    )

  def forward(self, user_ids, item_ids, genre_ids):
    user_emb = self.user_embed(user_ids)
    item_emb = self.item_embed(item_ids)
    genre_emb = self.genre_embed(genre_ids)
    x = torch.cat([user_emb, item_emb, genre_emb], dim=1)
    x = self.fc_layers(x)
    return x.view(-1)

# Instantiate the model
model=WideAndDeep(num_users=len(user_encoder.classes_),
num_items=len(item_encoder.classes_),
num_genres=len(genre_encoder.classes_), embed_dim=16)

# Define the loss function and optimizer
criterion = nn.MSELoss()
optimizer = optim.Adam(model.parameters(), lr=0.001)
```

FIGURE 15.1 A code snippet for a Wide and Deep recommender model using PyTorch on the MovieLens dataset.

Some well-known cases of model-agnostic strategies incorporate LIME and SHAP. These strategies function by perturbing the input data differently and observing the changes in the model's output. By analysing these changes, it can be understood how the model works and which features are most important for its predictions. Generally, model-agnostic strategies are an important tool for understanding and interpreting machine learning models, and they play a critical part in making AI more transparent and dependable. In the following examples, we will demonstrate how model-agnostic approaches can be implemented in our Wide and Deep example.

15.2.1.1 SHAP

We first create a SHAP descriptor for our model in the code below. Then, we select a random user and movie and extract the input attributes of the selected user and movie. Finally, we use the descriptor to calculate the SHAP values for the selected user and movie.

SHAP is a popular XAI technique that allows us to explain the predictions of a machine-learning model by dividing the output into the contributions of each input function. In this example, we used SHAP to explain the predicted rating for a given user and movie. The SHAP values represent the contribution of each input feature to the expected score. By using XAI techniques such as SHAP, we gain insight into how broadly and deeply our recommendation model makes its predictions. This can be useful in determining which features are most important for providing accurate recommendations and identifying potential model anomalies or limitations (Lundberg and Lee 2017) (Figure 15.2).

15.2.1.2 LIME

In this code, we first select a random user and movie and extract the input properties of the selected user and movie. Then, we create a LimeTabularExplainer that explains regression models using tabular data. We define a predictor function that accepts a matrix of inputs and returns the predicted scores for each row. Next, we use the explainer to calculate the predicted rating explanation given the input characteristics of the selected user and movie. Finally, we display Lime's explanation using the show_in_notebook method. LIME is another popular XAI technique that allows us to clarify the predictions of a machine learning model by training a local interpretable model on a subset of the inputs. In this example, we used LIME to explain the predicted rating for a given user and movie. The legend of Lime shows the effect of each input feature on the predicted score, along with the sign and magnitude of the contribution (Ribeiro, Singh, and Guestrin 2016).

Using XAI techniques such as LIME, we gain insight into how broadly and deeply our recommendation model makes its predictions. This insight can be useful for determining which features are most important for making accurate recommendations and for identifying potential model anomalies or limitations (Figure 15.3).

15.2.2 Gradient-Based Methods

Gradient-based methods are a class of XAI techniques that use the gradient of a model to calculate feature importance scores. These methods work by calculating

```
import shap

# Create a SHAP explainer
explainer = shap.Explainer(model)

# Select a random user and movie
user_id = 25
item_id = 150

# Get the input features for the selected user and movie

user_features=user_df.loc[user_df["user_id"]==user_id].drop("user_id",axis=1).values.flatten(
)

item_features = items_df.loc[items_df["item_id"] == item_id].drop(["item_id", "title",
"genres"], axis=1).values.flatten()
genre_id = items_df.loc[items_df["item_id"] == item_id, "genres"].values[0]

# Compute the SHAP values for the selected user and movie
shap_values=explainer.shap_values([torch.tensor(user_features), torch.tensor(item_features),
torch.tensor(genre_id)])

# Print the SHAP values
print(shap_values)
```

FIGURE 15.2 A code snippet for SHAP values computation for a Wide and Deep recommender model.

```
import lime.lime_tabular
import numpy as np

# Select a random user and movie
user_id = 25
item_id = 150

# Get the input features for the selected user and movie
user_features=user_df.loc[user_df["user_id"]==user_id].drop("user_id",
axis=1).values.flatten()

item_features = items_df.loc[items_df["item_id"] == item_id].drop(["item_id", "title",
"genres"], axis=1).values.flatten()

genre_id = items_df.loc[items_df["item_id"] == item_id, "genres"].values[0]

# Create a LimeTabularExplainer
explainer = lime.lime_tabular.LimeTabularExplainer(
  np.concatenate([user_df.drop("user_id", axis=1).values,
items_df.drop(["item_id", "title", "genres"], axis=1).values], axis=1),
  feature_names=user_df.columns.tolist() + items_df.drop(["item_id", "title", "genres"],
axis=1).columns.tolist() + ["genre"],
  class_names=["rating"],
  verbose=False,
  mode="regression"
)


# Compute the Lime explanation for the predicted rating
explanation = explainer.explain_instance(
  np.concatenate([user_features, item_features, np.array([genre_id])], axis=0),
  predict_fn,
  num_features=len(user_df.columns)+len(items_df.drop(["item_id", "title", "genres"],
axis=1).columns)+1
)
```

FIGURE 15.3 A code snippet for Lime explanation for Wide and Deep recommender model.

the gradient of a model's output concerning its input features, providing a measure of how much each feature influences the model's predictions (Simonyan, Vedaldi, and Zisserman 2013; Springenberg et al. 2014; Selvaraju 2016).

The main advantage of gradient-based methods is that they provide global and local explanations of model behaviour. Global annotations refer to the general behaviour of the model across its input space, while local annotations describe the model's behaviour for a specific input. In model gradient analyses, gradient-based methods can provide both types of explanations. Some popular examples of gradient-based methods are Integrated Gradients, Gradient*Input, and SmoothGrad. These methods differ in their approach to calculating feature importance scores, but they all depend to some extent on the gradient of the model (Bach et al. 2015).

One challenge of gradient-based methods is their sensitivity to small changes in the model or input data, leading to unstable or unreliable explanations. Researchers have proposed various modifications to these methods to address this challenge, such as regularization or smoothing techniques (Figure 15.4).

In general, gradient-based methods are a powerful tool for explaining the behaviour of complex machine learning models and play an important role in making artificial intelligence transparent and interpretable.

In this code, we first load the trained Wide and Deep model. We then define the IntegratedGradients method using the IntegratedGradients class of the Captum library (Adebayo et al. 2018). Next, we define an input tensor representing one pair of user objects. Finally, we calculate the object's importance score using the completeness property of the Integrated Gradients object, passing the input tensor and the target class (1 for positive evaluations). The output of the attribute method is a tensor with the same form as the input tensor, representing the feature importance scores for each input feature. These scores can be interpreted as a measure of how much each feature affects the model's prediction for a given input.

15.2.3 Custom Backpropagation

Custom backpropagation methods are a group of XAI methods that involve modifying the standard backpropagation algorithm to compute feature importance scores for a given input. These methods require the ability to examine the model and adapt to its composition and usually involve modifying the gradients computed in backpropagation in some way (Mirza and Osindero 2014; Sukhija et al. 2022).

One example of a common XAI backpropagation method is layer-based importance propagation (LRP), which is based on the idea of back-propagating importance scores (i.e., characteristic importance scores) in a model using a given rule-dependent activation function used in each layer. LRP is effective in explaining the predictions of neural networks in a variety of tasks (Bach et al. 2015).

Another example of a custom XAI backpropagation method is guided backpropagation, which modifies the backpropagation algorithm to propagate only positive gradients through the network, effectively highlighting features that positively affect the prediction. This method has proven to be effective in visualizing the parts of an image that are most important for a given classification.

```
import torch
import torch.nn.functional as F
from captum.attr import IntegratedGradients

# Load the trained model
model = WideAndDeep(num_users, num_items, embed_dim, hidden_dim, dropout)
model.load_state_dict(torch.load('wide_deep_model.pt'))

# Define the Integrated Gradients method
ig = IntegratedGradients(model)

# Define an input tensor
user_id = torch.tensor([0])
item_id = torch.tensor([0])
input_tensor = (user_id, item_id)

# Compute the feature importance scores
ig_attributions = ig.attribute(input_tensor, target=1)

# Print the feature importance scores
print('Integrated Gradients feature attributions:')
print(ig_attributions)
```

FIGURE 15.4 A code snippet for feature importance computation using integrated gradients for a Wide and Deep recommender model.

Adapted backpropagation XAI methods provide a flexible and efficient way to calculate feature importance scores for complex models. They are particularly well-suited to neural networks due to their compositional nature. However, these methods can be computationally expensive and require expertise to define appropriate rules or modifications to the backpropagation algorithm (Figure 15.5).

In this code, we first load the pre-trained Wide and Deep model and create a sample stream for clarification. We then use the research library to create parsers for LRP and guided backpropagation and use these parsers to compute explanations of the sample input. Finally, we print the five most important features of each explanatory method. Note that investigate library provides a convenient way to create custom LRP-based analysers, as it allows us to specify the LRP rule to use (lrp.alpha_1_beta_0 in this case).

Overall, this code is an example of how to calculate the explanations using standard XAI back propagation methods, such as LRP and guided backpropagation, for the broad and deep model above.

15.3 HYPERPARAMETER SELECTION

Hyperparameter determination could be vital in utilizing XAI methods like iNNvestigate, LIME, SHAP, and Captum viably. The hyperparameters determine how the XAI strategy works and can critically affect the quality of the clarifications produced. Below are a few consideration for hyperparameter selection for each strategy within the context of the outstanding recommender model.

For iNNvestigate, hyperparameters to consider incorporate the choice of investigation strategy (e.g., Layer-wise Pertinence Engendering, Slope, Profound Taylor Decay), which can affect the precision and interpretability of the clarifications

```
import torch
import numpy as np
import innvestigate
from innvestigate.utils.tests.networks import base as network_base
from innvestigate.analyzer import LRP
from innvestigate.analyzer import GuidedBackpropagation


# Define the model architecture and load pre-trained weights
model = WideAndDeep(num_users, num_items, num_genres, embedding_dim, hidden_dim)
model.load_state_dict(torch.load('model_weights.pt'))


# Create a sample input for explanation
sample_input = torch.tensor([1, 1, 1], dtype=torch.long)


# Compute LRP-based explanation
lrp_analyzer = innvestigate.create_analyzer("lrp.alpha_1_beta_0", model)
lrp_explanation = lrp_analyzer.analyze(sample_input)


# Compute Guided Backpropagation-based explanation
guided_backprop_analyzer = GuidedBackpropagation(model)
guided_backprop_explanation = guided_backprop_analyzer.analyze(sample_input)


# Print the top 5 most important features for each explanation method
num_top_features = 5
lrp_top_features = np.argsort(lrp_explanation)[-num_top_features:][::-1]
guided_backprop_top_features = np.argsort(guided_backprop_explanation)[-
num_top_features:][::-1]
print(f'Top {num_top_features} features using LRP: {lrp_top_features}')
print(f'Top {num_top_features} features using Guided Backpropagation:
{guided_backprop_top_features}')
```

FIGURE 15.5 A code snippet for feature importance computation using LRP and guided backpropagation for a Wide and Deep recommender model.

produced. Other vital hyperparameters incorporate the choice of reference input and the regularization parameter, which can affect the sensitivity and firmness of the clarifications generated.

For LIME and SHAP, vital hyperparameters to consider incorporate the choice of highlighting irritation strategy, the number of tests utilized for clarification, and the complexity of the neighbourhood surrogate demonstration utilized to approximate the behaviour of the original demonstration. These hyperparameters can affect the precision and soundness of the clarifications produced and the runtime required to create the explanations.

For Captum, vital hyperparameters to consider incorporate the choice of attribution strategy (e.g., Coordinate Slopes, DeepLift), the pattern input utilized to compute the attribution values, and the number of steps used to surmise the indispensibility over the input space. These hyperparameters can affect the accuracy and interpretability of the clarifications created and the runtime required to create them.

```
import numpy as np
from innvestigate.analyzer import create_analyzer
from sklearn.model_selection import ParameterSampler

# Define hyperparameters to optimize
params = {
  "method": ["lrp.z", "lrp.alpha_2_beta_1", "guided_backprop", "deep_taylor"],
  "reference_input": [np.zeros((1, 32)), np.random.rand(1, 32)],
  "tau": [1e-6, 1e-5, 1e-4, 1e-3, 1e-2],
}

# Define function to evaluate model with given hyperparameters
def evaluate_model(params):
  # Create iNNvestigate analyzer with given hyperparameters
  analyzer = create_analyzer(params["method"], model, **params)

  # Analyze sample input and output predictions
  analysis = analyzer.analyze(sample_input)
  prediction = model(sample_input)

  # Compute relevance-weighted output
  relevance = np.sum(analysis, axis=1)
  weighted_output = prediction * relevance

  # Compute accuracy and interpretability metrics
  accuracy = weighted_output[0, target_class]
  interpretability = np.sum(np.abs(relevance))

  return accuracy, interpretability

# Sample hyperparameters and evaluate model
n_iter = 100
random_search = ParameterSampler(params, n_iter=n_iter, random_state=123)
results = []
for i, param_set in enumerate(random_search):
  accuracy, interpretability = evaluate_model(param_set)
  results.append((accuracy, interpretability, param_set))
  print(f"Iteration {i+1}/{n_iter} - Accuracy: {accuracy:.4f}, Interpretability:
{interpretability:.4f}")

# Print best hyperparameters and results
best_result = sorted(results, key=lambda x: x[0], reverse=True)[0]
print(f"\nBest result: Accuracy: {best_result[0]:.4f}, Interpretability: {best_result[1]:.4f}")
print(f"Best parameters: {best_result[2]}")
```

FIGURE 15.6 A code snippet for hyperparameter selection using random search for the Wide and Deep model.

In common, the choice of hyperparameters for XAI strategies ought to be educated by the particular characteristics of the model and data being analysed, as well as the objectives of the investigation. It is valuable to regularly test with diverse hyperparameter settings and assess the quality and interpretability of the clarifications produced to choose the finest settings for a given application. Furthermore, it is imperative to carefully report the hyperparameters utilized in XAI examinations to guarantee reproducibility and straightforwardness (Figure 15.6).

15.4 CHALLENGES

Whereas LIME, SHAP, Captum, and iNNvestigate are capable model-agnostic XAI methods, there are a few challenges related to their use.

One major challenge with these methods is that they can be computationally costly, particularly with larger and more complex models. Lime and SHAP, for example, depend on testing and model retraining to generate clarifications, which can be time-consuming and resource-intensive. Applying these procedures in real-time or on large datasets could prove challenging.

Another challenge is that these strategies may need to provide clear and noteworthy clarifications continuously. Whereas they can distinguish the input highlights that are most critical for a given expectation, they may not give a clear understanding of how these features contribute to the expectation. This may make it troublesome for clients to interpret and act on the clarifications provided.

In the context of model interpretability methods, there is a risk of bias and other errors if these methods are not used judiciously. For instance, when constructing an explanation model, if the training data used is biased, that bias can propagate into the explanations generated by techniques like LIME and SHAP. Similarly, if the model is poorly calibrated or overfit to the training data, the explanations produced by these methods may not accurately reflect the true behavior of the model.

There are challenges related to deciphering the clarifications produced by these strategies. Whereas Lime and SHAP provide feature significance scores, determining how these scores ought to be deciphered can be difficult. Similarly, Captum and iNNvestigate can create heat maps and other visualizations to highlight important features. However, it can be problematic to perceive how these visualizations correspond to the fundamental model behaviour.

Although Lime, SHAP, Captum, and iNNvestigate are capable and valuable XAI strategies, they should be utilized with caution and in combination with other approaches to guarantee that their limitations and biases are legitimately accounted for.

15.5 FUTURE STUDIES

The field of interpretable artificial intelligence (XAI) is constantly evolving, and there are several promising areas for future research and development of XAI software. In this section, we discuss possible future research that can improve the functionality and usability of XAI software. Advanced XAI Techniques: While current XAI techniques such as LIME, SHAP, and gradient-based approaches have proven to be effective, there is still room for improvement. Future research can focus on developing better XAI techniques that provide more accurate, interpretable, and reliable explanations. This may include exploring new interpretation algorithms, incorporating domain information, or using advanced machine learning and statistical techniques (Ribeiro, Singh, and Guestrin 2016). Improvements in the visualization and user interfaces of XAI software can significantly improve the usability and adoption of XAI technologies. Future research may explore new visualization techniques to explain explanations more intuitively. User interfaces can be designed so that users can interact with explanations, explore specific properties, and gain more insight into the model's decision-making process (Strobelt et al., n.d.). Integrating human-centred design principles into XAI software development is critical to ensure that the explanations provided are relevant, useful, and understandable to end users. Future research could investigate user needs, preferences, and mental health patterns when interacting with XAI systems. This can include user research, feedback loops, and iterative design processes to create XAI software that meets the specific requirements and expectations of different user groups (Doshi and Kim 2017). Future studies may focus on objective evaluation criteria and benchmark data to compare XAI techniques. In addition, user studies can be conducted to assess the subjective quality and usefulness of the explanations, which can provide valuable information for further refinement and improvement. As XAI becomes more common, there is a need to address the ethical and legal implications of using XAI software. Future research should clarify frameworks and guidelines to ensure fairness, transparency, and accountability in XAI systems. This includes compliance with biases, privacy concerns, and regulations such as GDPR and data protection laws (Guidotti et al. 2018). The seamless integration of XAI software into the development pipelines of machine learning projects can facilitate its widespread adoption. Future research may focus on developing tools and frameworks that allow easy integration of XAI techniques into existing machine learning workflows, such as popular deep learning frameworks or automated ML platforms (Holzinger et al. 2018). XAI software should be tested and improved in different real-world applications and domains. Future research can explore the application of XAI software in critical areas such as healthcare, finance, autonomous systems, and cybersecurity. In summary, future XAI software research should strive for excellence in XAI techniques, improve usability through better visualization and user interfaces, consider human-centred design principles, develop evaluation metrics, and address ethical and legal aspects. Additionally, it should explore real-world applications and continuously develop and improve XAI software to meet the changing needs and challenges of explainable AI.

15.6 CONCLUSIONS

In summary, XAI is a rapidly growing field that offers many opportunities for software development. As machine learning and artificial intelligence grow in importance across industries, explaining and interpreting their predictions and decisions is critical. Software that can provide such explanations must build trust in AI models and ensure their ethical and responsible use.

This chapter discussed a few of the many XAI methods, including model-independent methods such as LIME and SHAP, gradient-based methods, and adaptive backpropagation methods such as LRP and guided backpropagation. An important consideration when implementing XAI software is the choice of hyperparameters. This requires choosing appropriate parameters for the XAI algorithm and hyperparameters for the underlying machine-learning model. This complex task requires careful consideration of your specific use case and dataset.

Despite the progress of XAI, there are still many challenges to overcome. These challenges include the trade-off between interpretability and accuracy, the need for large training data, and the difficulty of explaining deep neural networks. However, continued research and development of XAI software help meet these challenges and enable responsible and reliable use of artificial intelligence.

REFERENCES

Adebayo, Julius, Justin Gilmer, Michael Muelly, and Ian Goodfellow. 2018. "Sanity Checks for Saliency Maps." NIPS papers. https://papers.nips.cc/paper_files/paper/2018/hash/294a8ed24b1ad22ec2e7efea049b8737-Abstract.html.

Alber, Maximilian, Sebastian Lapuschkin, and Philipp Seegerer. 2019. "albermax/innvestigate: A toolbox to iNNvestigate neural networks' predictions!" GitHub. https://github.com/albermax/innvestigate.

Bach, Sebastian, Alexander Binder, Grégoire Montavon, Frederick Klauschen, Klaus-Robert Müller, and Wojciech Samek. 2015. "On Pixel-Wise Explanations for Non-Linear Classifier Decisions by Layer-Wise Relevance Propagation." PLOS. https://journals.plos.org/plosone/article?id=10.1371/journal.pone.0130140.

Cheng, Heng-Tze, Levent Koc, Jeremiah Harmsen, Tal Shaked, Tushar Chandra, Hrishi Aradhye, Glen Anderson, et al. 2016. "[1606.07792] Wide & Deep Learning for Recommender Systems." arXiv. https://arxiv.org/abs/1606.07792.

Doshi, Finale, and Been Kim. 2017. "[1702.08608] Towards A Rigorous Science of Interpretable Machine Learning." arXiv. https://arxiv.org/abs/1702.08608.

Guidotti, Riccardo, Anna Monreale, Salvatore Ruggieri, Franco Turini, Dino Pedreschi, and Fosca Giannotti. 2018. "[1802.01933] A Survey of Methods For Explaining Black Box Models." arXiv. https://arxiv.org/abs/1802.01933.

Gunning, David, and David W. Aha. 2019. "DARPA's Explainable Artificial Intelligence (XAI) Program | AI Magazine." AAAI Publications. https://ojs.aaai.org/aimagazine/index.php/aimagazine/article/view/2850.

F. Maxwell Harper and Joseph A. Konstan. 2015. The MovieLens Datasets: History and Context. *ACM Transactions on Interactive Intelligent Systems*, 5(4), Article 19 (January 2016), pp. 1–19. DOI: 10.1145/2827872

Holzinger, Andreas, Markus Plass, Michael Kickmeier-Rust, and Katharina Holzinger. 2018. "Interactive machine learning: experimental evidence for the human in the algorithmic loop." *Applied Intelligence* 49, 2401–2414.

Lipton, Zachary C. 2016. "[1606.03490] The Mythos of Model Interpretability." arXiv. https://arxiv.org/abs/1606.03490.

Lundberg, Scott, and Su-In Lee. 2017. "[1705.07874] A Unified Approach to Interpreting Model Predictions." arXiv. https://arxiv.org/abs/1705.07874.

Mirza, Mehdi, and Simon Osindero. 2014. "[1411.1784] Conditional Generative Adversarial Nets." arXiv. https://arxiv.org/abs/1411.1784.

Molnar, Christoph. 2020. "Interpretable Machine Learning." https://christophm.github.io/interpretable-ml-book/.

Paszke, Adam, Sam Gross, Francisco Massa, Adam Lerer, and James Bradbury. 2019. "[1912.01703] PyTorch: An Imperative Style, High-Performance Deep Learning Library." arXiv. https://arxiv.org/abs/1912.01703.

Ribeiro, Marco Tulio, Sameer Singh, and Carlos Guestrin. 2016. "[1602.04938] "Why Should I Trust You?": Explaining the Predictions of Any Classifier." arXiv. https://arxiv.org/abs/1602.04938.

Rudin, Cynthia. 2018. "[1811.10154] Stop Explaining Black Box Machine Learning Models for High Stakes Decisions and Use Interpretable Models Instead." arXiv. https://arxiv.org/abs/1811.10154.

Samek, Wojciech, Thomas Wiegand, and Klaus-Robert Müller. 2017. "[1708.08296] Explainable Artificial Intelligence: Understanding, Visualizing and Interpreting Deep Learning Models." arXiv. https://arxiv.org/abs/1708.08296.

Selvaraju, Ramprasaath R. 2016. "[1610.02391] Grad-CAM: Visual Explanations from Deep Networks via Gradient-based Localization." arXiv. https://arxiv.org/abs/1610.02391.

Simonyan, Karen, Andrea Vedaldi, and Andrew Zisserman. 2013. "[1312.6034] Deep Inside Convolutional Networks: Visualising Image Classification Models and Saliency Maps." arXiv. https://arxiv.org/abs/1312.6034.

Springenberg, Jost Tobias, Alexey Dosovitskiy, Thomas Brox, and Martin Riedmiller. 2014. "[1412.6806] Striving for Simplicity: The All Convolutional Net." arXiv. https://arxiv.org/abs/1412.6806.

H. Strobelt, S. Gehrmann, M. Behrisch, A. Perer, H. Pfister and A. M. Rush. 2019. "Seq2seq-Vis: A Visual Debugging Tool for Sequence-to-Sequence Models," *IEEE Transactions on Visualization and Computer Graphics*, vol. 25, no. 1, pp. 353–363, Jan. 2019, DOI: 10.1109/TVCG.2018.2865044.

Sukhija, Bhavya, Nathanael Köhler, Miguel Zamora, Simon Zimmermann, Sebastian Curi, Andreas Krause, and Stelian Coros. 2022. "[2204.04558] Gradient-Based Trajectory Optimization with Learned Dynamics." arXiv. https://arxiv.org/abs/2204.04558.

16 Interpretation and Visualization Techniques in AI Systems and Applications

Arka De, Sameeksha Saraf, Tusar Kanti Mishra, and B. K. Tripathy

16.1 INTRODUCTION

Artificially intelligent (AI) systems, in recent times, have revolutionized the interaction of humans with technology. These AI systems contain the potential to transform various industries, from Healthcare to Finance, Autonomous vehicles to Judicial Systems. However, the black box nature, which is the undefined working of these AI models and systems, poses a significant challenge in understanding their decision-making processes. The increase in complexity and sophistication of these AI systems makes it very important to come up with methods and approaches that can visualize, interpret, and provide insights into the inner workings of such models.

Interpretation and visualization are methods that help us understand the internal processes of decision-making and predicting capabilities of an AI system. By providing insights into how AI models arrive at conclusions, they also enhance the transparency and explainability of these systems. Interpretation involves disclosing various internal mechanisms such as identifying the features or factors influencing the model's output, as well as understanding the reasoning behind its choices. Visualization, on the other hand, is termed as the visual representation of the system's behavior, mostly in the form of intuitive visuals or graphical explanations that can be easily understood and interpreted by humans. It includes various techniques such as heatmaps, saliency maps, feature importance graphs, etc., which are capable of presenting information in a meaningful and comprehensive way.

The insufficiency in the interpretation and visualization of AI workings and decision-making raises many concerns about the outcomes of the models. It makes it difficult to understand why an AI system has reached a particular decision or prediction, especially when sensitive or consequential tasks and data are involved in the process. In fields like healthcare, where life and death situations are at stake, it is extremely necessary to understand the logic and reasoning behind the AI-based diagnosis as well as treatment. Similarly, in other fields as well, the ability to interpret the decision-making process of an AI system is very important for the trust and safety

DOI: 10.1201/9781003442509-16

of the user. Thus, interpretability and visualization of AI systems' workings can be considered as effective collaboration-enabling and trust-building factors between humans and AI.

This chapter presents an overall collection of methods and approaches involved in the process of interpreting and visualizing AI systems. Model-specific techniques involve understanding the internal workings of specific models, along with model-agnostic methods that aim to provide insights into the model without relying on any specific model architecture. These approaches involve different dimensions such as feature importance analysis, local explanations, visual interpretation, bias and fairness analysis, prototype-based explanations, as well as ensemble-based interpretation. Each of these methods has its own unique advantages and considerations, helping users to understand, interpret, and visualize according to their specific needs and context. The practical applications along with the methods of evaluating these interpretable models are also highlighted in this chapter, which helps to understand the different domains in which these techniques of interpreting and visualizations are successfully applied and also gives an understanding of how reliable these techniques can be.

Interpretation and visualization techniques are pivotal in addressing the nature of Black-box AI models and help foster transparency, accountability, and trust. This chapter provides a comprehensive understanding of these techniques along with their applications, evaluation, and comparisons with other similar terms present in this field. It also helps the reader to understand the present challenges faced and the expected future of these techniques.

16.2 INTERPRETABLE ARTIFICIAL INTELLIGENCE

Interpretability in AI refers to the ability to comprehend how machine learning models make their predictions or decisions. A model whose decision-making process can be easily understood and explained by humans without necessitating a deep understanding of the underlying algorithms or technical details can be successfully termed as an interpretable model. Interpretability is an important concept in AI because it fosters humans to trust and rely on AI systems more effectively by facilitating the comprehension of why a specific decision was made. Without interpretability, it can be difficult for humans to understand why a particular decision was made by an AI system, which can lead to mistrust, skepticism, and ultimately result in the rejection of AI systems altogether [1].

There are many techniques to achieve interpretability in AI, such as identifying important features, visualizing models, building surrogate models, and providing local and counterfactual explanations to understand the decision-making process, ensuring transparency and reliability of AI systems. Interpretability is especially critical in domains of healthcare, finance, and autonomous vehicles, where decisions made by AI systems can have significant real-world consequences. By ensuring that AI systems are interpretable, we can build more trustworthy and reliable AI systems that are better equipped to meet the needs of society [2].

Interpretability in AI helps build trust and confidence in AI systems by enabling humans to understand and explain how AI models arrive at their decisions. Without interpretability, AI decisions can seem like "black boxes", which is a machine learning model that makes decisions without providing any insight into how it arrived at those decisions, which humans can't understand. Moreover, interpretability is crucial for ensuring that AI decisions are fair and transparent. By understanding how AI models are making their decisions, we can identify any biases or discriminatory practices and address them appropriately. Interpretability is also necessary for regulatory compliance. Many industries are subjected to regulations that require them to explain their decision-making processes. Without interpretability, it can be challenging to comply with these regulations and demonstrate that AI decisions are made ethically and transparently.

16.2.1 Interpretable AI vs Explainable AI

Explainability and interpretability are two closely related but distinct concepts in the field of AI. While both make AI models more understandable and transparent, they differ in their goals and approaches. Although both explainability and interpretability are important for building trust and transparency in AI systems, they may have different implications for various domains or applications. Explainability is focused more on providing human-interpretable justifications for the model's decisions, while interpretability is focused on enabling humans to understand the internal workings of the model. Figure 16.1 presents a basic schematic representation of the relationship between explainability, interpretability, and visualization of AI [3].

Explainability focuses on providing clear and concise explanations for how an AI model makes its decisions. It uses techniques such as generating natural language explanations, providing visualizations, or highlighting the most relevant features that can influence the model's decision. The goal of explainability is to provide a way through which humans can understand and trust the model's decisions by providing a comprehensible justification or reason. For instance, if an AI model is used to predict

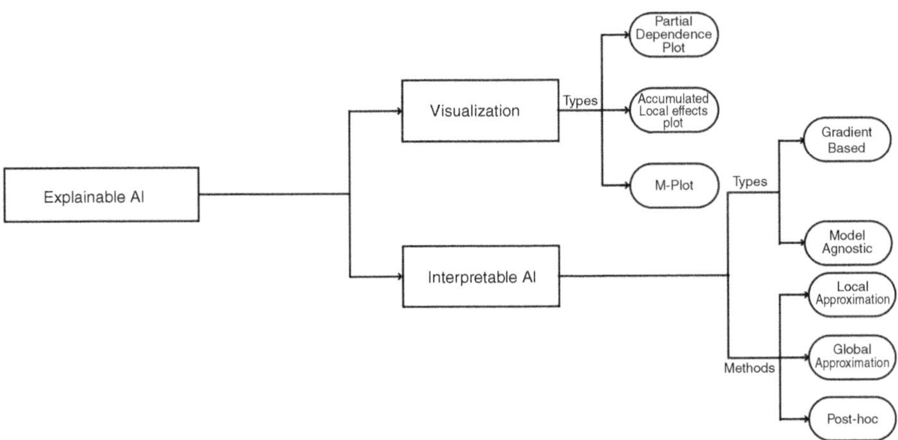

FIGURE 16.1 Schematic representation of relation between Explainable and Interpretable AI.

the risk of heart disease in an individual, an explanation may be a simple English sentence like "You can have heart disease because you are a smoker, have diabetes, and have a family history of heart disease". An explanation can also be a visualization that highlights the most relevant features or inputs that influenced the model's decision or a comparison with similar cases or previous decisions made by the model [4].

Interpretability, on the other hand, focuses on understanding how an AI model works and the conclusions it makes. It involves techniques such as sensitivity analysis, feature importance analysis, or decision boundary visualization, among others. The goal of interpretability is to ensure that humans can interpret and analyze the internal workings of the model, like identifying features most important for the model's decisions or understanding how the model generalizes to new data and interprets it. In the above instance taken, an interpretability technique such as feature importance analysis may reveal that the model's decision to predict a high risk for heart disease for the individual was highly driven by the patient's smoking status and family history, while diabetes had a relatively lower importance in its decision. Another interpretability technique such as decision boundary visualization may show how the model separates high-risk and low-risk patients based on age, gender, or medical history [5].

16.2.2 Interpretation of Black Box Models

Black box models are known for their complexity and the use of intricate, nonlinear relationships between inputs and outputs. Interpreting black box models, such as deep neural networks, involves the task of understanding and explaining how AI models make decisions that are difficult to understand. To comprehend how these models arrive at their predictions or decisions, various techniques are being used. For example, approach-agnostic models like model-agnostic methods can be applied to any type of model. These methods involve analyzing feature importance, partial dependence plots, or permutation feature importance to find the most influential features or inputs affecting the model's decision-making process. Similarly, model-specific methods are tailored to the specific characteristics and architecture of the black box model. These include features like activation maximization or saliency maps, which aim to visualize and understand how the model responds to changes in inputs and thus identify the factors driving its predictions. Interpreting black box models holds significant importance in ensuring the accountability and transparency of AI systems, particularly in domains like healthcare or finance. By gaining insights into the decision-making mechanisms of these models, we can increase trust and confidence in the outcomes they provide and ensure that they are making ethical and explainable decisions [6].

As in the black box model, internal workings are not easily accessible or interpretable to users, and the model behaves as an opaque system, taking inputs and producing outputs without revealing how the predictions or decisions were made. Other models provide a much greater level of transparency and interpretability in their workings. Two other models widely used are White models and grey models. In a white box model [7], the internal workings and structure are fully transparent and understandable. The model architecture, parameters, and decision-making process are explicitly known and can be easily accessed. Users can directly interpret how the

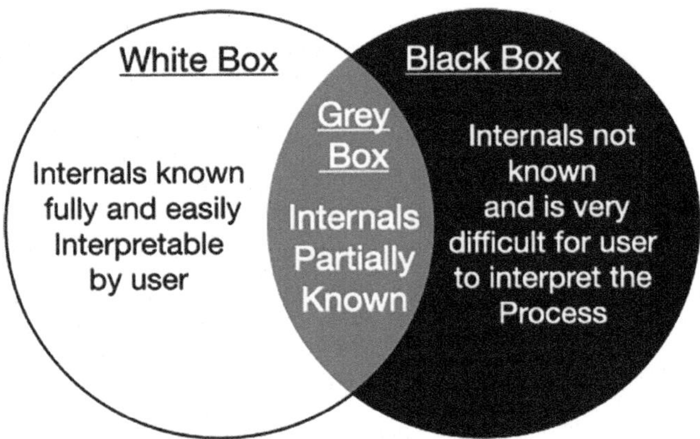

FIGURE 16.2 White box and black box models.

model arrived at its predictions or decisions. Examples of white box models include linear regression and decision trees. A grey box model [8], meanwhile, is a mixture of a white box model and a black box model. It has transparency and interpretability but is not fully transparent like a white box model. In a grey box model, users have partial knowledge about the model's internal workings or structure. They have access to certain information or insights that provide a limited understanding of the model's decision-making process. An example is a neural network with hidden layers, where the input and output layers are interpretable, but the intermediate layers are not fully transparent. Moreover, Figure 16.2 represents the overall lookout of White, Black, and Grey box models in the fields of AI.

16.3 IMPORTANCE OF VISUALIZING AI

The concept of Artificial Intelligence (AI) has sparked curiosity for many years now and holds immense potential to revolutionize many aspects of our lives. AI can transform the way we work, live, and interact with the world. However, as AI becomes a crucial part of our lives, the need to understand and trust its decision-making processes becomes more important. We can achieve this understanding and trust through the visualization of AI models. By visualizing AI, we can gain insights into how models function, the factors responsible for their predictions, decisions, and how they may be biased or unfair. Visualization techniques such as heatmaps, scatterplots, decision trees, and network graphs can help us identify patterns and relationships in data that may not be immediately obvious through statistical analysis. For example, a heatmap can show how different variables in a dataset are correlated, while a decision tree can reveal the factors that contribute to a particular decision. These visualizations can help us interpret the results of AI models, understand how they arrive at their decisions, and identify potential areas for improvement [9].

Moreover, visualizing AI can also help improve communication and collaboration between technical and non-technical stakeholders. Visual representations of AI models can help narrow down the gap between experts and laymen and, in turn, facilitate

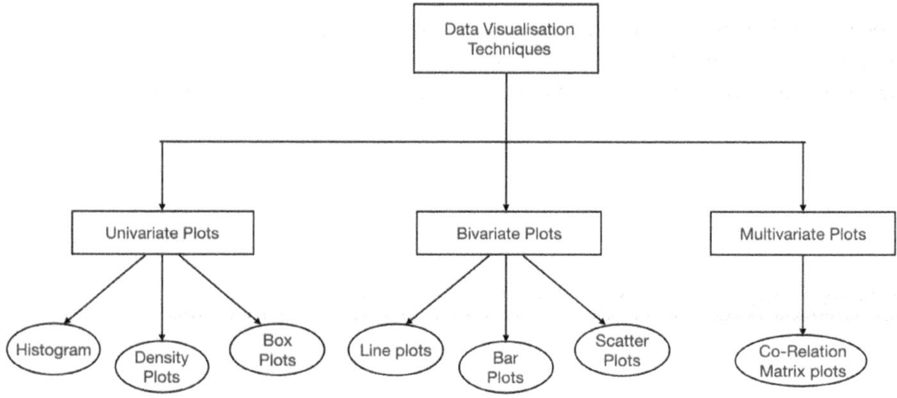

FIGURE 16.3 Data visualization in AI.

discussions about the benefits, limitations, and ethical implications of AI. Visualizing AI can even help us create more interpretable and explainable models in domains such as healthcare, criminal justice, and financial services, where transparency, accountability, and fairness play a major role. By using visualization techniques, we can ensure that the model's outputs are reliable and trustworthy. For instance, in the domain of healthcare, visualizing AI models can help clinicians understand how the model arrives at its diagnoses and treatment recommendations. This can help build trust between patients and doctors, as patients can better understand the reasoning behind their treatment plans. Likewise, in the domain of criminal justice, visualizing AI models can help us identify and address potential biases in the decision-making process. By visualizing the factors that contribute to a particular decision, we can ensure that the model's outputs are fair and unbiased. As AI becomes more prevalent in our daily lives, the importance of visualizing AI models will increase further. Currently, new and innovative visualization techniques can help us unlock the full potential of AI.

Data visualization techniques are methods used to present data in a visual format, making it easier to understand patterns, trends, and insights. There are various techniques, some of which include bar charts, line charts, pie charts, scatter plots, among others. Figure 16.3 gives a detailed perspective into the data visualization techniques used. The choice of technique depends on the data characteristics, the insights we want to convey, and the target audience. Data can be briefly divided into univariate, bivariate, and multivariate data that describe the number of variables or dimensions present in a dataset, based on which the technique is widely chosen [10].

16.4 CORRELATION AMONG INTERPRETATION AND VISUALIZATION

Interpretation and visualization are interconnected processes in data analysis. Visualization techniques provide a visual representation of data, thus aiding in pattern recognition, contextual understanding, and facilitating effective communication.

Meanwhile, interpretation uses visualizations to derive meaningful insights and understanding from the data. We can successfully say that they share a symbiotic relationship, which enhances the analysis process and enables a more comprehensive understanding of the information that needs to be processed.

Visualization plays a vital role in facilitating interpretation by providing a visual representation of data, making it easier for analysis and for users to grasp patterns, trends, and relationships. Visual representation of data can simplify complex data, which can thus be communicated more effectively, aiding in the interpretation process. Visualization techniques help in identifying patterns and trends within the data, which is the most crucial aspect of interpretation. On visual presentation, patterns not apparent in raw data or through statistical analysis alone become more visible. Visualizations such as line charts, scatter plots, or heatmaps can reveal correlations, clusters, or anomalies that can guide interpretation. Visualization provides the necessary context for interpretation. By presenting data in a visual format, analysts can observe the distribution, scale, and context of the data, which helps in understanding the significance and implications of the findings. Visualization helps to place data in a meaningful context and enables analysts to interpret results within the larger context of the problem or research question.

16.5 INTERPRETING INTELLIGENT SYSTEMS – METHODS

16.5.1 Model- Specific Interpretation

Model-specific interpretation methods mainly focus on understanding the decision-making processes of a specific artificial intelligence (AI) model. The internal workings of a particular model architecture are delved into to gain a better understanding of how the model arrives at its predictions or decisions. The model's structure and parameters are examined in model-specific interpretation methods to obtain valuable insights related to the underlying principles, mechanisms, and reasoning of the AI systems.

16.5.1.1 Decision Trees

Decision trees are one of the widely used model-specific interpretation methods. They are capable of providing a clear and intuitive representation of the decision logic employed by any model. Each decision based on a specific feature or attribute is represented by an internal node in the decision tree, whereas the leaf nodes represent the final prediction or outcome given by the model [11]. The visualization of the decision tree helps to understand the path that led to the specific decision and the relative importance of different features in the model's decision-making process. Figure 16.4 shows a similar approach to understanding the model's internal workings. Enhancing the interpretability of these decision trees is possible by directly training the tree or deriving it from other models.

16.5.1.2 Rule-Based Systems

Rule-based systems encode decision rules explicitly, making them highly interpretable. A specific set of rules is predefined in these systems that dictate the overall decision-making of the model. These rules can be based on specific feature thresholds or

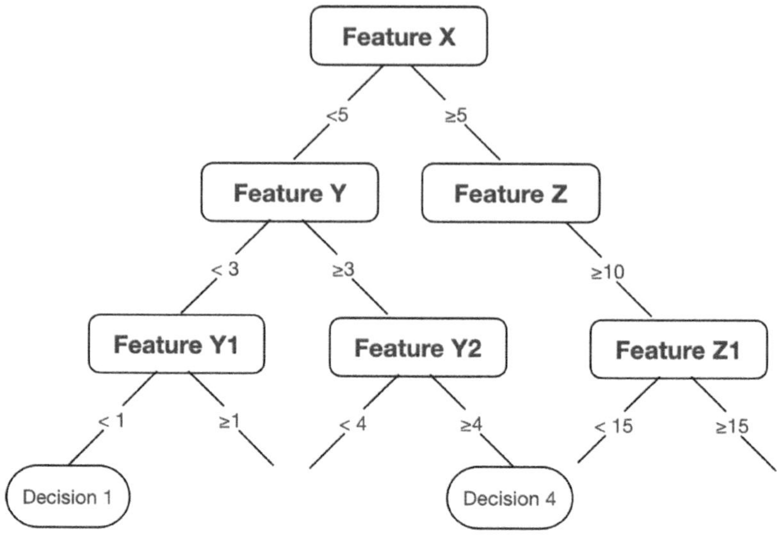

FIGURE 16.4 Interpretation using decision trees.

even logical conditions as well. The examination of these rules allows stakeholders to gain a clear understanding of the process taken up for reaching a specific decision as well as the factors that influence the model's prediction and classification capabilities. Rule-based systems are particularly useful when transparency and explainability are essential [12].

These rule-based systems work by extracting human-readable rules from the model that provide insights into the decision-making process and help in interpreting the decisions as well. These techniques aim to uncover the underlying logic and reasoning behind the model's predictions or decisions, making it easier for stakeholders to understand and validate its behavior [13]. These systems basically follow a certain sequence of steps to interpret the decision, which is started by choosing one of the rule extraction algorithms such as decision tree induction, RuleFit, Bayesian rule lists [14], or genetic algorithms [15]. Once an appropriate algorithm has been selected, the algorithm analyzes the trained AI model and the corresponding training data to identify patterns and decision rules that replicate the model's behavior. These identified rules and patterns are then validated to ensure the model's accuracy and consistency with the original AI model. Thus, after the rules are identified and validated, they can be interpreted and used to explain the AI model's decision-making process. Users can analyze the rules to understand the model's working and understand the key features, conditions, and their influence on the model's prediction.

16.5.2 Model Agnostic Interpretation

Model-agnostic interpretation methods play a crucial role in understanding the decision-making process of artificial intelligence (AI) models. Unlike model-specific interpretation methods that focus on a particular model architecture, regardless of the complexity or underlying algorithms, model-agnostic techniques are capable of providing insights into any AI model. This method offers a generalizable approach to interpretability, enabling stakeholders to gain transparency and trust in AI systems without requiring in-depth knowledge of the model's internal workings. In this section, we will explore various model-agnostic interpretation methods and their applications in understanding and validating AI models.

16.5.2.1 Feature Importance

Feature importance analysis is one of the most fundamental model-agnostic interpretation approaches. This technique mainly aims to identify the most influential features in the model's workings. Quantification of the contribution of each feature helps understand the factors that have the greatest impact on the model's prediction [16]. Feature importance metrics such as permutation importance, SHAP values, or LIME (Local Interpretable Model-agnostic Explanations) can be utilized to estimate the significance of individual features. These methods provide an understanding of the relative importance of the features, helping us identify which input the model relies on the most for the predictions.

16.5.2.2 Global Surrogate Models

Global surrogate models are a powerful model-agnostic technique used to approximate the behavior of complex AI models with more interpretable models, such as decision trees or linear models. These surrogate models capture the overall decision-making process of the AI model without revealing its intricate internal architecture. Training a surrogate model on the AI model's inputs and outputs allows a transparent and understandable representation of the model's behavior. Surrogate models offer interpretability benefits by providing explicit decision rules or feature importance estimates that can be easily understood and validated [1].

16.5.2.3 Shapley Values

Shapley values, based on cooperative game theory, provide a model-agnostic approach to attributing contributions to individual features for a given prediction. These values distribute the prediction's outcome among the features, considering all possible coalitions of features. By calculating the Shapley values, stakeholders can understand the impact of each feature on the model's prediction, accounting for interactions and dependencies among features. Shapley values offer a unified framework for interpreting any type of model, allowing for fair and consistent feature attribution across different predictions [17].

16.5.2.4 Counterfactual Explanations

Counterfactual explanations provide alternative input scenarios to explain why a specific prediction was made. These methods generate new instances that are similar to the original input but result in a different prediction. By presenting counterfactual examples, users can understand the key changes required in the input to influence the model's decision. Counterfactual explanations are valuable in understanding the model's sensitivity to specific input variations and identifying potential biases or limitations [18].

16.6 VISUALIZATION TECHNIQUES

Data visualization techniques are the methods and approaches used to visually represent data in a meaningful and informative way. These techniques aim to enhance data understanding, identify patterns and trends, and communicate insights effectively to the audience. They transform raw data into visual representations such as charts, graphs, maps, and diagrams, making it easier for individuals to explore, analyze, and interpret complex information. Visualization techniques comprise a wide range of tools, methods, and visual representations that can be applied to different types of data. Some common visualization techniques are listed in this section [19].

16.6.1 Charts and Graphs

These techniques are effective for displaying numerical and categorical data, comparing values, and showing trends over time. For instance, bar charts represent data using rectangular bars of lengths proportional to the values they represent. Line charts, on the other hand, display data points connected by line segments. They are suitable for showing trends over time or continuous data. Scatter plots, meanwhile, represent individual data points as dots on a 2-D plane. They are useful for visualizing relationships between two variables. Similarly, pie charts display data as slices of a circular pie, with each slice representing a proportion of the whole. They are useful for illustrating parts of a whole or comparing categorical data. These were some widely used examples of charts and graphs; there are many other ways like area charts, histograms, and box plots, among many others [20].

16.6.2 Maps and Geospatial Visualization

These techniques utilize geographic information systems (GIS) to represent data on maps, allowing for the exploration of spatial relationships and patterns. For example, heatmaps use color to represent values in a matrix or a table. They are useful for visualizing patterns and relationships in large datasets. Likewise, tree maps display hierarchical data using nested rectangles, where the size of each rectangle represents a proportionate value. They are thus useful for representing hierarchical structures. Other examples include choropleth maps, point maps, and thematic maps, among many others [21].

16.6.3 Networks and Graphs

These techniques represent relationships and connections between entities using vertices, which act as nodes, and edges, which become the link. Network graphs can be used to visualize social networks, organizational structures, and interconnected systems. For example, in Force-directed Layout, nodes and links are positioned based on attractive and repulsive forces between them. In this arrangement, related nodes are closer together, and unrelated nodes are farther apart. Similarly, in Hierarchical Layout, nodes are organized in a hierarchical structure, often based on levels or categories. It is useful for visualizing organizational structures or any network with

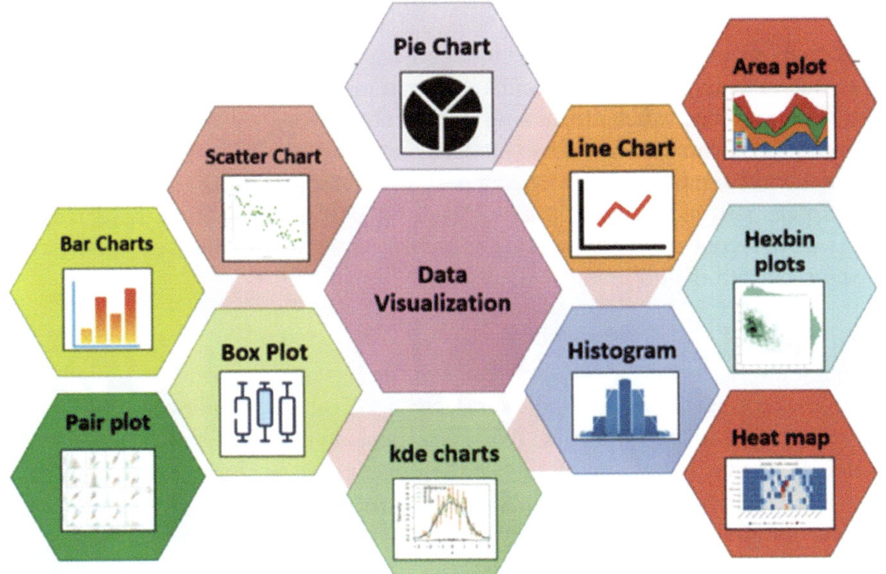

FIGURE 16.5 Different forms of data visualization.

a clear hierarchy as nodes are positioned in levels or layers, and the links connect nodes between different levels [22].

There are several other techniques like Word Clouds and Text Visualization, which visually display text data, highlighting the frequency or importance of words through varying sizes or colors. Interactive Visualization involves the use of interactive elements to allow users to explore and interact with the data. Interactive visualizations enable dynamic data exploration. These were a few commonly used visualization techniques. The choice of technique should depend on the nature of the data and the target audience, ensuring that the visual representation enhances understanding and facilitates data-driven decision-making. Figure 16.5 presents a few of the commonly used data visualization techniques [23].

16.7 EVALUATION OF INTERPRETABLE AI

Interpretable AI models are designed to provide human-understandable explanations related to the decisions made by the model or the actions taken—that is, why the decision was made or the action was taken. These explanations of the decision-making approach of the model allow users to understand and trust the reasoning behind their outputs. Thus, evaluating the interpretability of such models is of utmost importance in the development and deployment of AI systems, ensuring their effectiveness in various domains such as healthcare and finance. The process of assessing and measuring the interpretability and explainability of AI models is termed as the evaluation of interpretable AI models.

16.7.1 Why Evaluate?

The development of interpretable AI models has gained significant attention in recent times due to its huge potential to enhance the transparency, accountability, and user trust in AI systems, which are lacking in current technology. However, just ensuring the interpretive capability of these models is not enough. It is equally important to ensure that these models are producing desirable outputs with correct interpretations and explanations as well, and this can be done only by evaluating their interpretability through rigorous assessment processes. Lipton [24] has discussed in detail the need for evaluating interpretable AI systems.

16.7.1.1 Trust and Transparency

Interpretable AI models make their behavior understandable to users by providing the logic and explanations behind their decisions. Evaluations allow us to determine the reliability and trustworthiness of these explanations and logics given by the model. Assessing the interpretability of these models also helps us gain insights into the comprehensibility and consistency of the given logic. This builds confidence in the decision-making process and fosters trust between the users and the AI systems.

16.7.1.2 Accountability and Bias Mitigation

Evaluation of interpretable AI systems helps identify biases and unfairness in the decision-making process by analyzing the explanations and determining if biases of any kind are being propagated or if there is any discrepancy in the treatment of different groups. Evaluations provide an opportunity to identify and rectify biases and discriminatory patterns of any kind and help ensure the accountability and fairness of the system.

16.7.1.3 Ethical Considerations and Regulations

Addressing ethical considerations and compliance with regulatory requirements are crucial achievements of evaluating an interpretable AI system, as it allows us to examine the model's explanations for potential ethical concerns such as privacy violations and propagation of sensitive information. Evaluations of these systems proactively identify and mitigate these issues and ensure that the AI model adheres to legal and ethical standards. A prime example of this would be the use of interpretable AI for the selection of embryos during IVF, due to its capability of detecting even minute abnormalities and providing more objective judgment in the selection of embryos [25].

16.7.1.4 Model Improvement and Iterations

The process of evaluation also serves as a feedback loop for model improvement and iterations by rigorously assessing the interpretation skills of the AI system. It also provides us with crucial insight data such as the strengths, weaknesses, and limitations of the system. This information helps to advance the model design and its engineering, as well as its explanation generation techniques.

16.7.2 Evaluation Methodologies

The systematic approaches and frameworks used to assess and validate the performance and effectiveness of interpretable AI models are still in the very early stages of development. Researchers are still figuring out different algorithms and metrics to judge and evaluate the explanations provided for the decisions made by an AI system.

One of the evaluation taxonomies proposed by Velez and Kim [26] categorizes the process into three distinct groups: Application Grounded Evaluation, Human Grounded Evaluation, and Functionality Grounded Evaluation. Application Grounded Evaluation involves real humans performing real tasks. According to the authors, if the application of a model is well-defined for a specific task, then the best way to evaluate the model is to compare its performance with real humans doing the same tasks, such as doctors diagnosing a specific disease. Human Grounded Evaluation involves real humans performing simplified tasks. When the tasks performed are relatively more complex and domain experts are difficult to find, this type of evaluation can be employed. Velez and Kim suggest evaluating the model based on the quality of the explanations and interpretations provided by the model for its decision-making process. Functionally grounded evaluation involves no human interaction and instead uses proxy tasks for judging the quality of explanations provided.

Gilpin et al., in their study [27], have proposed another taxonomy for the evaluation of interpretable models. The proposed taxonomy also consists of three levels, namely: processing, representation, and explanation producing. The first level, processing models or emulation-based models, includes the use of proxy methods to judge the completeness of the models by assessing their performance on substitute tasks. Representation systems are categorized as systems capable of judging the completeness of substitute tasks and detection of biases. Explanation-producing systems are evaluated according to the match of user expectations and thus include human evaluation and detection of biases.

16.7.2.1 Human Centric Evaluation

Human-centric evaluation of interpretable AI refers to the assessment and validation of AI models and their interpretability methods by incorporating the feedback and perspectives of human users. This evaluation approach recognizes the critical role that humans play in interpreting and making decisions based on AI-generated explanations. Human-centric evaluation involves conducting user studies, surveys, interviews, or focus groups to gather qualitative and quantitative feedback from individuals who interact with the AI models and their explanations. These evaluations aim to assess the user experience, comprehension, trust, and satisfaction with the provided explanations. By involving human users in the evaluation process, researchers can gain insights into the effectiveness, usability, and impact of interpretability methods on user understanding and decision-making. Human-centric evaluation not only provides valuable feedback for improving the interpretability of AI models but also helps build user trust, transparency, and acceptance of AI systems, leading to more ethical and accountable deployment of interpretable AI solutions.

The use of human-centric evaluation has been widely employed in many studies and different scenarios with the involvement of various AI models. Yu et al. [28] demonstrate one of the similar experiments conducted by Yu et al., which included the evaluation of a music rules generating AI model namely MUS-ROVER by 23 students. The aim of this study was to understand the level of interpretability of the rules generated by the AI model. Another similar study was conducted by Kim et al. [29].

A systematic human evaluation technique namely HIVE (Human Interpretability of Visual Explanations) was proposed to assess the human-centered evaluation of visual interpretability methods.

16.7.2.2 Quantitative Evaluation

The use of numerical or measurable metrics for assessing the performance, characteristics, or effectiveness of an AI system is referred to as its quantitative evaluation. This type of evaluation involves the use of quantitative measures to analyze specific aspects or properties of the subject being evaluated. In the field of interpretable AI models, quantitative evaluation entails the use of numerical metrics to assess various aspects of interpretability, such as the quality of explanations, the accuracy of the generated insights, and the effectiveness of the interpretability techniques. Quantitative evaluation is important because it offers a systematic and objective way to measure and benchmark the performance of interpretable AI models. It allows for rigorous analysis, repeatability, and comparability of results across different experiments or studies. By employing quantitative evaluation, researchers can make data-driven decisions, identify trends, validate hypotheses, and ensure the reliability and effectiveness of the interpretability techniques employed.

Though the specific metrics used to evaluate the models vary from one model to another, some of the commonly used metrics are discussed below and a pictorial representation has been presented in Figure 16.6.

16.7.2.2.1 Complexity Metrics

Complexity metrics are quantitative measures used to assess the complexity of interpretable AI models and their generated explanations. These metrics provide insights into the cognitive load required to understand the explanations and the level of complexity involved in the interpretability process. Complexity metrics can include factors such as explanation length, lexical complexity, and sentence structure complexity.

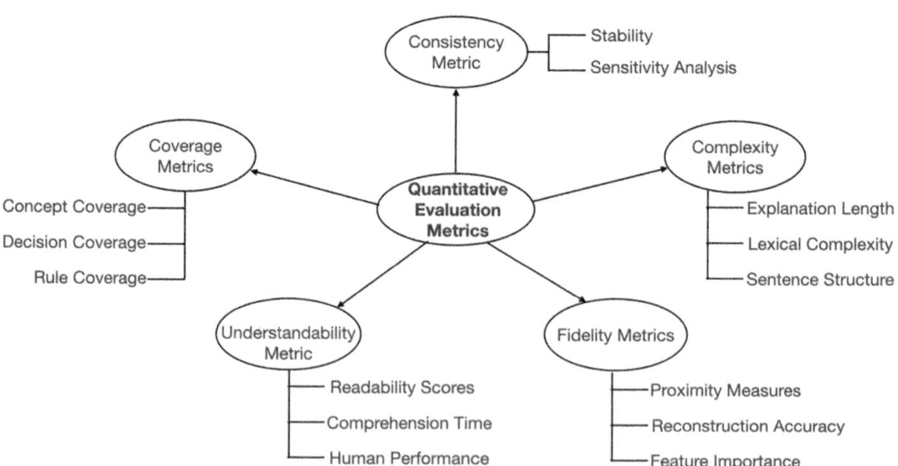

FIGURE 16.6 Commonly used Quantitative Evaluation metrics.

Explanation length measures the number of words or characters in the generated explanations, with longer explanations indicating higher complexity. Lexical complexity evaluates the difficulty of the vocabulary and linguistic structures used in the explanations, considering factors such as word choice, sentence complexity, or technical jargon. Sentence structure complexity assesses the syntactic complexity of the sentences, considering features like sentence length, grammatical constructs, or use of subordination. By employing complexity metrics, researchers can gain insights into the level of cognitive effort required for users to comprehend the explanations and identify opportunities for simplification and improvement in the interpretability of AI models [30].

16.7.2.2.2 Fidelity Metrics

Fidelity metrics are measures used to assess the degree of agreement or similarity between the explanations generated by interpretation methods and the underlying decision-making process of the AI model. Fidelity metrics aim to evaluate how faithfully the explanations capture the model's internal mechanisms and reasoning [31]. These metrics help assess the descriptive accuracy and reliability of the interpretation methods in providing insights into the AI model's behavior. Fidelity metrics can be based on various criteria, such as feature importance consistency, prediction consistency, or local approximation accuracy. Feature importance consistency measures the degree to which the identified important features in the explanations align with the actual influential features in the model's decision. Prediction consistency evaluates the extent to which the explanations consistently align with the model's predictions across different instances. Local approximation accuracy assesses how well the interpretation methods approximate the model's decision locally in the vicinity of specific instances. Fidelity metrics enable researchers and practitioners to quantify the faithfulness of the explanations, identify potential discrepancies or limitations, and guide the improvement of interpretation methods to provide more reliable and accurate insights into the AI model's decision-making process.

16.7.2.2.3 Understandability Metrics

Understandability metrics are quantitative measures used to assess the degree to which the explanations generated by interpretation methods are understandable and comprehensible to human users. These metrics aim to evaluate the clarity, simplicity, and ease of understanding of the provided explanations [32]. Understandability metrics play a crucial role in ensuring that the interpretability and explainability of AI models are accessible and usable to a wide range of users, including domain experts and non-experts. Common understandability metrics include readability scores, linguistic simplicity measures, and user comprehension tests. Readability scores assess the difficulty level of the explanations by considering factors such as sentence length, word complexity, and syntactic complexity. Linguistic simplicity measures evaluate the use of plain language, avoidance of technical jargon, and the clarity of explanations. User comprehension tests involve presenting explanations to users and assessing their understanding and comprehension through quizzes, comprehension tasks, or feedback surveys. By employing understandability metrics, researchers and practitioners can assess the effectiveness of the interpretation methods in providing clear and understandable explanations, identify areas of improvement, and ensure that the interpretability of AI models meets the needs and expectations of the intended users.

16.7.2.2.4 Coverage Metrics

Coverage metrics are quantitative measures used to evaluate the extent to which the generated explanations capture all relevant aspects of the AI model's decision-making process. These metrics aim to assess the comprehensiveness and completeness of the provided explanations, ensuring that they cover important factors and considerations. Coverage metrics can include factors such as the proportion of relevant features or factors mentioned in the explanations, the coverage of different decision classes or scenarios, or the diversity of explanation types. For example, a coverage metric may evaluate the percentage of influential features mentioned in the explanations compared to the total number of features. Another metric may assess the coverage of various decision outcomes or classes in the explanations. Additionally, the diversity of explanation types can be evaluated to ensure that different aspects of the decision process are adequately represented. By employing coverage metrics, researchers and practitioners can quantitatively assess the breadth and inclusiveness of the explanations, identify potential gaps or biases, and strive for comprehensive interpretability in AI models.

16.7.2.2.5 Consistency metrics

Consistency metrics are quantitative measures used to evaluate the degree of agreement or coherence in the explanations generated by interpretation methods. The main aim of using these metrics is to measure the stability and robustness of the interpretations across different instances or perturbations [33]. Consistency metrics can include factors such as inter-instance consistency and stability under perturbations. Inter-instance consistency is the measure of similarity or agreement between the explanations generated for different instances of the same input. It checks for the consistency in the interpretation methods to provide similar insights for similar inputs. Stability under perturbations evaluates the robustness of the explanations when the input or the model undergoes minor changes or perturbations. It assesses whether the explanations remain consistent and do not drastically change in response to small variations. Consistency metrics are crucial in ensuring that the interpretations are reliable, providing consistent insights regardless of slight variations in the input or model conditions. By employing consistency metrics, researchers and practitioners can quantitatively assess the stability and reliability of the interpretation methods, identify potential inconsistencies or biases, and enhance the robustness of interpretability in AI models.

16.8 APPLICATION OF VISUALIZATION AND INTERPRETABLE AI

16.8.1 HEALTHCARE SYSTEMS

The healthcare domain presents a critical and complex landscape where the interpretability and transparency of AI models are of utmost importance. Visualization techniques, combined with interpretable AI, have the potential to revolutionize healthcare by providing actionable insights, facilitating decision-making, and promoting trust in AI-driven healthcare systems. This section explores the application of visualization and interpretable AI in healthcare, highlighting its benefits and key areas of implementation.

16.8.1.1 Medical Image Interpretation

Visualization plays a vital role in medical image interpretation, enabling healthcare professionals to analyze and understand complex imaging data. Techniques such as heatmaps, overlays, and 3D reconstructions can highlight regions of interest, abnormalities, or lesions, assisting radiologists and clinicians in diagnosing diseases such as cancer, cardiovascular conditions, or neurological disorders. Interactive visualization tools enhance collaboration, allowing experts to discuss and annotate medical images in real time, leading to more accurate and efficient diagnoses.

16.8.1.2 Clinical Decision Support Systems

Interpretable AI models, combined with visualizations, can provide clinicians with decision support by explaining the reasoning behind the AI system's recommendations. Visual representations of model predictions, such as graphs, charts, or timelines, can help clinicians understand the factors influencing the decision, identify high-risk patients, or assess the effectiveness of treatment options. This facilitates shared decision-making between the AI system and healthcare professionals, leading to more personalized and informed patient care.

16.8.1.3 Predictive Analytics and Risk Assessment

Visualization techniques enable healthcare providers to comprehend and communicate predictive analytics and risk assessment models. Visualizations of patient risk scores, survival curves, or risk heatmaps can aid in identifying individuals at high risk for certain diseases, such as diabetes or cardiovascular events. These visual representations empower clinicians to proactively intervene, monitor patients, and implement preventive measures, ultimately improving patient outcomes and reducing healthcare costs.

16.8.1.4 Explainable Electronic Health Records (EHR)

Interpretable AI models combined with visualization techniques can unlock valuable insights from electronic health records (EHR). By visualizing patient records, including lab results, medications, and vital signs, clinicians can gain a holistic view of the patient's health status and identify trends or patterns that may contribute to disease progression. This promotes early detection, personalized treatment planning, and more efficient utilization of healthcare resources.

16.8.2 Judicial Systems

The integration of visualization and interpretable AI techniques in judicial systems has the potential to revolutionize the legal field by enhancing transparency, improving decision-making processes, and promoting accountability [34]. Here, we explore the applications and benefits of visualization and interpretable AI in judicial systems.

16.8.2.1 Legal Decision Support

Visualization techniques combined with interpretable AI can provide valuable insights to legal professionals, aiding in legal decision-making processes. Visual representations of case data, legal precedents, and relevant statutes can assist judges, lawyers, and legal researchers in identifying patterns, correlations, and inconsistencies. This helps in formulating legal arguments, evaluating the strength of evidence, and making well-informed decisions.

16.8.2.2 Explainability of AI-Driven Decisions

As AI algorithms are increasingly employed in legal systems, the need for explainable AI becomes crucial. Visualization techniques can help elucidate the decision-making process of AI models, allowing legal professionals to understand and justify AI-driven decisions. Visualizations can highlight the features, factors, or legal principles that contribute to the model's decision, providing transparency and accountability in legal proceedings.

16.8.2.3 Legal Analytics and Predictive Modeling

Visualization and interpretable AI can be utilized to analyze large volumes of legal data, including court cases, statutes, and legal texts. By visualizing legal analytics, such as trends in case outcomes, judge's decisions, or legal precedent networks, legal professionals can gain valuable insights for case assessment, risk prediction, and strategy development. Visual representations enable the identification of relevant factors and can aid in making data-driven decisions.

16.8.2.4 Case Outcome Explanation

Interpretable AI can provide explanations for case outcomes, helping legal professionals and litigants understand the factors that influenced the decision. Visualizations can highlight the key evidence, legal principles, or precedents that played a role in the outcome, making the decision-making process more transparent and understandable. This can contribute to fairer and more just legal proceedings.

16.8.2.5 Legal Research and Document Analysis

Visualization techniques can assist legal researchers in analyzing and extracting relevant information from extensive legal documents and case law. By visualizing relationships between legal concepts, entities, or citations, researchers can navigate complex legal information more effectively, identify connections, and derive meaningful insights. This promotes efficient legal research, accelerates the discovery of relevant case law, and enhances the preparation of legal arguments.

16.8.3 Financial Systems

Visualizing financial data and interpreting the decisions taken by artificially intelligent financial systems hold the potential to revolutionize the way financial data is analyzed, understood, and utilized. Financial institutions can gain deeper insights, improve their decision-making, and enhance transparency in decision-making by fully embracing visualization and interpretation techniques. Here, we explore the key applications and benefits of visualization and interpretable AI in financial systems.

16.8.3.1 Data Visualization and Exploration

Visualization techniques enable financial professionals to explore and analyze complex financial data more effectively. By visually representing financial data such as market trends, portfolio performance, or customer behavior, analysts can identify patterns, correlations, and anomalies. Visualizations aid in understanding complex relationships, uncovering hidden insights, and making data-driven decisions. This enhances financial risk management, asset allocation strategies, and investment analysis [35].

16.8.3.2 Explainability of AI Models

Interpretable AI models are crucial in financial systems, where transparency and accountability are paramount. Visualization techniques can be used to explain the reasoning behind the predictions or decisions made by AI models. Visualizations provide insights into the features, factors, or rules that influence the model's outputs, enabling financial professionals to understand and justify the model's recommendations. This helps in regulatory compliance, risk assessment, and auditability of AI-driven financial systems.

16.8.3.3 Fraud Detection and Risk Management

Visualization and interpretation of AI play a critical role in detecting fraudulent activities and managing financial risks. Visualization of transaction data, network relationships, and behavioral patterns enables financial institutions to identify potential fraud indicators as well as anomalies. Interpretable AI models can provide explanations for fraud detection alerts, enabling investigators to understand the underlying factors and take appropriate actions [36]. Visualizations facilitate the identification of high-risk transactions, suspicious patterns, and emerging threats, enhancing overall risk management. They also help in reducing the number of false alarms and provide more accurate results.

16.8.3.4 Portfolio Optimization and Asset Allocation

Visualization techniques combined with interpretable AI models can assist in portfolio optimization and asset allocation strategies. Visual representations of historical performance, risk profiles, and market dynamics help financial professionals evaluate and adjust their investment portfolios. Interpretable AI models provide insights into the factors driving portfolio performance and recommend adjustments based on risk preferences, market conditions, and financial goals. Visualizations aid in understanding the impact of different asset allocation scenarios, facilitating informed investment decisions.

16.9 CHALLENGES IN VISUALIZATION AND INTERPRETATION OF AI

Though visualization and interpretation of AI offer great promise in enhancing transparency and understanding of AI systems, they also come with their own set of challenges. Overcoming these challenges is crucial to ensure the effective application and interpretation of AI models. Here, we discuss some of the key challenges in visualizing and interpreting AI.

16.9.1 Lack of Ground Truth

With recent developments in this field and the increasing interest of researchers in interpretable AI, it is essential to have an accurate understanding of what interpretability entails and the criteria an AI model must satisfy to be declared interpretable. Interpretable AI models aim to provide explanations for their decisions or predictions. However, in many cases, there is no ground truth or definitive answer to compare against. This makes it challenging to evaluate the correctness or accuracy of the explanations generated by interpretable AI models. Developing robust evaluation methodologies that can assess the quality and fidelity of the explanations remains a significant challenge [1].

16.9.2 Complexity and High-Dimensional Data

AI models primarily operate on high-dimensional data, which can pose challenges for visualization. It is quite difficult to represent and comprehend complex data structures and their relationships in a visually intuitive manner. The challenge of oversimplification or loss of critical information also persists in the case of complex AI models, where the visualization technique may or may not be able to capture the full complexity of the model. Thus, identifying techniques capable of effectively visualizing and interpreting high-dimensional data in a meaningful way remains an ongoing challenge.

16.9.3 Balancing Interpretability and Performance

There is often a trade-off between model interpretability and performance. Highly interpretable models may sacrifice predictive accuracy or generalizability. Conversely, complex models capable of delivering high performance may lack interpretability [37]. Thus, it becomes very important to strike the right balance between interpretability and performance, which makes it a challenge that requires careful consideration and exploration of different model architectures and techniques.

16.9.4 Ethical and Legal Implications

Ethical and legal concerns are raised in the application of AI in sensitive domains such as healthcare or finance. The ability to visualize and interpret the decisions made by AI can help address these concerns by providing transparency and accountability. Challenges may still persist regarding the assurance of fairness, transparency, and compliance of these AI systems. Thus, developing ethical guidelines, addressing biases, and incorporating legal considerations [38] is crucial for deploying AI responsibly.

16.10 FUTURE OF INTERPRETABLE AI

The future of interpretable AI holds immense potential as researchers and practitioners continue to address many challenges and enhance the transparency and interpretability of AI systems. One key aspect of the future of interpretable AI lies in the development of advanced model architectures and algorithms capable of striking a balance between interpretability and performance. Extensive research can be carried out in fields such as novel techniques including rule-based models, sparse linear models, or Bayesian networks, which can provide interpretable representations while maintaining competitive accuracy. Additional focus is needed on developing standardized evaluation metrics and methodologies to assess the interpretability and robustness of AI models. The integration of ethical considerations, privacy-preserving techniques, and legal frameworks will also play a vital role in shaping the future of interpretable AI and ensuring responsible and accountable AI deployment across various domains. The future of interpretable AI holds great promise in enabling transparency, trust, and understanding in AI systems, paving the way for more reliable and socially beneficial AI applications.

16.11 CONCLUSION

Interpretation and visualization techniques play a crucial role in advancing AI systems by addressing the challenges posed by black box models and fostering transparency and trust. Interpretable AI and explainable AI approaches provide valuable insights into the inner workings of complex models, while visualization techniques effectively communicate AI-driven insights in a manner that is understandable to humans. The evaluation of interpretability is paramount, and metrics and user feedback serve as valuable tools in refining and improving these techniques. By continuously enhancing interpretability, we can ensure that AI systems meet the highest standards of performance and transparency. Interpretable AI not only facilitates collaboration between humans and intelligent systems but also meets the legal and ethical requirements for explainability. It helps to identify and mitigate biases, promoting fairness and accountability in AI decision-making processes. Looking ahead, future research should focus on developing model-specific interpretability methods and advancing visualization techniques to accommodate the complexity and scale of modern AI systems. Collaboration among academia, industry, and policymakers is crucial in defining standards and guidelines for the responsible adoption of interpretable AI. Ultimately, these techniques enable the development of transparent and beneficial AI systems while upholding ethical principles. By empowering users with understandable explanations and intuitive visualizations, we can build trust, foster responsible AI deployment, and harness the full potential of AI for the benefit of society.

REFERENCES

1. Molnar, C., Casalicchio, G., & Bischl, B. (2020). Interpretable machine learning - a brief history, state-of-the-art, and challenges. In *ECML PKDD 2020 Workshops* (pp. 417–431). Springer International Publishing. doi:10.1007/978-3-030-65965-3_28.
2. Bakar, N. A., Abu-Siada, A., & Islam, S. (2014). A review of dissolved gas analysis measurement and interpretation techniques. *IEEE Electrical Insulation Magazine*, *30*(3), 39–49.

3. Samek, W., Wiegand, T., & Müller, K. R. (2017). Explainable artificial intelligence: Understanding, visualizing and interpreting deep learning models. arXiv preprint arXiv:1708.08296.
4. Heuillet, A., Couthouis, F., & Díaz-Rodríguez, N. (2020). Explainability in deep reinforcement learning. arXiv preprint arXiv:2008.06693.
5. Li, X., Xiong, H., Li, X., Wu, X., Zhang, X., Liu, J., Bian, J., & Dou, D. (2022). Interpretable deep learning: Interpretation, interpretability, trustworthiness, and beyond. arXiv preprint arXiv:2103.10689.
6. Guidotti, R., Monreale, A., Ruggieri, S., Turini, F., Giannotti, F., & Pedreschi, D. (2018). A survey of methods for explaining black box models. *ACM Computing Surveys (CSUR), 51*(5), 1–42.
7. Loyola-Gonzalez, O. (2019). Black-box vs. white-box: Understanding their advantages and weaknesses from a practical point of view. *IEEE Access, 7,* 154096–154113.
8. Kroll, A. (2000). Grey-box models: Concepts and application. *New Frontiers in Computational Intelligence and Its Applications, 57,* 42–51.
9. Wu, A., Wang, Y., Shu, X., Moritz, D., Cui, W., Zhang, H., … & Qu, H. (2021). AI4VIS: Survey on artificial intelligence approaches for data visualization. *IEEE Transactions on Visualization and Computer Graphics.* doi: 10.1109/TVCG.2021.3099002.
10. Cawthon, N., & Moere, A. V. (2007, July). The effect of aesthetic on the usability of data visualization. In *2007 11th International Conference Information Visualization (IV'07)* (pp. 637–648). IEEE.
11. Moshkovitz, M., Yang, Y. Y., & Chaudhuri, K. (2021, July). Connecting interpretability and robustness in decision trees through separation. In *International Conference on Machine Learning* (pp. 7839–7849). PMLR.
12. Kliegr, T., Bahník, Š., & Fürnkranz, J. (2021). A review of possible effects of cognitive biases on interpretation of rule-based machine learning models. *Artificial Intelligence, 295,* 103458.
13. Palade, V., Neagu, D. C., & Patton, R. J. (2001). Interpretation of trained neural networks by rule extraction. In *Computational Intelligence. Theory and Applications: International Conference,* 7th Fuzzy Days Dortmund, Germany, October 1–3, 2001, Proceedings 7 (pp. 152–161). Springer, Berlin Heidelberg.
14. Letham, B., Rudin, C., McCormick, T. H., & Madigan, D. (2015). Interpretable classifiers using rules and bayesian analysis: Building a better stroke prediction model. *The Annals of Applied Statistics, 9*(3), 1350–1371.
15. Yedjour, D., & Benyettou, A. (2018). Symbolic interpretation of artificial neural networks based on multiobjective genetic algorithms and association rules mining. *Applied Soft Computing, 72,* 177–188.
16. Altmann, A., Toloşi, L., Sander, O., & Lengauer, T. (2010). Permutation importance: A corrected feature importance measure. *Bioinformatics, 26*(10), 1340–1347.
17. Messalas, A., Kanellopoulos, Y., & Makris, C. (2019, July). Model-agnostic interpretability with shapley values. In *2019 10th International Conference on Information, Intelligence, Systems and Applications (IISA)* (pp. 1–7). IEEE.
18. Sharma, S., Henderson, J., & Ghosh, J. (2019). Certifai: Counterfactual explanations for robustness, transparency, interpretability, and fairness of artificial intelligence models. arXiv preprint arXiv:1905.07857.
19. Sadiku, M., Shadare, A. E., Musa, S. M., Akujuobi, C. M., & Perry, R. (2016). Data visualization. *International Journal of Engineering Research and Advanced Technology (IJERAT), 2*(12), 11–16.
20. Few, S., & Edge, P. (2007). Data visualization: past, present, and future. *IBM Cognos Innovation Center,* 1–12.

21. Sopan, A., Noh, A. S. I., Karol, S., Rosenfeld, P., Lee, G., & Shneiderman, B. (2012). Community health map: A geospatial and multivariate data visualization tool for public health datasets. *Government Information Quarterly, 29*(2), 223–234.
22. Conti, G. (2007). *Security Data Visualization: Graphical Techniques for Network Analysis.* No Starch Press, San Francisco, CA.
23. Chi, M. T., Lin, S. S., Chen, S. Y., Lin, C. H., & Lee, T. Y. (2015). Morphable word clouds for time-varying text data visualization. *IEEE Transactions on Visualization and Computer Graphics, 21*(12), 1415–1426.
24. Lipton, Z. C. (2017). The mythos of model interpretability. arXiv preprint arXiv:1606.03490.
25. Afnan, M. A. M., Liu, Y., Conitzer, V., Rudin, C., Mishra, A., Savulescu, J., & Afnan, M. (2021). Interpretable, not black-box, artificial intelligence should be used for embryo selection. *Human Reproduction Open, 2021*(4), hoab040.
26. Doshi-Velez, F., & Kim, B. (2017). Towards a rigorous science of interpretable machine learning. arXiv preprint arXiv:1702.08608.
27. Gilpin, L. H., Bau, D., Yuan, B. Z., Bajwa, A., Specter, M., & Kagal, L. (2019). Explaining explanations: An overview of interpretability of machine learning. arXiv preprint arXiv:1806.00069.
28. Yu, H., Taube, H., Evans, J. A., & Varshney, L. R. (2020). Human evaluation of interpretability: The case of AI-generated music knowledge. arXiv preprint arXiv:2004.06894.
29. Kim, S. S. Y., Meister, N., Ramaswamy, V. V., Fong, R., & Russakovsky, O. (2022). HIVE: Evaluating the human interpretability of visual explanations. arXiv preprint arXiv:2112.03184.
30. Molnar, C., Casalicchio, G., & Bischl, B. (2020). Quantifying model complexity via functional decomposition for better post-hoc interpretability. In *Machine Learning and Knowledge Discovery in Databases* (pp. 193–204). Springer International Publishing. doi: 10.1007/978-3-030-43823-4_17.
31. Markus, A. F., Kors, J. A., & Rijnbeek, P. R. (2021). The role of explainability in creating trustworthy artificial intelligence for health care: A comprehensive survey of the terminology, design choices, and evaluation strategies. *Journal of Biomedical Informatics, 113*, 103655. doi: 10.1016/j.jbi.2020.103655. PMID: 33309898.
32. Nothdurft, F., Richter, F., & Minker, W. (2014). Probabilistic Human-Computer Trust Handling. In *Proceedings of the 15th Annual Meeting of the Special Interest Group on Discourse and Dialogue (SIGDIAL)* (pp. 51–59). Philadelphia, PA: Association for Computational Linguistics.
33. Honegger, M. (2018). Shedding light on black box machine learning algorithms: Development of an axiomatic framework to assess the quality of methods that explain individual predictions. arXiv preprint arXiv:1808.05054.
34. Zhong, H., Wang, Y., Tu, C., Zhang, T., Liu, Z., & Sun, M. (2020). Iteratively questioning and answering for interpretable legal judgment prediction. *AAAI, 34*(01), 1250–1257.
35. Marghescu, D. (2007). Multi-dimensional data visualization techniques for exploring financial performance data.
36. Weber, P., Carl, K. V., & Hinz, O. (2023). Applications of explainable artificial intelligence in finance-a systematic review of finance, information systems, and computer science literature. *Management Review Quarterly*, 1–41. doi: 10.1007/s11301-023-00320-0.
37. Guo, W. (2019). Explainable Artificial Intelligence (XAI) for 6G: Improving Trust between Human and Machine. arXiv preprint arXiv:1911.04542.
38. Graziani, M., Dutkiewicz, L., Calvaresi, D., et al. (2023). A global taxonomy of interpretable AI: Unifying the terminology for the technical and social sciences. *Artificial Intelligence Review, 56*, 3473–3504. doi: 10.1007/s10462-022-10256-8.

17 A Study on Transparent Recommender Systems

V. Lakshmi Chetana and Hari Seetha

17.1 INTRODUCTION

For the past few decades, the usage of AI-based systems has increased enormously for decision-making in various applications like recommender systems, healthcare, computer vision, and image processing, etc. Most of the models that are built using Machine Learning/Deep Learning are considered to be 'black-box' by many researchers in the literature because these models are complex, non-linear and extremely difficult to be interpreted and explained to the layman. To make these models transparent, scrutable, and explainable, transparent AI is the only solution. Transparent AI is the future generation AI that revolutionizes the business decision-making process across industries. This is an emerging area and the future of AI that combines machine learning, statistics, and cognitive science. Transparent AI is considered critical because it explains how and why the AI model has come to a specific conclusion i.e., in other words, it explains AI-based systems' decisions to their users. Figure 17.1 shows the black-box versus white-box AI models. Transparent AI is a paradigm where we develop a model by keeping the computation behind its decision open. With this transparency, people will understand the reason behind the inevitable decision made by the AI system. The major benefit of transparent AI is that

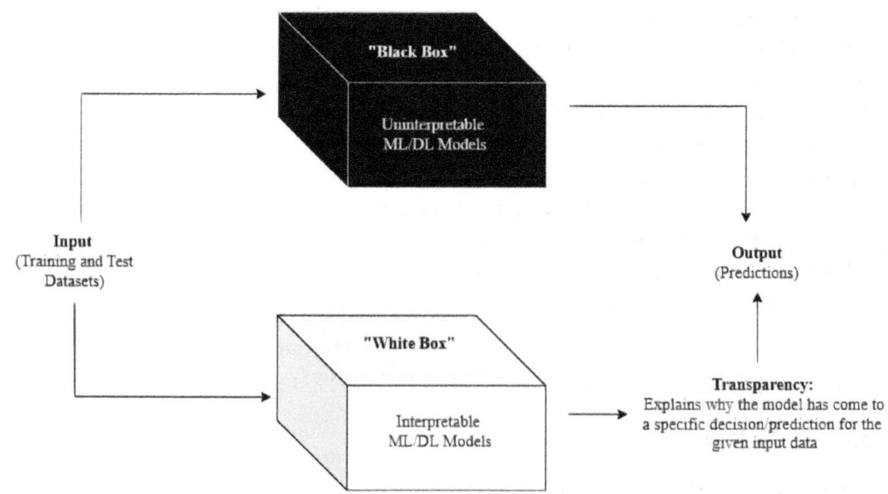

FIGURE 17.1 Black-box Vs White-box models.

DOI: 10.1201/9781003442509-17

it removes the fear of losing control over the AI system, as the transparent AI-based model is developed without hiding the logic behind its decisions. This will help in avoiding disasters and realizing the positive potential of AI systems in making significant contributions to society.

Transparent AI models are broadly categorized into *intrinsic* and *post-hoc* models. The intrinsic models (which can be synonymously called white-box models) are the models that are simple and clearly understood from the beginning. These models can be termed interpretable models, providing explanations for the predictions from the outset. Examples of intrinsic models are linear regression, logistic regression, KNN, decision trees, and Bayesian models [1]. The post-hoc models are methods that become transparent, understandable, and explainable after the model is completely constructed. In this method, we attempt to interpret the black-box models using any of the transparency techniques like Shapley values, feature importance scores, partial dependency plots, etc., which will be explained in Section 17.2.1. These models are too complex for humans to interpret. A few examples of post-hoc models are SVM, Neural networks, and Random Forest.

The post-hoc models can be further categorized based on the two criteria – scope and model. Based on the model criteria, the models can be categorized into model-agnostic or model-specific. The first approach tries to give interpretations of any machine learning model, no matter how much complex it is, while the latter gives interpretations only to some specific machine learning models. Based on the scope, we categorize the model into local and global. Local explainers interpret the models' prediction based on a specific data point. Global explainers interpret the overall behavior of a model based on the entire dataset and identify the bias in the model. The taxonomy of transparent AI models is shown in Figure 17.2.

The organization of the chapter is as follows: Section 17.2 explains the methods of transparency and tools to make the machine and deep learning models transparent. This section also elaborates on five commonly used tools like SHAP, LIME, WIT, interpretML, and DALEX, and presents the evaluation metrics used to assess the transparent AI models. Section 17.3 presents the transparent recommender systems, their benefits and challenges, and presents the recent literature on transparent recommender systems. Section 17.4 concludes and presents the future scope of transparent recommender systems.

17.2 TRANSPARENT AI TECHNIQUES AND TOOLS

Nowadays, the need for understanding the predictions made by complex AI models has increased. Some powerful tools and platforms for understanding and interpreting AI models are listed in Table 17.1. These frameworks are used to enhance the transparency, understanding, interpretability, scrutability, and explainability of ML/DL models, which helps in better decision-making.

17.2.1 Model Transparency Methods

The transparency tools consist of interpretability methods that gives explanations of the model predictions and make the model understandable to the user. This section

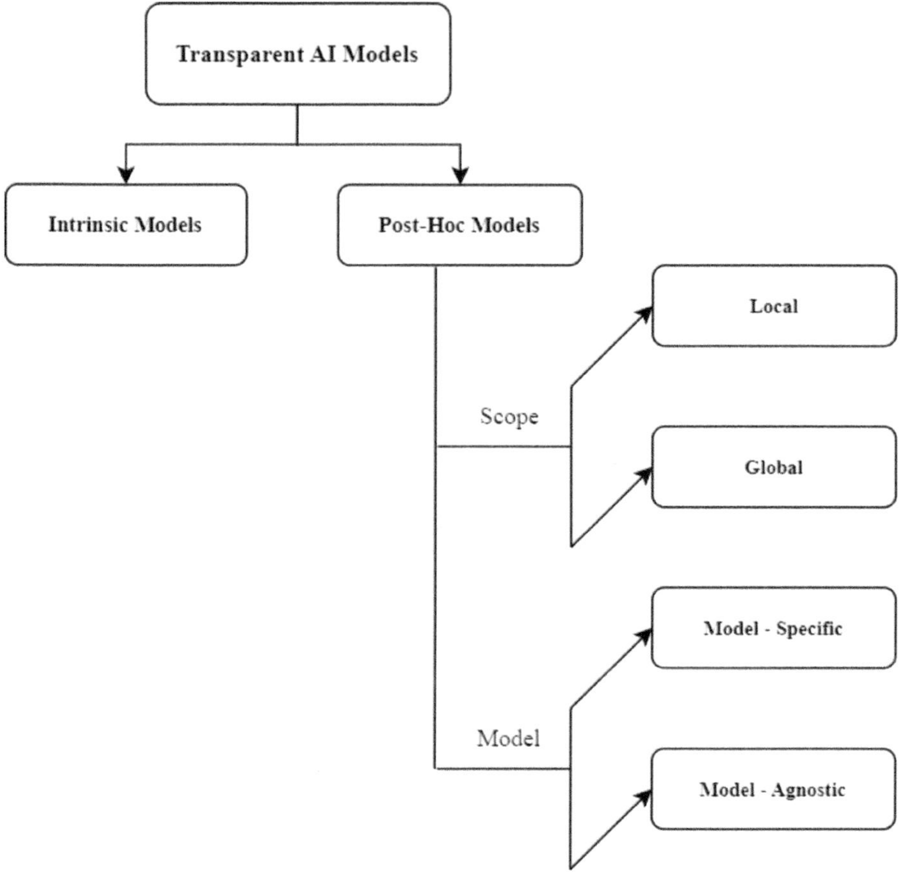

FIGURE 17.2 Taxonomy of transparent AI models.

presents the various transparency methods to make the AI model transparent. These methods can also be called explainers as they render the model transparent. A few of the interpretability techniques available in the literature are:

- **Shapley values:** These values determine each features' contribution to the model's prediction.
- **Partial dependence plots (PDP):** These plots demonstrate how the model's predictions differ as a feature's value is changed.
- **Individual Conditional Expectation (ICE):** It is a method of explaining the predictions of machine learning models. ICE works by generating a line for each instance in the dataset that displays how the prediction of the instance alters as a feature changes.
- **Feature significance values:** These values indicate how significant each feature is for the model's prediction. The better the value, the more relevant the feature is [14].

TABLE 17.1
Opensource Frameworks for Transparent AI

S. No	Name of the Framework	Documentation Link	Github Link
1	SHAP [2]	https://shap.readthedocs.io/en/latest/	https://github.com/slundberg/shap
2	LIME [3]	https://lime-ml.readthedocs.io/en/latest/	https://github.com/marcotcr/lime
3	ELI5[4]	https://eli5.readthedocs.io/en/latest/	https://github.com/eli5-org/eli5
4	What-if Tool [5]	https://cloud.google.com/ai-platform/prediction/docs/using-what-if-tool#install_the_what-if_tool	https://pair-code.github.io/what-if-tool/
5	AIX 360 [6]	https://aix360.readthedocs.io/en/latest/	https://github.com/Trusted-AI/AIX360
6	interpretML [7]	https://interpret.ml/docs/getting-started.html	https://github.com/interpretml/interpret
7	iModels [8]	https://csinva.io/imodels/	https://github.com/csinva/imodels
8	Dalex [9]	https://dalex.drwhy.ai/python/api/	https://github.com/ModelOriented/DALEX/tree/master/python/dalex
9	Explain [10]	https://docs.seldon.io/projects/alibi/en/stable/	https://github.com/SeldonIO/alibi
10	Captum [11]	https://captum.ai/	https://github.com/pytorch/captum
11	iNNvestigate [12]	https://innvestigate.readthedocs.io/en/latest/	https://github.com/albermax/innvestigate
12	Tf-explain	https://tf-explain. readthedocs.io/en/latest/	https://github.com/sicara/tf-explain
13	DeepLift [13]	https://pypi.org/project/deeplift/	https://github.com/kundajelab/deeplift

- **Text explanations:** These explanations are generated by explaining the linear model that was created to approximate the decision boundary of the original model.
- **Local surrogate models:** These models are used to approximate the decision boundary of the original model around the instance being explained.
- **Integrated gradients:** These values indicate how much the model's prediction changes when the value of a feature is changed.
- **Accumulated Local Effects (ALE):** ALE works by creating a local linear approximation of the model's decision boundary around the instance being explained. This approximation is then used to generate a set of explanations that are interpretable to humans. The explanations generated by ALE are in the form of feature significance values. These values indicate how significant each feature is for the model's prediction. The better the value, the more relevant the feature is [14].

- **Contrastive Explanation Method (CEM):** CEM makes the model transparent by first determining which features are crucial for the model's prediction. These features are then used to generate a contrastive explanation, which is a collection of features that are required by the model to make an accurate classification prediction.

17.2.2 Model Transparency Tools

There are myriad tools to make AI models transparent, interpretable, scrutable, explainable, and understandable. A few of the tools are provided in Table 17.1. This section explains in detail those that are most commonly used to make the models transparent.

17.2.2.1 SHapley Additive exPlanations (SHAP) [2]:

It is a framework based on game theory to make the machine learning model transparent. Using the traditional Shapley values of game theory and associated expansions, it establishes a link between optimum credit distribution and local explanations.

Some of the benefits of SHAP are

- SHAP values are feature-level explanations, which can be more useful than global explanations such as accuracy or precision.
- SHAP values are local explanations, which means that they can be used to explain the prediction of a single instance.
- SHAP values are sensitive to the model's input features, which means that these values determine the most significant features for a given prediction.
- SHAP values are easy to understand and interpret.

Some of the limitations of SHAP are

- SHAP may be computationally expensive.
- SHAP may not always be reliable, accurate, and interpretable in the case of complex models.
- SHAP is not a substitute for understanding the model.

The different types of explainers available in SHAP are Tree Explainers, Linear Explainers, Kernel Explainers, and Sampling Explainers. Tree explainers determine each feature's importance in the prediction of tree-based models like decision trees, random forest, XGBoost, etc. Linear explainers determine each feature's importance in the prediction of linear models like linear regression and logistic regression. Kernel explainers determine each feature's importance in the prediction of non-linear models like SVM and neural networks. Sampling explainers determine each feature's importance in complex models apart from the above-mentioned models. They work by sampling a subset of the data and calculating the Shapley values for the features in the subset.

17.2.2.2 Local Interpretable Model Agnostic Explanations (LIME) [3]

It is a framework for making any machine learning model transparent.

Some of the benefits of LIME are

- LIME is a model-agnostic framework, which explains the predictions of any machine learning model, regardless of the model's type or complexity.
- LIME is a local explanation method, which explains the models' prediction of a single instance.
- LIME is easy to understand and interpret.

Some of the limitations of using LIME are

- LIME can be computationally expensive to use.
- LIME's explanations may not be accurate for complex models.
- LIME's explanations may not be complete, as they only explain the local behavior of the model.

The different types of explainers available in LIME are feature importance explainers and counterfactual explainers. Feature importance explainers identify the features that are most important in prediction, while counterfactual explainers are used to identify why the model made a particular prediction.

17.2.2.3 What-If Tool (WIT) [5]

The What-If Tool is a feature in the Google Cloud AI Platform that allows you to explore the predictions of a machine learning model. It provides different explainers that help us understand how the model works and why it makes the predictions it does.

Some of the benefits of the What-if Tool are

- This tool is easy to use. It does not require any coding or technical expertise.
- This tool provides a visual explanation of the model's predictions, making it easy to understand how the model works and why it makes the predictions that it does.
- This tool provides a list of significant features that contributed to the prediction. This information is used to identify the features that are most important in the model prediction.

Some of the limitations of using the What-if Tool are

- It is not always accurate.
- It can be difficult to interpret and may not be reliable in some cases.
- It is only applicable to models that can be explained in terms of feature contributions.
- It is computationally expensive to calculate, especially for large models.

Some of the explainers available in this tool are feature importance explainers, partial dependence plots, individual conditional expectation and counterfactual explainers. Feature importance explainers identify the features that are most important in prediction. Some examples of feature importance explainers are SHAP explainers that calculate the SHAPley values for each feature in the model, which represent the contribution of each feature to the model's prediction. Partial dependence plots show how the model's prediction changes as a feature value changes. Individual conditional expectation plots show how the prediction of a model changes as a feature value changes, while holding all other features constant. Counterfactual explainers are used to identify why the model made a particular prediction.

17.2.2.4 interpretML [6]

This tool provides a unified Python API for machine learning interpretability. It provides various explainability techniques, such as local explainers (LIME, SHAP), global explainers (ALE, PDP), and model debugging tools.

Some of the benefits of InterpretML are

- InterpretML is easy to use. It does not require any coding or technical expertise.
- InterpretML provides a variety of explainability tools to understand how a model works, determine the most significant features for a given prediction, and debug a model.

Some of the limitations of using InterpretML are

- The explanations generated by the explainer can be complex and difficult to understand.
- It is not applicable to all machine learning models.
- It is computationally expensive to calculate, especially for large models.
- The explanations may not be reliable in all cases.

Some of the explainers available in this tool are Feature importance explainers, Counterfactual explainers, Local explainers, and Global explainers. Feature importance explainers are used to identify the features that are most important in prediction. These explainers also give information on how to improve the model's accuracy or debug the model. Counterfactual explainers are used to identify why the model made a particular prediction. Local explainers explain the model prediction based on a specific data point. Global explainers explain the overall behavior of a model based on the entire dataset and identify the bias in the model.

17.2.2.5 Dalex [8]

It is a framework developed in R for making machine learning models transparent. It is a powerful tool that can be used to benefit a wide range of stakeholders. It is a valuable tool for anyone who is working with machine learning models.

Some of the benefits of DALEX are

- DALEX can be used to identify problems with models, such as bias or overfitting, which in turn helps in improving the accuracy of the model.
- DALEX can be used to compare different models and select the most interpretable model.
- DALEX can be used to identify and address potential biases in models.

Some of the limitations of using DALEX are

- DALEX cannot explain all aspects of a model's behavior. For example, it cannot explain the role of non-linearities in the model.
- The explanations that DALEX provides can sometimes be difficult to interpret. This is especially true for complex models because the explanations are based on complex mathematical models.
- DALEX can be computationally expensive to use. This is because it needs to compute the explainability methods that it provides.

DALEX provides several explainers like Feature importance explainers, Counterfactual explainers, Local explainers, and Global explainers. Various feature importance explainers available in DALEX are SHAP values, Partial dependence plots, and Individual conditional expectation plots. Different counterfactual explainers are LIME and CEM (Contrastive Explanation Method). ICE (Interactive Individual Conditional Expectation) plots, which shows how the model's prediction changes with the change in the value of a feature of a specific data point, and SHapley values are the local explainers. ALE (Accumulated Local Effects) plots and PDP plots are the global explainers.

The summary of the most familiar open-source frameworks/tools for Transparent AI is given in Table 17.2.

17.2.3 Evaluation Metrics of Transparent AI Models

There are several evaluation metrics to assess the transparency and interpretability of a machine learning model. These metrics can be used to measure the accuracy, completeness, and understandability of the explanations given by the model.

Some of the most common evaluation metrics for transparency and interpretability include:

- **Accuracy:** This metric measures the accuracy of the explanations given by the model. It is calculated by measuring the proportion of correct explanations.
- **Completeness:** This metric measures the completeness of the explanations generated by the model. It is calculated by measuring the proportion of explanations that cover all of the important features.
- **Understandability:** This metric measures the understandability of the explanations generated by the model. It is calculated by measuring the proportion of explanations that are easy for humans to understand.

TABLE 17.2
Summary of Transparent AI Tools

S. No	Transparency Tool	Transparency Method	Model	Scope
1	LIME	LIME	Model-Agnostic	Local
2	SHAP	Tree Explainers	Model-Specific	Local
		Linear Explainers	Model-Specific	Local
		Kernel Explainers	Model-Agnostic	Local
3	What-if Tool	Feature importance	Model-Agnostic	Local
		ICE	Model-Agnostic	Local
		PDP	Model-Agnostic	Global
4	interpretML	SHAP	Model-Agnostic	Local
		LIME	Model-Agnostic	Local
		Partial Dependence	Model-Agnostic	Global
5	DALEX	Feature importance	Model-Agnostic	Global
		ALE	Model-Agnostic	Global
		PDP	Model-Agnostic	Global
		LIME	Model-Agnostic	Local
		SHAP	Model-Agnostic	Local
		ICE	Model-Agnostic	Local
		CEM	Model-Agnostic	Global

- **Fidelity:** This metric measures how closely the explanations generated by the model match the actual decision-making process of the model.
- **Relevance:** This metric measures how relevant the explanations provided by the model are to the user's needs.

17.3 TRANSPARENT AI IN RECOMMENDER SYSTEMS

Recommender systems have become an essential part of our daily lives. These systems provide us with personalized and relevant recommendations for products, services, and content based on our historical behavior and preferences. However, the decision-making processes behind these recommendation systems are often perceived as "black box" tools with little to no transparency. This lack of transparency causes distrust in the recommendations and reduces users' confidence in the system. To address this issue, the concept of transparent AI has emerged as an important topic in recommender systems. A transparent recommender system provides explanations of its recommendations to the user. This can be done in a variety of ways, such as providing users with information about the features that were used to make the recommendation or providing users with a way to see how their own preferences influenced the recommendation. Transparent AI in recommender systems is a way to make the algorithms that power these systems more understandable to users. This can be done by providing information about how the algorithms work, what data they use, and how they make decisions. Transparency helps users better understand the recommendations they receive and to make more informed decisions about which products or services to use. For example, Netflix allows its users to see the ratings

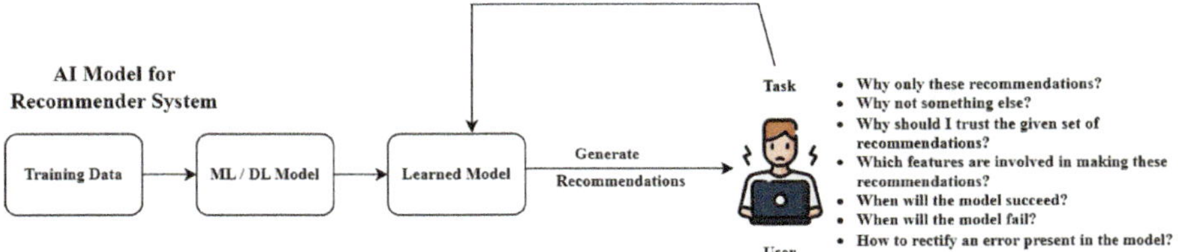

FIGURE 17.3 Basic AI model for recommender system.

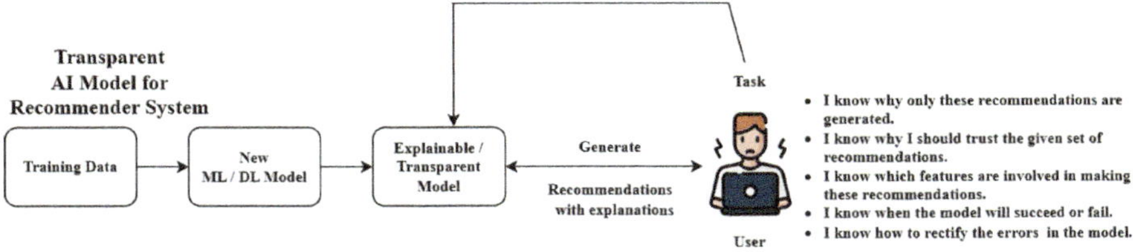

FIGURE 17.4 Transparent AI model for recommender system.

and reviews of other users before they make a recommendation. Amazon provides its users with information about the factors that influenced their recommendations, such as the products they have previously purchased, and Spotify allows users to see the songs that other users have listened to before they make a recommendation. The AI model for a recommender system is shown in Figure 17.3, and the transparent AI model for a recommendation system is shown in Figure 17.4.

17.3.1 Benefits of using Transparent AI in Recommender Systems

Several benefits to transparent AI in recommender systems are as follows:

- First, it can help to build trust between users and the systems that are making recommendations. When users understand how the algorithms work, they are less likely to feel that they are being manipulated or deceived.
- Second, transparency can help to improve the accuracy of recommendations. When users are able to provide feedback on the recommendations they receive, the algorithms can learn to make better decisions in the future.
- Third, transparency can help to identify and address bias in recommendation systems. By understanding how the algorithms work, it is possible to identify and remove biases that may be leading to unfair or inaccurate recommendations.

17.3.2 Challenges of Using Transparent AI in Recommender Systems

A few of the challenges to transparent AI in recommender systems are as follows:

- First, it can be difficult to explain how complex algorithms work.
- Second, users may not be interested in learning about the technical details of how recommendation systems work.
- Finally, transparency can make it more difficult for companies to protect their intellectual property.

17.3.3 Literature Review on Transparent AI Recommender Systems

Transparent Recommender Systems aims to address the challenges associated with traditional black-box recommendation algorithms. Traditional recommender systems are often criticized for their lack of transparency, as they do not provide any clear rationale for the recommendations they make. This lack of transparency can lead to user distrust and reluctance to engage with the recommender systems. However, transparent recommender systems have been developed to overcome these challenges by providing users with clear reasons for the recommendations made. Transparency is evaluated by whether or not the explanations can reveal the internal workings of the recommender models, allowing users to understand and trust how recommendations have been generated.

The state-of-the-art work done in the field of transparent recommender systems is presented in Table 17.3. This field is still in its early stages, but there has been a growing interest in this area in recent years. There are several different approaches to making recommendation systems more transparent, and there is no single approach that is universally accepted. However, there are a number of promising methods that have been developed, and there is a growing body of research on this topic. This section provides the state-of-the-art research done in making the recommendation system models transparent, interpretable, understandable, explainable, trustworthy, etc.

17.4 CONCLUSIONS AND FUTURE SCOPE

Transparent recommender systems are a promising new technology that has the potential to improve the user experience in a variety of ways. By providing users with more information about how recommendations are made, transparent recommender systems can help users make more informed decisions about what to consume. Additionally, transparent recommender systems can help build trust between users and recommender systems, which can lead to increased user satisfaction. The future of transparent recommender systems is promising. As research in this area continues, we can expect to see more transparent systems that are able to explain their decisions to users. This will help ensure that these systems are used in a fair and equitable way, and it will also help improve the user experience. In addition to providing users with information about how recommendations are made, transparent recommender systems can also help users understand their own preferences. By providing users with feedback about their past behavior, transparent recommender systems can help users learn more about what they like and dislike. This can be helpful for users who are trying to discover new content or who are trying to make better decisions about the content they consume. Overall, the future of transparent recommender systems is bright. These systems have the potential to improve the user experience, build trust

TABLE 17.3
Literature on Transparent Recommender Systems

Author	Title of the paper	Description	Limitations
Koras et al. [15]	Interpretable deep recommender system model for prediction of kinase inhibitor efficacy across cancer cell lines	The authors proposed a novel approach for predicting the efficacy of kinase inhibitors in cancer cell lines. • The approach, called DEERS, is a deep neural network recommender system that uses side information about drugs and cancer cell lines to make predictions. • DEERS was evaluated on a dataset of 1,000 cancer cell lines and 100 kinase inhibitors, and it was shown to outperform simpler matrix factorization models. • DEERS was also shown to be interpretable, meaning that it is possible to understand why it makes the predictions that it does.	• The study was conducted on a relatively small dataset of cancer cell lines and kinase inhibitors. • The study did not evaluate DEERS on a clinical dataset. • The study did not assess the clinical utility of DEERS.

(Continued)

TABLE 17.3 (*Continued*)
Literature on Transparent Recommender Systems

Author	Title of the paper	Description	Limitations
Balog et al. [16]	Transparent, scrutable and explainable user models for personalized recommendation.	• This paper presents a new set-based recommendation technique that permits the user model to be explicitly presented to users in natural language. • This allows users to understand recommendations made and improve the recommendations dynamically. • The approach is comparable to traditional collaborative filtering techniques in a standard static setting, but it allows users to efficiently improve recommendations. • Further, it makes it easier for the model to be validated and adjusted, building user trust and understanding.	• The approach is based on a set-based model, which may not be suitable for all types of recommendation systems. • The approach has only been evaluated on a limited number of datasets, and it is not yet clear how it would perform in other settings. • The approach is more complex than traditional collaborative filtering techniques, and it may be more difficult to implement and deploy.
Yu et al. [17]	Contextual-boosted deep neural collaborative filtering model for interpretable recommendation	• This paper proposes a new model, CDNC, for item recommendation. CDNC leverages both user ratings and item introductions to learn better user and item representations. • This allows CDNC to deal with the cold-start problem and provide interpretable recommendations. • CDNC can provide interpretable recommendations by explaining why it recommends a new item. • CDNC is performed by four modules: initial user representation learning, interactive-learning-based user representation learning, item representation learning, and feature fusion and prediction.	• CDNC is a relatively new model, and more research is needed to evaluate its performance on a wider range of datasets. • CDNC is computationally expensive to train, and it may not be suitable for real-time recommendation systems.

(*Continued*)

TABLE 17.3 (*Continued*)
Literature on Transparent Recommender Systems

Author	Title of the paper	Description	Limitations
Sinha et al. [18]	The Role of Transparency in Recommender Systems	• This paper explores the role of transparency in recommender systems. • There are a number of ways to provide transparency in recommender systems, including providing explanations, allowing users to see the underlying data, and allowing users to customize the system. • The authors argue that transparency can help to increase user trust in the system and lead to increased use. They conducted a user study of five music recommender systems and found that users who were given explanations for the recommendations were more likely to trust the system and use it again.	• The study was conducted with a small number of users. • The study focused on music recommender systems, and the results may not generalize to other types of recommender systems. • The study did not investigate the impact of different levels of transparency on user trust and confidence.
Sonboli et al. [19]	Fairness and Transparency in Recommendation: The Users' Perspective	• This paper explores the importance of fairness and transparency in recommender systems. • The authors argue that users have a right to understand how their data is being used to make recommendations, and that recommender systems should be designed to be fair and equitable. • They propose a number of design principles for fairness-aware recommender systems and discuss the challenges of implementing these principles in practice.	• The paper focuses on the user perspective and does not consider the perspectives of other stakeholders, such as content providers. • The paper does not provide a detailed implementation of any of the proposed design principles.

(Continued)

TABLE 17.3 (*Continued*)
Literature on Transparent Recommender Systems

Author	Title of the paper	Description	Limitations
Clara Siepmann et al. [20]	Trust and Transparency in Recommender Systems	• Trust and transparency are important factors in recommender systems (RS). There is a link between trust and transparency, and users are more likely to trust systems that are transparent. • Trust is an important factor in RS, and it can be defined as the willingness of a user to rely on a system to provide accurate and relevant recommendations. • Transparency is the degree to which a system reveals its inner workings to users. • In this paper, the authors discuss the different perspectives on trust and different ways to evaluate it. • It also reviews the relationships between trust and transparency, as well as mental models, and investigates different strategies to achieve transparency in RS such as explanation, exploration, and exploranation (i.e., a combination of exploration and explanation). • The paper identifies a need for further studies to explore these concepts as well as the relationships between them.	• The paper focuses on trust and transparency in RS, and it does not consider other factors that may influence user trust, such as accuracy and relevance. • The paper does not provide a comprehensive overview of the different strategies to achieve transparency in RS. • The paper is based on a review of the literature, and it does not present any new empirical findings.

(*Continued*)

TABLE 17.3 (*Continued*)
Literature on Transparent Recommender Systems

Author	Title of the paper	Description	Limitations
Zhang et al. [21]	Explainable Recommendation: A Survey and New Perspectives Suggested	• This paper provides a comprehensive overview of the research on explainable recommendations. • The paper begins by discussing the importance of explainable recommendations and the challenges involved in developing explainable recommendation systems. • The paper then reviews a variety of approaches to explainable recommendations, including post-hoc explanations, model-based explanations, and user-centered explanations. • The paper concludes by discussing the future of explainable recommendations and the challenges that need to be addressed.	• The research on explainable recommendations is still in its early stages. • There is no single approach to an explainable recommendation that is universally effective. • The development of explainable recommendation systems can be challenging and time-consuming.

(*Continued*)

TABLE 17.3 (*Continued*)
Literature on Transparent Recommender Systems

Author	Title of the paper	Description	Limitations
Herlocker et al. [22]	Explaining collaborative filtering recommendations.	• This paper discusses the importance of providing explanations for collaborative filtering recommendations automatically, called, Automatic Collaborative Filtering (ACF). The paper argues that explanations can help users to understand why they are being recommended certain items, and can also help to improve the trust that users have in the recommender system. • The paper discusses a number of different ways to provide explanations for collaborative filtering recommendations. These include: 1 *Item-based explanations:* These explanations show the user how similar items have been rated by other users. 2 *User-based explanations:* These explanations show the user how similar users have rated the item. 3 *Content-based explanations: These* explanations show the user how the item matches the user's interests.	• The cost of providing explanations can be significant, especially for large ACF systems. • The difficulty of providing accurate explanations can be challenging, especially for complex ACF systems. • The potential for explanations to be misleading, if they are not carefully designed.

(*Continued*)

TABLE 17.3 (*Continued*)
Literature on Transparent Recommender Systems

Author	Title of the paper	Description	Limitations
Mika [23]	Toward a Transparent Recommender System	• This paper proposes a framework for developing transparent RSs. • The framework is based on the idea of using knowledge graphs to represent the relationships between items. • This allows the system to explain its recommendations by providing users with a list of items that are similar to the ones that they have been recommended	• The paper does not provide a detailed implementation of the proposed framework. • The paper does not evaluate the effectiveness of the proposed framework in a real-world setting. • There are a number of challenges to developing transparent RSs, including the size and complexity of knowledge graphs, and the identification of relevant items for recommendations.

and confidence, and help users understand their own preferences. As research in this area continues, we can expect to see even more innovative and effective transparent recommender systems.

REFERENCES

1. M. S. Khan, M. Nayebpour, M. H. Li, H. El-Amine, N. Koizumi, and J. L. Olds, "Explainable AI: A neurally-inspired decision stack framework," *Biomimetics*, vol. 7, no. 3, p. 127, 2022.
2. S. M. Lundberg, and S.-I. Lee, "A unified approach to interpreting model predictions," in *Proceedings of the 31st International Conference on Neural Information Processing Systems*, Long Beach, California, USA, pp. 4768–4777, 2017.
3. M. T. Ribeiro, S. Singh, and C. Guestrin, "'Why Should I Trust You?' Explaining the predictions of any classifier," in *NAACL-HLT 2016-2016 Conference of the North American Chapter of the Association for Computational Linguistics: Human Language Technologies Proc. Demonstr. Sess.*, pp. 97–101, 2016.
4. A. Fan, Y. Jernite, E. Perez, D. Grangier, J. Weston, and M. Auli, "ELi5: Long form question answering," in *ACL 2019-57th Annual Meeting of the Association for Computational Linguistics; Proceedings of the Conference*, pp. 3558–3567, 2020.
5. J. Wexler, M. Pushkarna, T. Bolukbasi, M. Wattenberg, F. Viegas, and J. Wilson, "The what-if tool: Interactive probing of machine learning models," *IEEE Trans. Vis. Comput. Graph.*, vol. 26, no. 1, pp. 56–65, 2020.
6. V. Arya, et al., "One Explanation Does Not Fit All: A Toolkit and Taxonomy of AI Explainability Techniques," 2019. arXiv preprint arXiv:1909.03012.
7. H. Nori, S. Jenkins, P. Koch, and R. Caruana, "InterpretML: A Unified Framework for Machine Learning Interpretability," pp. 1–8, 2019. arXiv preprint arXiv:1909.09223.
8. C. Singh, K. Nasseri, Y. Tan, T. Tang, and B. Yu, "Imodels: A Python package for fitting interpretable models," *J. Open Source Softw.*, vol. 6, no. 61, p. 3192, 2021.
9. H. Baniecki, W. Kretowicz, P. Piatyszek, J. Wisniewski, and P. Biecek, "dalex: Responsible machine learning with interactive explainability and fairness in python," *J. Mach. Learn. Res.*, vol. 22, pp. 1–7, 2021.
10. J. Klaise, A. Van Looveren, G. Vacanti, and A. Coca, "Alibi explain: Algorithms for explaining machine learning models," *J. Mach. Learn. Res.*, vol. 22, pp. 1–7, 2021.
11. N. Kokhlikyan, et al., "Captum: A unified and generic model interpretability library for PyTorch," pp. 1–11, 2020. CoRR abs/2009.07896.
12. M. Alber, et al., "INNvestigate neural networks!" *J. Mach. Learn. Res.*, vol. 20, pp. 1–8, 2019.
13. A. Shrikumar, P. Greenside, and A. Kundaje, "Learning important features through propagating activation differences," in *Proceedings of the 34th International Conference on Machine Learning - Volume 70*, Sydney, NSW, Australia pp. 3145–3153, 2017.
14. U. Kamath, and J. Liu, *Explainable Artificial Intelligence: An Introduction to Interpretable Machine Learning*. Springer International Publishing, Cham, 2021.
15. K. Koras, E. Kizling, D. Juraeva, E. Staub, and E. Szczurek, "Interpretable deep recommender system model for prediction of kinase inhibitor efficacy across cancer cell lines," *Sci. Rep.*, vol. 11, no. 1, pp. 1–16, 2021.
16. K. Balog, F. Radlinski, and S. Arakelyan, "Transparent, scrutable and explainable user models for personalized recommendation," in *Proceedings of 42nd International ACM SIGIR Conference on Research and Development in Information Retrieval*, Paris, France, pp. 265–274, 2019.

17. S. Yu, M. Yang, Q. Qu, and Y. Shen, "Contextual-boosted deep neural collaborative filtering model for interpretable recommendation," *Expert Syst. Appl.*, vol. 136, pp. 365–375, 2019.
18. R. Sinha, and K. Swearingen, "The role of transparency in recommender systems," in CHI '02 Extended Abstracts on *Human Factors* in *Computing Systems,* Minneapolis, Minnesota, USA, pp. 830–831, 2002.
19. N. Sonboli, J. J. Smith, R. Burke, and C. Fiesler, *Fairness and Transparency in Recommendation: The Users' Perspective*, vol. 1, no. 1. Association for Computing Machinery, 2021. doi: 10.1145/3450613.3456835.
20. M. A. C. Clara Siepmann, "Trust and transparency in recommender systems," in *ACM CHI 2023 Workshop Human-Centered Perspectives in Explainable AI (HCXAI)*, Hamburg, Germany, April 25–30, 2023.
21. Y. Zhang, and X. Chen, "Explainable recommendation: A survey and new perspectives," *Found. Trends Inf. Retr.*, vol. 14, no. 1, pp. 1–101, 2020.
22. J. L. Herlocker, J. A. Konstan, and J. Riedl, "Explaining collaborative filtering recommendations," in *Proceedings of 2000 ACM Conference on Computer Supported Cooperative Work*, Philadelphia, Pennsylvania, USA, pp. 241–250, 2000.
23. G. P. Mika, "Toward a transparent recommender system," *Proceedings of the 14th International Rule Challenge 4th Doctoral Consortium and 6th Industry Track@ RuleML+RR*, vol. 2644, pp. 111–119, 2020.

Index

For Product Safety Concerns and Information please contact our EU representative GPSR@taylorandfrancis.com Taylor & Francis Verlag GmbH, Kaufingerstraße 24, 80331 München, Germany

Batch number: 10399582

Printed by Printforce, the Netherlands